GEOLOGY AND MINERALOGY RESEARCH DEVELOPMENTS

PEROVSKITE

CRYSTALLOGRAPHY, CHEMISTRY AND CATALYTIC PERFORMANCE

GEOLOGY AND MINERALOGY RESEARCH DEVELOPMENTS

Additional books in this series can be found on Nova's website under the Series tab.

Additional e-books in this series can be found on Nova's website under the e-book tab.

MATERIALS SCIENCE AND TECHNOLOGIES

Additional books in this series can be found on Nova's website under the Series tab.

Additional e-books in this series can be found on Nova's website under the e-book tab.

GEOLOGY AND MINERALOGY RESEARCH DEVELOPMENTS

PEROVSKITE

CRYSTALLOGRAPHY, CHEMISTRY AND CATALYTIC PERFORMANCE

JINGHUA ZHANG
AND
HUAN LI
EDITORS

nova publishers

New York

For permission to use material from this book please contact us:
Telephone 631-231-7269; Fax 631-231-8175
Web Site: http://www.novapublishers.com

Library of Congress Cataloging-in-Publication Data

ISBN: 978-1-62417-800-9

Published by Nova Science Publishers, Inc. † New York

CONTENTS

PREFACE

In this book, the authors present current research in the study of the crystallography, chemistry and catalytic performance of perovskites. Topics discussed include the defect structure and defect-induced expansion of perovskite oxides; perovskite-based catalysts for transformation of natural gas and oxygenates into syngas; Bi containing multiferroic perovskite oxide thin films; perovskites as catalysts for environmental remediation; microwave-assisted synthesis and characterization of perovskite oxides; perovskite and lead based ceramic materials; photocatalytic properties of perovskite-type layered oxides; structure of perovskite electron-ionic conductors; and distorted perovskites.

Chapter 1 - This chapter reviews results of research aimed at design of catalysts of methane and biofuels transformation into syngas based upon perovskite-like oxides (doped ferrites –nickelates/(cobaltites) and manganese – chromates of La and Pr) and their nanocomposites with fluorites (YSZ, GDC) promoted by nanoparticles of Ni—based alloys (Ni-Fe, Ni-Ru, Co-Fe). The structural and surface characteristics of complex perovskite-like precursors ($LnFe_{1-x-y}Ni_xRu_yO_{3-z}$ etc) and nanocomposites generated due to segregation of metal alloy nanoparticles from perovskites under reducing action of H_2, CH_4, and EtOH or reaction feed (CH_4+ CO_2; EtOH+ H_2O, etc) were studied by XRD, TEM with EDX, XPS. The oxygen bonding strength, mobility and reactivity were characterized by O_2 TPD, H_2/CH_4/EtOH TPR followed by temperature-programmed reoxidation with H_2O Catalytic properties were estimated in reactions of CH_4 dry reforming and ethanol steam reforming at short contact times. The most promising active components were supported on heat-conducting monolithic substrates and tested in concentrated feeds. Catalytic activity and coking stability of these catalysts were found to be controlled by the efficiency of reagents activation (fuel molecules –on metal sites, oxidants –on oxide support sites) and the surface/near surface oxygen diffusion to metal-support interface where activated C-H-O fragments interact with oxygen atoms generating syngas. Strong metal –oxide support interaction provides stability of metal nanoparticles to sintering in real operation conditions.

Chapter 2 - In this work multiferroic thin films of Bi containing perovskite oxides are discussed, where the driving force for ferroelectricity are the Bi^{3+} lone-pair electrons. First, a brief introduction of Bi containing multiferroic perovskite oxides will be presented to describe the mechanisms for establishing magnetic and ferroelectric order in these compounds and some recent developments in this field of research are reported. The second section addresses experimental aspects of epitaxial thin film growth of $BiMnO_3$, $(Bi_{0.9}La_{0.1})_2NiMnO_6$ and $BiFeO_3$ thin films by pulsed laser deposition. The third section is dedicated to the physical properties of such films in terms of the structural characterization and the magnetic and ferroelectric properties and their correlations in form of magnetoelectric coupling (MEC) effects.

Chapter 3 - The current awareness about environmental and health protection has increased the regulations of the chemical industrial processes. Consequently, the use of heterogeneous catalysts for pollutant abatement from vehicle emissions and industrial effluent is indispensable in order to attenuate the global warming and other environmental effects.

Currently, most of the heterogeneous catalysts are based on noble metals, due to the high activity; however, their main disadvantages are related to volatility, sulfur poisoning and cost. Therefore, much effort has been devoted to the substitution of noble metal catalysts with mixtures of selected metal oxides.

Transition metal perovskites, with the general formula ABO_3, are well known as very good oxidation catalysts due to thermal stability, poisoning resistance and low cost. Moreover, the oxidation on perovskite-type catalysts, the combustion is carried out much below the flame temperature, substantially reducing the formation of harmful species such as CO and NO_x. In addition, perovskite-like oxides can be tailored to create a wide family of catalysts by varying either the A-site or the B-site metal ion, or both. Indeed, the catalytic activity of a perovskite can be modified by inserting proper transition metal ions. Partial substitution at the A-site with another metal cation, doping, can strongly affect the catalytic activity, due to stabilization of unusual oxidation states of the B cation and simultaneous formation of structural defects created by such a substitution. Structural defects are responsible not only for part of the catalytic activity, but also for oxygen mobility within the crystal lattice of the solid due to non-stoichiometry created by the substitution. The presence of ionic vacancies affects the catalytic activity by favouring (or not) reactant adsorption from the gas phase. Thus, an oxygen vacancy in the surface layer is not only the place where oxygen can enter into the solid state structure, but also the site with a missing oxygen ion surrounded by a maximum number of cations represents the most acidic site, much more acidic than the regular surface cations.

In this chapter, the synthesis, characterization and catalytic activity of pure and doped perovskites for the abatement of some of the most hazardous pollutants delivered to the atmosphere as volatile organic compounds (VOCs) and particulate soot, is addressed. In particular, the performance of pure and doped lanthanum ferrites and cobaltites in the catalytic oxidation of some VOCs as toluene, ethanol, acetyl acetate and n-hexane is discussed. Finally, the catalytic activity of alkali metal niobates in the soot combustion and the relation between defects and catalytic activity of calcium doped lanthanum ferrites in the combustion of methane is also discussed.

Chapter 4 - The use of microwave irradiation is a promising alternative heat source for the synthesis of inorganic materials such as perovskite oxides. The method offers massive energy and time savings as compared to the traditional ceramic method. In this work the authors review the basic principles of the microwave heating mechanism based on interactions between dipoles in the material and the electromagnetic microwave.

Furthermore, the authors comment on and classify all different sub-types of microwave synthesis such as solid-state microwave and microwave assisted hydrothermal synthesis. In the experimental part of this work the authors present a large range of materials that were synthesized in our laboratories by one of the microwave techniques, where such materials include superconducting, ferromagnetic, ferroelectric, dielectric and multiferroic perovskite systems.

Chapter 5 - Perovskites based upon substituted $LaBO_{3-\delta}$, where B is a 3d transition metal, are the state-of-the-art materials for a variety of different high and intermediate temperature devices such as solid oxide fuel cells (SOFCs) and mixed ionic and electronic conducting (MIEC) membranes. The unique feature of the oxides is their ability to undergo both thermal expansion and that induced by the defects of oxygen nonstoichiometry in the oxide lattice. The latter is chemical or defect-induced expansion.

There are different possible mechanisms to explain the isothermal expansion observed. If formation of oxygen vacancies in oxide lattice at low oxygen partial pressures/high temperature is accompanied by the reduction of 3d metal then one of them may be based on a change of mean size of the reducible cation due to substitution of "large" $B^{(n-1)+}$ for smaller B^{n+} as a result of the reduction.

The priority purpose of this chapter is to explain the defect-induced expansion for substituted chromites, manganites and cobaltites on the basis of their defect structure using both the experimental data available in the literature and those recently obtained by the authors.

Theoretical values of the chemical expansion were computed on the basis of a relative change of the mean ionic radius. In order to enable such computation modeling analysis of the defect structure of the perovskites studied was carried out. As a result, most adequate models were found. The experimental data and those calculated on the basis of model proposed were shown to be in good agreement for all oxides investigated. Change of a spin state of cobalt in different oxidation states was found to contribute significantly in an expansion value of cobalt containing perovskites.

Chapter 6 - The history, crystallographic signature of perovskite materials along with physical properties has been discussed. Basically lead-based perovskite material (PT, PZT) has so many technological applications in electronic industries. The property of the lead based material can also be enhanced by substituting different transition materials within the perovskite. The main objectives of the present review articles are to understand the crystal structure and electrical transport properties of PT and PZT based ceramic perovskite.

Chapter 7 - The chapter presents a comparative study of photocatalytic activity of layered perovskite-type oxides based on original results. Series of Ruddlesden-Popper ($A_2Ln_2Ti_3O_{10}$ and $ANdTiO_4$, where A = H, Li, Na, K; Ln = La, Nd) and Dion-Jacobson ($ANdTa_2O_7$, where A = H, Li, Na, K, Rb, Cs and $ANdNb_2O_7$, where A = Rb, Cs) phases where synthesized and characterized by XRD, SEM, UV-VIS diffuse-reflectance spectroscopy, BET. Photocatalytic activity was determined from the hydrogen evolution rate in aqueous suspensions under UV-irradiation in the presence of isopropyl alcohol as a sacrificial electron donor. Dependence of photocatalytic activity of complex oxides on their characteristics, such as BET surface area, particle morphology, band gap energy and water intercalation ability is discussed.

Chapter 8 - Nowadays one of the most urgent problems facing the researchers in the field of solid state chemistry is the problem of obtaining materials for energy-saving technologies. The so called solid oxide fuel cells require the materials for cathodes and electrolytes. Lanthanum gallate with perovskite structure doped with strontium and magnesium is a popular subject to use as an electrolyte, since it is conducting oxygen through the vacancies in the oxygen sublattice resulting from heterovalent doping. However this doped lanthanum gallate appears to be nonstable at high temperatures, at which these systems operate. On doping lanthanum gallate with strontium and magnesium and also with a transition element such systems become stable enough, even though some electron conductivity appears, this is good for cathodes but seems to be bad for electrolytes.

Several problems arise in this context, which led us to study the electron structure of these systems. First: what is the reason for the stabilization of the whole system on doping with transition elements in addition to an alkaline-earth or magnesium. The second: which element is the best for electrolytes and which for cathodes. The third: what must be the quantitative relationships between transition element and an element of the second group of Periodic Table. Up to now these problems were solved empirically, the most commonly used ratio M:Sr or M:Mg was 5:1. There are other problems, such as, why strontium but not calcium or barium is used for SOFC.

The explanations for these questions seem to lie in the electron structure of such systems, which means the state of transition element atoms and interatomic interactions in these systems. Transition elements are paramagnetic, and the whole systems are solid solutions. Therefore their electron structure can be studied by the authors method of magnetic dilution. This means a systematic study of magnetic susceptibility of a wide series of concentrations in the systems $La_{1-y}Sr_yM_xGa_{1-x}O_{3-\delta}$ or $LaM_xGa_{1-x-y}Mg_yO_{3-\delta}$ with fixed ratio $x:y$, starting with solid solutions $LaMO_3$ in $LaGaO_3$ and then passing to $x:y = 5:1$, $x:y = 2:1$ and doping with both Sr and Mg – M:Sr:Mg = 5:1:1.

The examination of magnetic characteristics resulted in the following conclusions: 1) no oxidation of paramagnetic elements is observed in the solid solution upon doping with a bivalent element; 2) in the solid solutions of lanthanum gallate doped with a transition and a diamagnetic elements we see the formation of large (about 1-2 nm in size) clusters; 3) these clusters are not destroyed by the dilution and necessarily include also strontium or magnesium and accompanying vacancies in the oxygen sublattice.

Therefore the reason for the structure stabilization upon doping lanthanum gallate with a diamagnetic and paramagnetic elements is the formation of nanosize clusters, containing both elements and vacancies in the oxygen sublattice. Clustering increases as the content of bivalent element increases.

The study of the impedance spectra of these systems showed that an increase in the bivalent element concentration results in a decrease in conductivity in doped lanthanum gallate. A comparison between several d-elements and calculation of the bond structure allowed nickel to be recommended as a doping element for cathodes whereas cobalt and chromium – for electrolytes. A scheme is suggested to account for for the fact that the ratio M:Sr(Mg)=5:1 appears to be the best for optimizing the conductivity of doped lanthanum gallate.

The general conclusion is that clustering of paramagnetic elements in lanthanum gallate is an important factor for their conductivity. Clustering is necessary for maintaining the stability of the structure, but an increase in clustering following an increase in the bivalent atom concentration results in a decrease in conductivity owing to the fact that the field of negatively charged cluster hampers the mobility of vacancies.

Chapter 9 - The ability to rationally design materials for use in fundamental studies or applications is a key motivation in the field of synthetic chemistry. For this, one needs a basic 'building block' which is flexible for incorporation of organic as well as inorganic counter part.

The cubic AMX_3 perovskite structure will be a 'role' model for the same. AMX_3 pervoskite consists of an extended three-dimensional (3-D) network of corner-sharing MX_6 octahedra, where M is generally a divalent metal and X a halide. With the increase in size of 'A', an organic counter part, one can generate layered perovskite consists where MX_4^{2-} layers of corner-sharing metal halide octahedra alternate with organic cation bilayers (for monofunctional cations) or monolayers (for bifunctional cations). Here two metal centers are connected by halide, which in turn are indulge in hydrogen bonding with the oraganic cations.

That means apart from ionic bonding, hydrogen bonding in designing these type compounds play equally important role, a topic covered in this chapter. These so called, organic-inorganic hybrid perovskites, self-assemble at ambient temperature solutions and are generally found stable up to the decomposition point (typically >250 °C). But the authors and others have observed very interesting thermal decomposition behavior in these compounds will be addressed in detail.

Finally, the chapter will discuss technologically important properties explored by these layered perovskites, which include magnetism, thermochromism and structural phase-transitions.

In: Perovskite
Editors: Jinghua Zhang and Huan Li
ISBN: 978-1-62417-800-9

Chapter 1

PEROVSKITE-BASED CATALYSTS FOR TRANSFORMATION OF NATURAL GAS AND OXYGENATES INTO SYNGAS

***V. A. Sadykov*[1,2], *S. N. Pavlova*[1], *G. M. Alikina*[1], *N. N. Sazonova*[1], *N. V. Mezentseva*[1,2], *M. V. Arapova*[1], *V. A. Rogov*[1,2], *T. A. Krieger*[1], *A. V. Ishchenko*[1], *R. V. Gulyaev*[1], *A. V. Zadesenets*[2,3], *A. C. Roger*[4], *C. E. Chan–Thaw*[4] *and O. L. Smorygo*[5]**

[1]Boreskov Institute of Catalysis SB RAS, Novosibirsk, Russia
[2]Novosibirsk State University, Novosibirsk, Russsia
[3]Institute of Inorganic Chemistry SB RAS, Novosibirsk, Russia
[4]Laboratory of Materials, Surfaces and Processes, University of Strasbourg, France
[5]Institute of Powder Metallurgy, Minsk, Belarus

ABSTRACT

This chapter reviews results of research aimed at design of catalysts of methane and biofuels transformation into syngas based upon perovskite-like oxides (doped ferrites – nickelates/(cobaltites) and manganese – chromates of La and Pr) and their nanocomposites with fluorites (YSZ, GDC) promoted by nanoparticles of Ni—based alloys (Ni-Fe, Ni-Ru, Co-Fe). The structural and surface characteristics of complex perovskite-like precursors ($LnFe_{1-x-y}Ni_xRu_yO_{3-z}$ etc) and nanocomposites generated due to segregation of metal alloy nanoparticles from perovskites under reducing action of H_2, CH_4, and EtOH or reaction feed (CH_4+ CO_2; EtOH+ H_2O, etc) were studied by XRD, TEM with EDX, XPS. The oxygen bonding strength, mobility and reactivity were characterized by O_2 TPD, H_2/CH_4/EtOH TPR followed by temperature-programmed reoxidation with H_2O Catalytic properties were estimated in reactions of CH_4 dry reforming and ethanol steam reforming at short contact times. The most promising active components were supported on heat-conducting monolithic substrates and tested in concentrated feeds. Catalytic activity and coking stability of these catalysts were found to be controlled by the efficiency of reagents activation (fuel molecules –on metal sites, oxidants –on oxide support sites) and the surface/near surface oxygen diffusion to metal-support interface where activated C-H-O fragments interact with oxygen atoms generating syngas. Strong metal –oxide support interaction provides stability of metal nanoparticles to sintering in real operation conditions.

Keywords: Perovskites, nanocomposites, catalysis, CH_4 dry reforming, ethanol steam reforming, structure, reactivity, coking stability

1. Introduction

Generation of synthesis gas via routes alternative to traditional processes such as steam reforming of methane has recently attracted considerable attention due to both environmental and commercial reasons [1-2]. Transformation of natural gas using carbon dioxide and reforming of oxygenates derived from fast pyrolysis of biomass are the most promising processes [3-14]. Dry reforming transforms cheap undesirable greenhouse gases and is particularly important in the case of biogas or gas fields containing a significant amount of methane and carbon dioxide. The synthesis gas produced in this reaction having the H_2/CO ratio close to unity is a favorable feedstock for further chemical processes such as the Fischer–Tropsch synthesis and production of oxygenates followed by the liquid fuel synthesis [3-8]. Oxygenates derived from fast pyrolysis of biomass is a viable alternative to fossil sources for syngas generation via catalytic steam reforming (SR) [9-14].

The active catalysts for methane dry reforming (MDR) and steam reforming of oxygenates are usually based on the supported transition metals (Ni, Co, Cu, Cr) [15-26] or noble metals (Rh, Ru, Ir, Pt) [27-35]. The major problem of both processes hindering their industrial application is the coking of catalysts and, as a result, their deactivation especially in the case of Ni-containing catalysts [15-26]. Different approaches have been considered up to now to minimize the coke deposition on catalysts [4-13]. Among them an attractive option is the use of perovskite-like oxides ABO_3 as the catalyst precursors when formation of highly dispersed active metal (Ni, Ru, Co, their alloys etc) fixed in oxide matrix occurs under reaction conditions as a result of perovskite reduction [36-38]. The structural and thermal stability as well as catalytic activity of such systems can be adjusted by substitution with various cations in A and/or B sites [18-68]. For Ni-containing perovskite-type oxides of general formula $A_{1-x}A`_xNi_{1-y}B`_yO_3$ (A, A` - La, Ca, Sr, Ce, Pr; B` - Fe, Co, Ru; x, y - the substitution degree) as precursors of MDR catalysts, a high dispersion of metals depending on the nature of A, A` and/or B` cations as well as participation of mobile oxygen of oxide matrix in the reaction thus preventing the catalyst coking and increasing its activity and stability have been demonstrated [18-60]. For example, in the case of $La_{1-x}A_xNiO_3$ (A=Ce, Pr) perovskites, $La_{0.9}Pr_{0.1}NiO_3$ demonstrated the highest catalytic activity and resistance to coking in MDR explained by a higher dispersion of Ni in the reduced catalyst as well as redox properties of praseodymium oxide [55]. For $Ln_{1-x}Ca_xRu_{0.8}Ni_{0.2}O_3$ (Ln = La, Sm, Nd) catalysts, change of the nature and composition of cations in A-site was shown to strongly affect stability and selectivity of catalysts [58]. A partial substitution of Ni with Co [46-49], Fe [50-54] and/or Ru [56-62] leading to formation of Ni-Me alloys under reducing MDR reaction conditions improves catalyst activity and stability. Thus, for the $LaNi_{1-x}Ru_xO_3$ ($x=0 \div 1$) catalysts, partial substitution of Ni with Ru hampers coking [57]. In our study of $LnFe_{0.7-x}Ru_xNi_{0.3}0_{.3}O_{3-\delta}$ (Ln = La, Pr, Sm, $x = 0 \div 0.1$) perovskites [54-55] under realistic conditions of MDR, their transformation into composites comprised of Ni-Fe-(Ru) alloy particles and LnO_x epitaxially bound with remaining Ln-Fe-O perovskite particles was demonstrated. This microstructure plays an important role in MDR through activation of CO_2 on the oxide sites and transfer of active oxygen-containing species to alloy particles where they interact with CH_x fragments producing syngas. The nature of Ln cation affects both the composition of Ni-Fe-(Ru) alloy and oxygen mobility/reactivity in perovskite.

In the case of oxygenates reforming, perovskites as the catalyst precursors were reported only in few works [63-69]. Thus, $LaMO_3$ (M= Ni, Co, Fe, Mn) perovskites have been studied in authothermal reforming of ethanol [63]. The Ca-substituted $La_{1-x}Ca_xFe_{1-y}Ni_yO_3$ perovskite oxide supported nickel catalysts have shown very good stability and activity in steam reforming of ethanol [66].

This chapter overviews results of research within the national and International research projects aimed at design of nanocomposite active components of catalysts for dry reforming of methane and steam reforming of biofuels into syngas for synfuels production. A lot of attention was paid to systematic studies of the effect of chemical composition of perovskites and their nanocomposites with doped zirconia or ceria solid solutions on their real structure, oxygen mobility and reactivity considered being the most important factors controlling their performance. Factors determining activation/deactivation of catalysts due to reducing pretreatment or under reaction media effect as well as their catalytic properties were elucidated.

Nanocomposites possessing promising performance and coking stability in target reactions were supported on heat-conducting substrates and successfully tested in realistic conditions in pilot -scale reactors of CH_4 or oxygenates transformation into syngas.

2. Experimental

2.1. Synthesis

Perovskites of general formula $A_{1-x}A`_xNi_{1-y}B_y$ (A=La, Pr, Sm; A`=Sr, Pr, Ce; x=0-3; B= Fe, Mn, Ru, y= 0-0.7) and mixed manganite-chromite $La_{0.8}Pr_{0.2}Mn_{0.2}Cr_{0.8}O_3$ were synthesized by Pechini route using metal (Me) nitrates, $Ru(OH)Cl_3$, citric acid (CA), ethylene glycol (EG) and ethylenediamine (ED) as reagents [70-71]. The molar ratio of CA:EG:ED:Me was 3.75:11.25:3.75:1. CA and metal nitrates were dissolved in ethylene glycol at 80°C and in distilled water at room temperature, respectively. The prepared solutions were mixed together at room temperature under stirring followed by addition of ED. The mixed solution was stirred for 60 min and then heated at 70°C for 24 h allowing for the gel formation. The gel was calcined under air at 600°C for 2 h, then obtained solid was ground and annealed at 900°C.

The perovskites $LaCo_xFe_{1-x}O_3$ were prepared by thermal decomposition of mixed La, Co and Fe propionates. This preparation method is derived from the sol-gel method and based on the oligomerisation of propionate precursors in propionic acid [72]. The starting materials for La, Co and Fe were La acetate, Co acetate and Fe powder, which all led to propionate precursors in propionic acid. The three boiling solutions were mixed and propionic acid was evaporated until a resin was obtained. The resin was calcined under air at 700°C for 6 hours.

The amount of cobalt versus iron cations in the B sites of the perovskite was varied between x = 0 ($LaFeO_3$) and x = 1 ($LaCoO_3$).

The (100-x) wt.%$LnFeNi_{0.3}$+x wt.%$Ce_{0.9}Gd_{0.1}O_{2-\delta}$ (GDC) (x = 5 - 50wt.%) composites were prepared using a one-pot method and ultrasonic dispersion approach. In a one-pot method the composite was prepared similarly to perovskite samples by adding a corresponding quantity of cerium and gadolinium nitrates. In another approach, the mixed powders of GDC calcined at 500°C and $LnFeNi_{0.3}$ calcined at 700°C were ultrasonically dispersed in a solvent (distilled water, acetone, isopropanol) with addition of 2 wt% of a surfactant using a Ika T25 ULTRA-TURRAX basic (IKA, Germany) generator to prepare a composite. The dispersed mixture was filtered, dried and calcined under air.

The nanocomposites comprised of perovskite $La_{0.8}Pr_{0.2}Mn_{0.2}Cr_{0.8}O_3$ in combination with NiO and YSZ ($Y_{0.08}Zr_{0.92}O_{2-\delta}$) were synthesized using earlier described one-pot Pechini method [32, 73, 74]. Here, YSZ powder prepared via Pechini route and calcined at 600°C (specific surface area 70 m^2/g) was dispersed in mixed water solution of metal nitrates with addition of citric acid and ethylene glycol. After evaporation, the solid residue was decomposed in air at 500°C and calcined at 800°C for 2 h. Ru was supported on powdered nanocomposites by insipient wetness impregnation followed by drying and calcinations at 800°C.

2wt.% Ru and 1.15wt.% Ni were supported on $La_{0.8}Pr_{0.2}Mn_{0.2}Cr_{0.8}O_3$ powder from acetone solution with complex $[RuNO(NO_2)_3\mu\text{-}NO_2\ \mu\text{-}OHNi(H_2O)_4]$ by the incipient wetness impregnation followed by drying and calcination under air at 300°C.

To prepare heat-conducting metal substrates for structured catalysts of oxygenates reforming, open-cell nickel foams were manufactured by the nickel electroplating of the polyurethane foam samples (thickness 5 mm, the cell density 60 ppi) followed by sintering in the dissociated ammonia atmosphere at 1100°C for 1 h. The foam samples were then deformed by a uniaxial compression to 1 mm thickness modifying the cell morphology and decreasing porosity from 95.5% to 50 – 80%. Deformed foams were subjected to the pack aluminizing and then annealed at 1000 °C for 1 h under air to form a thin corundum layer over the foam cell walls for a better adhesion of catalytic layers [73].

To support active components on such substrates, powders of nanocomposite complex oxides were ultrasonically dispersed in isopropyl alcohol with addition of polyvinyl butyral as a binder to make slurry. Thin layers of composites were supported on substrates using these slurries followed by drying and calcination at 1100°C after each supporting step until loading of 4–7 wt.% was achieved. Precious metals (Pt, Ru) or $LaNi(Pt)O_3$ (up to 10 wt. %) were supported by the incipient wetness impregnation followed by drying and calcination under air at 800°C.

A stack comprised of 12 Ni-Al-foam plates and 11 sheets of Fecralloy gauzes (volume 34x34x34 mm^3) loaded with 4% (50 $La_{0.8}Pr_{0.2}Mn_{0.2}Cr_{0.8}O_3$ + 30 NiO + 20 YSZ) +0.7Ru was prepared using procedures described above [73].

For comparison, SiC monolithic substrate with square channels 1x1 mm and wall thickness ~ 0.1 mm usually applied for design of diesel particulate filters was used for preparation of structured catalysts of ethanol reforming using the same procedure as described above.

2.2. Characterization

2.2.1. Structural Features and Surface Properties

The samples of perovskites and composites were characterized by XRD, BET, TEM with EDX, H_2 and CH_4 temperature-programmed reduction. X-ray diffraction patterns were obtained with an ARLX`TRA diffractometer (Thermo, Switzerland) using CuKα radiation in 2θ scanning range 15–80° with step 0.05°. The structural refinement has been carried out using a PCW software (version 2.4). Qualitative phase analysis has been carried out by using PDF-2–ICDD files and ICSD/retrieve database. The X-ray particle sizes were estimated by using the Scherrer equation. BET specific surface area (SSA, m^2/g) was determined from the data on Ar thermodesorption. The TEM micrographs were obtained with a JEM-2010 instrument (lattice resolution 1.4Å and acceleration voltage 200 kV). Local elemental analysis was performed with EDX method (a Phoenix Spectrometer). The X-ray photoelectron spectra (XPS) were recorded on a photoelectron spectrometer KRATOS ES 300 with Mg K_α radiation (1253.6 eV) at pressure $1x10^{-8}$ Torr. Preliminarily all samples were ground up in an agate mortar and deposited on a conductive carbon tape.

2.2.2. Oxygen Mobility and Reactivity

The surface/bulk oxygen mobility for samples pretreated in O_2 at 500°C was characterized by the temperature-programmed desorption (TPD) in He flow. The redox properties of perovskites were studied using temperature-programmed reduction (TPR) by H_2 (10% H_2 in Ar, the feed rate 2.5 L/h, the temperature ramp from 25 to 900°C at 10°/min) or CH_4 (1% CH_4 in He, the feed rate 10 L/h, the temperature ramp from 25 to 880°C at 5°/min). The experiments were carried out in kinetic installations equipped with GC Tcvet-500 and PEM-2M analyzer (IR absorbance and electrochemical gas sensors). In the case of repeated red-ox cycles, samples were reoxidized in the flow of O_2 for 1 h at 500°C.

2.2.3. Catalytic Activity

Methane dry reforming (MDR) was studied in the temperature-programmed mode (1% CH_4 + 1% CO_2 in He, contact time 0.005 s) as well as in steady-state conditions (10%CH_4 + 10% CO_2 in He, contact time 0.01-0.015 s) at temperatures up to 800–850°C in a flow installation. The long-term testing was performed in the reaction mixture 45% CO_2 + 45% natural gas +10% N_2 at 800 °C and contact time 0.24 s. Before testing, the catalysts were activated in the reaction feed at 850 °C for 2 hours. The concentration of reactants and products were analyzed by using GC Tcvet-500. Temperature-programmed oxidation (TPO) in the feed of 1% O_2 in He was used to evaluate an amount of carbon formed over catalysts during long-term testing.

To estimate the reactivity of perovskites toward oxygenates reforming, for every sample a set of serial experiments were conducted using a flow quartz reactor (internal diameter 2.5 mm) packed with 0.3 g catalyst (0.25 - 0.5 mm fraction) diluted with a quartz sand. First, the sample pretreated in O_2 at 500 °C was subjected to temperature-programmed reduction (TPR) in the feed of 1% C_2H_5OH in He (the rate 10 L/h, the temperature ramp from 25 to 880°C at 5°/min) then the reactor was fast cooled to the room temperature and temperature-programmed reoxidation (TPO) in the feed of 1% H_2O in He (the feed rate 10 L/h, the temperature ramp from 25 to 880°C at 5°/min) was conducted. The catalytic activity of reoxidized samples was estimated in the ethanol steam reforming (ESR) using a feed 0.5vol.% C_2H_5OH + 2.vol.% H_2O in He at contact time 36 ms at 400 – 800 °C. The unreacted ethanol and liquid products were frozen in the trap at -40°C. The gaseous products (H_2, CO and CO_2) were continuously monitored by using on-line IR absorbance gas analyzer PEM-2M, TCD and an electrochemical H_2 sensor with the data acquisition and processing through a PC.

Catalytic experiments at high ethanol concentrations were carried out in U-shaped tubular quartz flow reactor (4.5 mm i.d.) at atmospheric pressure. Usually 0.18 g of catalyst (fraction 0.5-0.25 mm) diluted in a 1:10 weight ratio with quartz sand was taken. Reaction was performed at 600 – 800 °C using feed with volume composition $EtOH:H_2O:N_2$ = 1:4:5 fed into the reactor with the total flow rate 9 l/h. Before testing the catalysts were pretreated for 1 hour at 850 °C in the flow of 10 % H_2 in N_2 or in the reaction mixture. Three on-line gas chromatographs (GC) "LHM-8" equipped with thermal conductivity detectors and a flame ionization detector respectively were used for the analysis of reactants and products. Hydrocarbons (CH_4, C_2H_6, C_2H_4, etc.) and oxygenates (EtOH, CH_3OH, acetone, diethyl ether, etc) were analyzed using a Porapak T column; N_2, O_2, CH_4 and CO were analyzed with a molecular sieve column, and H_2, N_2, CO, CO_2, CH_4 – with an active carbon "SYT" column. Ar and He were used as carrier gases.

The ethanol steam reforming reaction over $LaCo_xFe_{1-x}O_3$ was carried out in a fixed bed reactor, at atmospheric pressure, using 160 mg of catalyst, heated at 2 C min^{-1} from room temperature to 550°C in an Ar/N_2 mixture (total flow 35 $mL.min^{-1}$, 4:1 mole ratio, GHSV~5000 h^{-1}), without any reduction treatment. After 1 hour at 550 °C, the ethanol/water mixture (water:ethanol =6:1 molar ratio) was introduced using the same Ar/N_2 mixture as a carrier gas (total GHSV 26000 h^{-1}, ethanol concentration 4.3%), and the reactive test was carried out at 550°C for 3 hours. The gas products were analysed on-line by micro-gas chromatography to identify the principal products and by-products (column Poraplot Q; column molecular sieve 5 Å).

Testing of monolithic catalysts in ethanol or glycerol (oxy)steam reforming using concentrated feeds was carried out using installation equipped with current-heated monolithic evaporation/mixing unit earlier described in details [74].

3. Structural and Textural Features of Catalysts

Characteristics of initial perovskites and their composites are presented in Tables 1-4. The BET specific surface areas of samples are in the range of 4-12 m^2/g which are comparable with the usual values obtained for such systems [50-53, 75].

According to XRD data (Table 1-3, Figures 1-3), the phase composition of initial perovskites depends on their composition.

Thus, $LnFe_{0.7}Ni_{0.3}O_3$ (Ln = La, Pr, Sr) samples are single phase orthorhombic perovskites, the lattice parameters being close to known values [76, 77] while $SmFe_{0.7}Ni_{0.3}O_3$ and $La_{0.9}Ce_{0.1}Fe_{0.7}Ni_{0.3}O_3$ samples contain a trace admixture of NiO. The observed differences in the lattice parameters are due to the difference in the size of Ln cations [78] (Table 1). For $SmFe_{0.7}Ni_{0.3}O_3$, splitting of reflections is observed due to the cooperative lattice distortion (Figure 1). A similar phenomenon was earlier observed for $SmFeO_3$

perovskite and explained by a strong distortion of Sm-oxygen polyhedra, distortion of the Fe-O octahedra being small [74]. The average X-ray particle size for all perovskites was estimated to be ~ 100 nm.

The phase composition and structural peculiarities of mixed ferrites –nickelates (cobaltites) depend on the Fe content (Fig, 1, Table 2). Thus, samples $La_{0.8}Sr_{0.2}Fe_{1-x}Ni_xO_3$ with x=0.1-0.3 calcined at 900 °C are the single phase perovskite of orthorhombic structure (Pnma space group) [79-83]. For $LSFN_{0.4}$, noticeable reflections of tetragonal phase (T) and NiO are observed, and the structure of the perovskite is rhombohedral (Table 2). For $LaCo_xFe_{1-x}O_3$ system, the perovskite structure for x < 0.5 corresponds to that of $LaFeO_3$ (orthorhombic), while being isomorphic to $LaCoO_3$ (rhombohedral) for x ≥ 0.5 (Figure 1b).

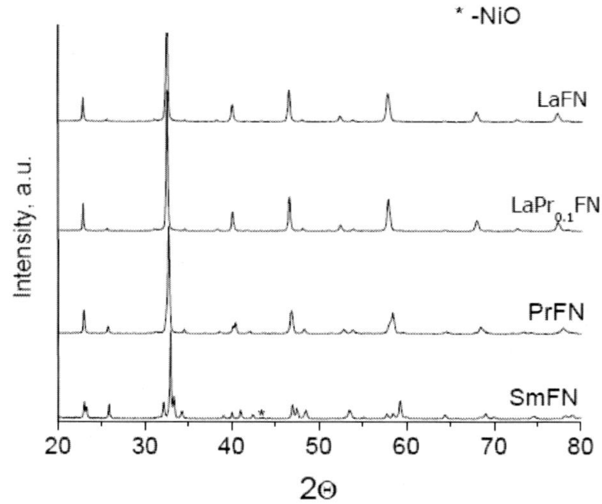

(a)

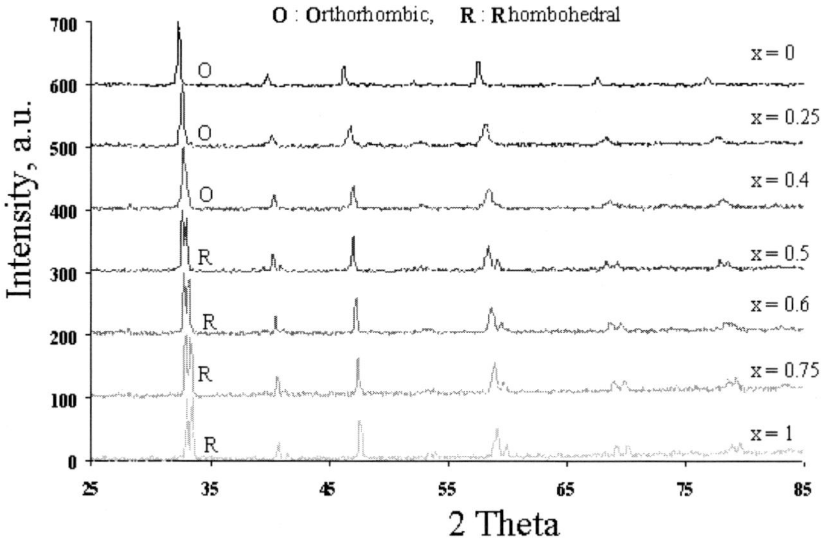

(b)

Figure 1. (Continued).

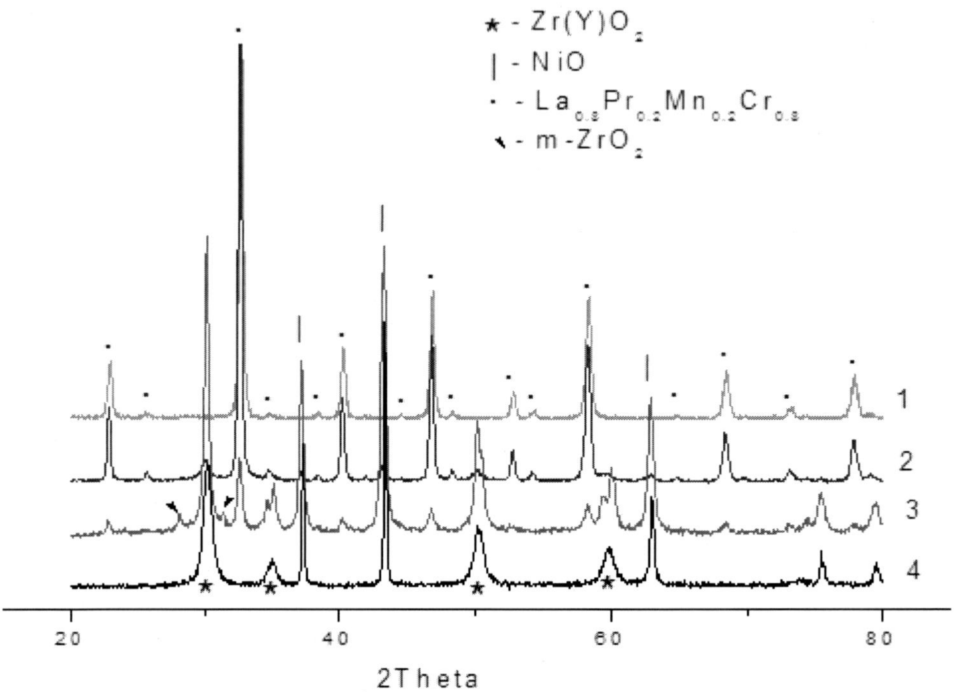

(c)

Figure 1. XRD patterns of as-prepared catalysts:
a - $LnFe_{0.7}Ni_{0.3}O_{3-\delta}$ perovskites calcined at 900 °C. b -$LaCo_xFe_{1-x}O_3$ perovskites calcined at 700 °C; c - $La_{0.8}Pr_{0.2}Mn_{0.2}Cr_{0.8}O_3$ (1) and 60%NiO + 40%YSZ (4) calcined at 900 °C and their nanocomposites comprised of 80% $La_{0.8}Pr_{0.2}Mn_{0.2}Cr_{0.8}O_3$+ 10%NiO + 10%YSZ (2) and 10% $La_{0.8}Pr_{0.2}Mn_{0.2}Cr_{0.8}O_3$+ 45% NiO + 45%YSZ (3).

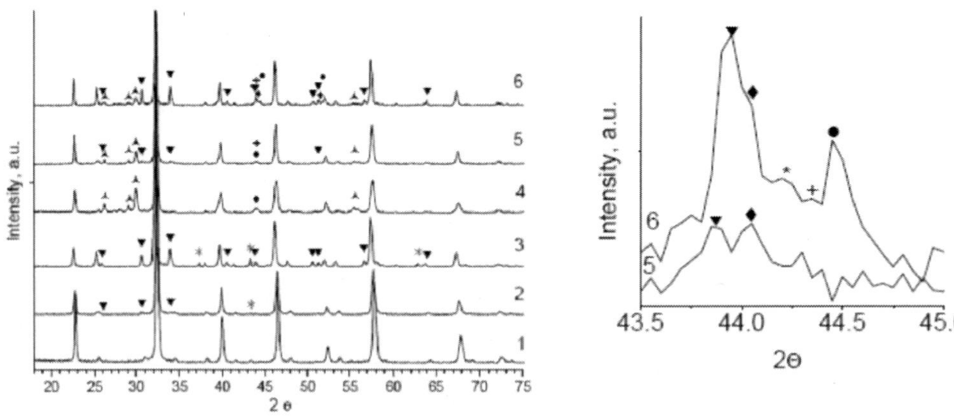

Figure 2. XRD patterns of $LaFe_{0.7-x}Ru_xNi_{0.3}O_{3-\delta}$: as prepared (1-3); reduced in H_2 (4-6); x=0 (1, 4); x=0.05 (2, 5); x= 0.1 (3, 6). NiO (✱), LaOCl (▼), La_2O_3 (⅄), Ni-Fe alloy (♦), Ni-Fe-Ru alloy (*), Ru (+), Ru-Ni alloy (●). Reduction conditions: 5% H_2 in He, 800°C, 2 h. Adapted from refr. [55].

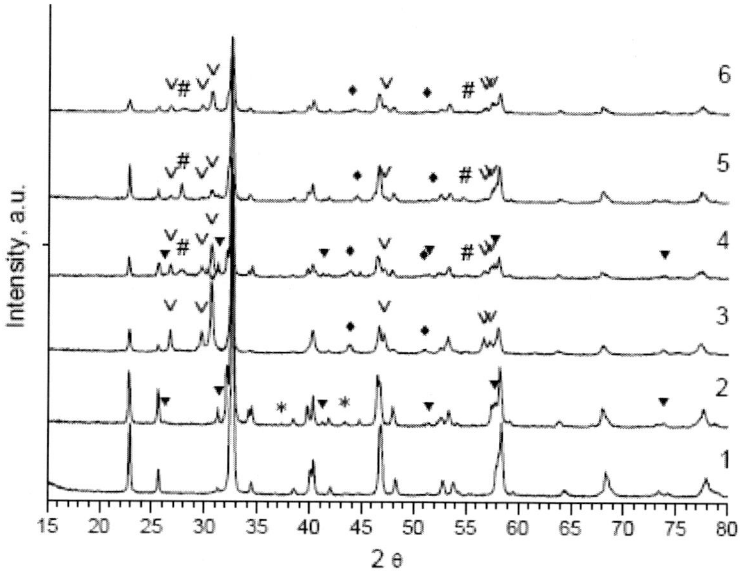

Figure 3. XRD patterns of $PrFe_{0.7}Ni_{0.3}O_{3-\delta}$ (1, 3, 5) and $PrFe_{0.6}Ru_{0.1}Ni_{0.3}O_{3-\delta}$ (2, 4, 6): as prepared (1, 2), reduced in H_2 (3, 4) and used in MDR (5, 6). NiO (✱), PrOCl (▼), Pr_2O_3 (v), Pr_xO_y (#), Ni-Fe alloy (♦). Reduction conditions: 5% H_2 in He, 800°C, 2 h. MDR conditions: 800°C in the feed 50 % CO_2 + 50% natural gas, contact time 0.2 s. Adapted from refr. [55].

Table 1. Unit cell parameters[1], secondary phases[1] and BET specific surface area [2] for perovskites $LnFe_{0.7-x}Ni_{0.3}Ru_xO_{3-\delta}$ (Ln=La, Pr, Sm, x=0.05, 0.1) calcined at 900 °C

Perovskite	Ln ionic radius[3] (Å)	*a* (Å)	*b* (Å)	*c* (Å)	*V* (Å^3)	Secondary phase	SSA, m^2/g
LaFN	1.36 (La^{3+})	5.511	7.808	5.537	59.56	-	7
$La_{0.8}Sr_{0.2}$FN		5.530	7.775	5.486	58.97	-	12
$La_{0.9}Pr_{0.1}$FN		5.500	7.809	5.462	58.65	-	4
$La_{0.9}Ce_{0.1}$FN		5.519	7.801	5.545	59.68	CeO_2, NiO	10
PrFN	1.31(Pr^{3+})	5.522	7.742	5.462	58.38	-	4
SmFN	1.24(Sm^{3+})	5.572	7.657	5.376	57.34	NiO	5
$LaFNRu_{0.05}$		5.529	7.817	5.547	59.94	LaOCl, NiO	7
$PrFNRu_{0.05}$		5.574	7.779	5.480	59.40	PrOCl, NiO	7
$PrFNRu_{0.1}$		5.584	7.777	5.478	59.47	PrOCl, NiO	4

[1] - The structural refinement has been carried out using a PCW software (version 2.4). Qualitative phase analysis has been carried out by using PDF-2–ICDD files and ICSD/retrieve database; [2]- from the data on Ar thermodesorption; [3]- by [78].

The partial substitution of Fe with Ru ($LnFe_{0.7-x}Ni_{0.3}Ru_xO_{3-\delta}$; Ln=La, Pr; x=0.05, 0.1) leads to appearance of NiO and praseodymium or lanthanum oxychloride [JCPDF 88-0064] phase and distortion of perovskite structure (Table 1, Figures 2, 3). The formation of oxychlorides is caused by using $Ru(OH)Cl_3$ as a source of Ru that is confirmed by the increase of the oxychloride reflections intensity at increasing Ru concentration being more pronounced for La-containing samples. Simultaneously, especially in the case of $LaFe_{0.7-}$

${}_xNi_{0.3}Ru_xO_{3-\delta}$ samples, the noticeable growth of NiO reflections is observed (Figure 2). The shift of all perovskite reflections towards lower angles shows the unit cell expansion due to partial segregation of nickel from perovskite and insertion of larger Ru^{3+} cations into the structure (0.068 nm versus 0.06 and 0.0645 nm for Ni^{3+} and Fe^{3+} [84]).

Table 2. Specific surface area, phase composition and structural characteristics of $La_{0.8}Sr_{0.2}Fe_{1-x}Ni_xO_{3-\delta}$ perovskites calcined at 900 °C

Perovskite	*a* (Å)	*b* (Å)	*c* (Å)	*V* (Å³)	Secondary phase	SSA, m²/g
$LSFNi_{0.1}$	5.520	7.810	5.550	5.520	-	11
$LSFNi_{0.2}$	5.534	7.815	5.505	5.534	-	12
$LSFNi_{0.3}$	5.530	7.775	5.486	5.530	-	12
$LSFNi_{0.4}$	5.502	13.364	5.502	5.502	P, T, NiO	12

Table 3. X-ray size and lattice parameter of Fe-Ni alloy for perovskites $LnFe_{0.7}Ni_{0.3}O_{3-\delta}$ with different Ln cations reduced at 800 °C in H_2

Ln	La	LaPr	Pr
D_{alloy} (Å)	175	115	160
Lattice parameter of alloy (Å)	3.567	3.560	3.579

After reduction at 600 °C in H_2, perovskites $LnFe_{0.7}Ni_{0.3}O_3$ (Ln = La, Pr) retain their structure but some shift of all reflections towards lower angles is observed showing the unit cell expansion due to reduction of Fe^{4+} and Ni^{3+} ions to larger Fe^{3+} and Ni^{2+} cations [77]. The XRD patterns of perovskites reduced in H_2 at 800 °C are presented in Figure 4.

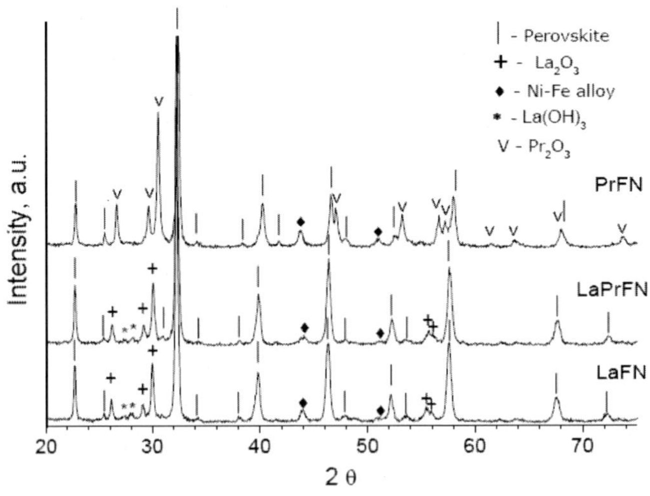

Figure 4. XRD patterns of perovskites $LnFe_{0.7}Ni_{0.3}O_{3-\delta}$ reduced at 800°C in H_2. Adapted from refr. [54].

For all samples, the observed phases are Ni-Fe alloy, remaining perovskites with changed parameters and La_2O_3, Pr_2O_3 or Sm_2O_3 depending on the composition of starting samples. The lattice parameter of Ni-Fe alloy (Table 3) changes with the Ln cation nature but in all cases it is intermediate between those of pure metal Ni face-centered cubic lattice - 3.5240 Å [JCPDF 65-2865] and $Fe_{0.64}Ni_{0.36}$ alloy - 3.5922 Å [JCPDF 47-1405]. For the $LaFNi_{0.3}$ and $LaPrFNi_{0.3}$ precursors, the Ni-Fe alloy composition is close (Table 3) while for PrFN, a larger parameter indicates that the alloy is highly enriched by Fe. The X-ray alloy particle size is smaller for Pr-containing samples.

XRD patterns of Ru-containing perovskites $LnFe_{0.7-x}Ni_{0.3}Ru_xO_{3-\delta}$ reduced in H_2 at 800 °C (Figures 2, 3) show the reflections in the range of $2\theta \sim$ 44-44.5° corresponding to Ni-Fe(Ru)-alloy [54] and metal Ru [59], as well as those of orthorhombic perovskite, rare-earth oxides (La_2O_3 [JCPDF 83-1344] or Pr_2O_3, Pr_xO_y) and lanthanum or praseodimium oxychloride. The position and intensity of perovskite reflections are practically the same in the initial and reduced samples showing, in agreement with literature data [58-60], that reduction of Ru-containing perovskites is hampered as compared with Ru-free samples. However, the presence of La_2O_3(Pr_2O_3, Pr_xO_y) reflections indicates some decomposition of perovskites considering the invariance of La(OH)Cl_2 reflections and only some decrease of Pr(OH)Cl_2 reflection intensity after reduction. The reflections of metal phase are broad and of a low intensity that makes difficult their analysis. However, some information on the composition of metal phase could be deduced. The reflections of Ni-Fe alloy shift to higher angles with the increase of Ru concentration due to its incorporation into alloy. Thus, in the case of $LaFe_{0.65}Ni_{0.3}Ru_{0.05}O_{3-\delta}$ sample, the position of alloy reflection is rather close to that for Ru-free Ni-Fe alloy (2θ = 44°) showing only a small content of Ru in alloy. In the XRD pattern of $LaFe_{0.6}Ni_{0.3}Ru_{0.1}O_{3-\delta}$ sample, the reflection at 2θ = 44.2° indicates the presence of Ni-Fe alloy enriched with Ru, the reflection at 44.35° corresponds to pure metal Ru and the reflection at 44.5° corresponds to Ni-Ru alloy. The presence of metal Ru in this sample implies that a part of ruthenium in the oxidized sample is not incorporated into the perovskite structure, though in the XRD pattern of the initial $LaFe_{0.6}Ni_{0.3}Ru_{0.1}O_{3-\delta}$ perovskite any reflections of individual ruthenium oxide are not detected.

XRD data for $LnFe_{0.7}Ni_{0.3}O_{3-\delta}$, GDC and their composites prepared via a one-pot method calcined at 900°C are presented in Table 4 and Figures 5-6.

Table 4. Phase composition and BET for composites $xLnFe_{0.7}Ni_{0.3}O_{3-\delta}$ +yGDC (Ln= Pr, La, Sm, x = 50 - 95%, y = 5 - 50%) prepared via one-pot method and calcined at 900 °C

Sample	Phase composition	SSA (m^2/g)
95%PrFN+5%GDC	P, NiO, CeO_2	6
90%PrFN+10%GDC	P, NiO, CeO_2	6.6
80%PrFN+20%GDC	P, NiO, CeO_2	7
50%PrFN+50%GDC	P, NiO, CeO_2	4
50%PFN+50%GDC(USD)*	P, CeO_2	6
90%LaFN+10%GDC	P, CeO_2	8
90%LaPrFN+10%GDC	P, CeO_2	8
90%SmFN+10%GDC	P, NiO, CeO_2	8

* - Composite was prepared by ultrasonic dispersion of perovskite and GDC powders.

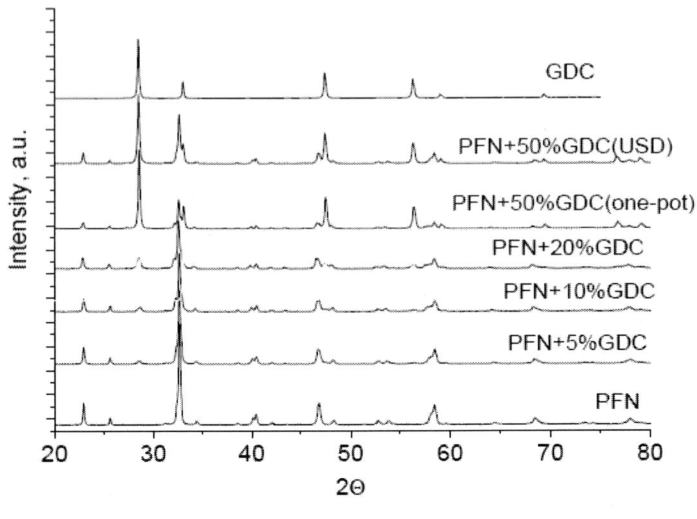

Figure 5. XRD patterns of $PrFe_{0.7}Ni_{0.3}O_{3-\delta}$, GDC and their composites prepared via one-pot method or ultrasonic dispersion and calcined at 900 °C.

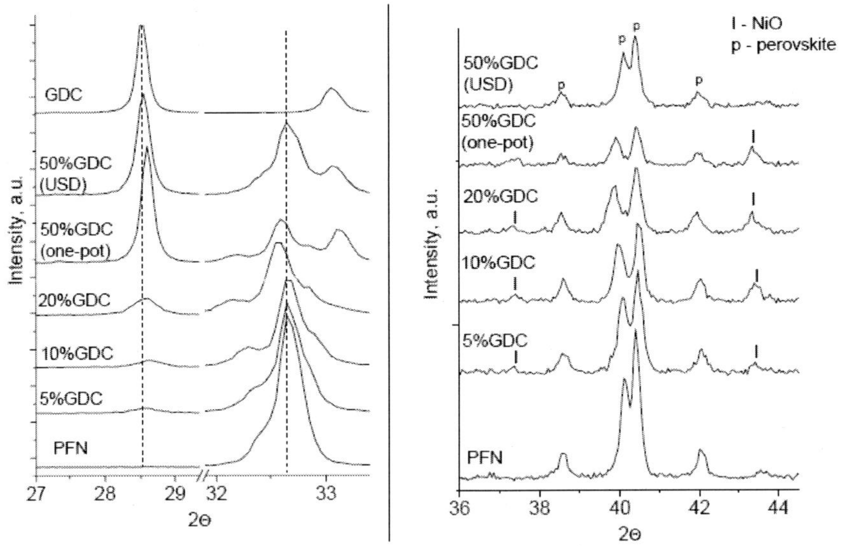

Figure 6. Enlarged 27 - 33° and 36 - 44° 2θ regions of XRD patterns of $PrFe_{0.7}Ni_{0.3}O_{3-\delta}$, GDC and their composites prepared via one-pot method or ultrasonic dispersion and calcined at 900 °C.

The phase composition of samples depends on the Ln-cation nature. Thus, for La-containing systems, only perovskite and cubic oxide phase are present while NiO is also observed in the case of Pr- and Sm-containing samples (Table 4).

For all (100-x) wt.%$PrFe_{0.7}Ni_{0.3}O_{3-\delta}$+x wt.%GDC composites, the reflections of perovskite, cubic GDC phase and NiO are observed (Figures 5-6). Reflections of GDC rise in intensity and are shifted with increasing GDC concentration. Simultaneously, the intensity of perovskite reflections decreases while their shift is noticeable only for composites containing 20 - 50% GDC (Figures 5-6).

The changes in the reflection positions are determined by cations interchange between perovskite and fluorite oxide phases in the course of a one-pot synthesis. For the sample 50%$PrFe_{0.7}Ni_{0.3}O_{3-\delta}$+50%GDC prepared via ultrasonic dispersion, the position of perovskite and GDC reflections are not changed and reflections of NiO are absent showing a weaker interaction between phases in this composite [85].

For $La_{0.8}Pr_{0.2}Mn_{0.2}Cr_{0.8}O_3$+NiO+YSZ nanocomposite prepared via modification of Pechini route, interaction between YSZ and perovskite phase is reflected also in appearance of reflections corresponding to monoclinic ZrO_2 phase (Figure 1c). Hence, leaching of small Y cations from YSZ and their incorporation into perovskite structure apparently occur [85].

4. Reactivity of Catalysts

4.1. Temperature-programmed Reduction by H_2

The H_2 TPR profiles of $LnFNi_{0.3}$ perovskites are presented in Figure 7. In all cases, three main peaks are observed in the patterns: the intense low-temperature peak with the maximum at 317 – 354 °C, the low intensity peak at 386 - 445 °C and an unresolved intense high-temperature peak above 600 °C.

The first low-temperature peak is assigned to reduction of Ni^{3+} to Ni^{2+}, and the second one corresponds to reduction of Fe^{4+} cations to Fe^{3+} state within the perovskite lattice [50-54]. The high-temperature peak corresponds to the reduction of perovskite with its partial destruction leading to formation of Ni-Fe alloy and corresponding oxides (Figure 4). In going from LaFN to SmFN, the low-temperature peaks are shifted to higher temperature while the high-temperature peaks are shifted to lower temperatures.

The trend of increasing the temperature of a partial Ni^{3+} (Fe^{4+}) cations reduction in this order agrees with the data for $LaFeO_3$, $NdFeO_3$ and $SmFeO_3$ reported in [75, 77]. Such a trend has been explained by the decrease of the Fe-O bond distances from $LaFeO_3$ to $SmFeO_3$ estimated from the cell size and, thus, by the increase of the Fe-O bond strength. Since for $LnFe_{0.7}Ni_{0.3}O_3$ the cell volume also decreases in the order from La to Sm (Table 1), it is reasonable to suppose the same reason for more difficult reduction of Ni^{3+}(Fe^{4+}) cations for SmFN perovskite.

The shift of high-temperature peak to lower temperatures is determined by the decrease of perovskite lattice stability in going from LaFN to SmFN. LaFN and PrFN having less disordered structures are more stable to reduction as compared with SmFN perovskite (Figure 4) with the most distorted Sm-O polyhedra [77].

The spectra profiles and peak positions depend on the Ni concentration. Thus, for $La_{0.8}Sr_{0.2}Fe_{1-x}Ni_xO_{3-\delta}$ the low-temperature peak shifts to lower temperature at increasing Ni concentration from 10 to 30at.% (Figure 7b).

At x = 0.4 shift of this peak to higher temperature is observed which is determined by the presence of NiO (Table 2). Similar spectra were earlier obtained for $La_{1-x}Sr_xNiO_3$ [41-45] and $LaNi_xFe_{1-x}O_3$ [50-54] perovskites.

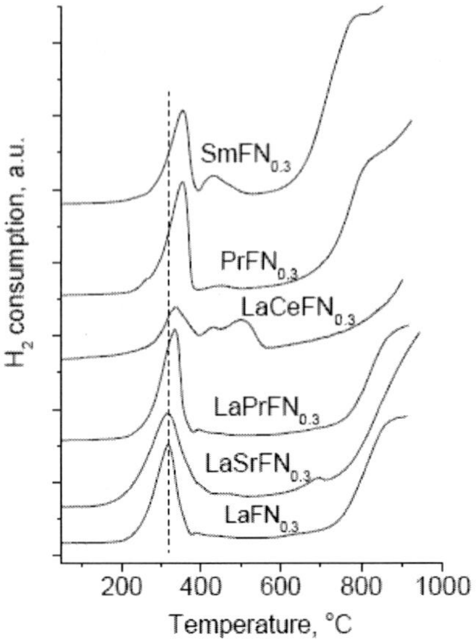

(a)

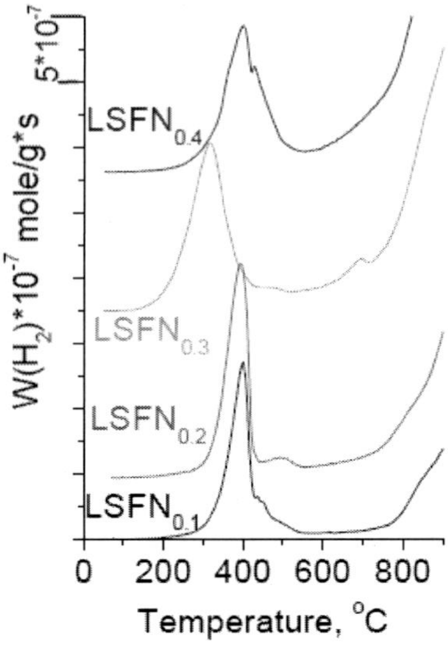

(b)

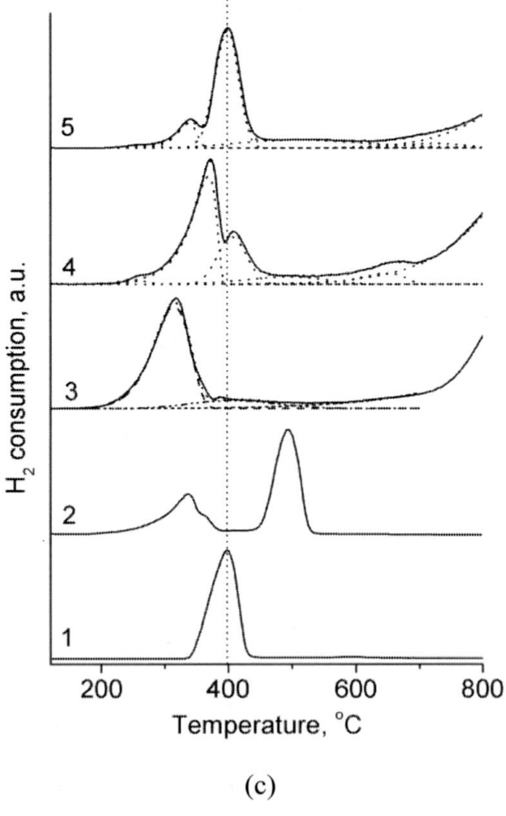

(c)

Figure 7. H_2-TPR profiles for (a) $LnFe_{0.7}Ni_{0.3}O_{3-\delta}$ perovskites calcined at 900 °C, (b) $La_{0.8}Sr_{0.2}Fe_{1-x}Ni_xO_{3-\delta}$ perovskites and (c) NiO (1), $LaNiO_3$ (2), $LaFe_{0.7}Ni_{0.3}O_{3-\delta}$ (3), $LaFe_{0.65}Ru_{0.05}Ni_{0.3}O_{3-\delta}$ (4), $LaFe_{0.6}Ru_{0.1}Ni_{0.3}O_{3-\delta}$ (5). Adapted from refr. [55].

H_2-TPR patterns for NiO, $LaNiO_3$ and complex nickelates-ferrites are presented in Figure 7c. A single peak with a maximum at 400°C is observed in the H_2 TPR pattern of NiO whereas $LaNiO_3$ shows two peaks: the first one centered near 337° is assigned to the reduction of Ni^{3+} to Ni^{2+} and the second peak at 495° corresponds to the reduction of Ni^{2+} to metallic nickel [47, 53, 60]. For $LaFe_{0.7}Ni_{0.3}O_{3-\delta}$, in agreement with known data [50-53], the first peak shifts to lower temperatures and the second one (corresponding to a partial reduction of perovskite) shifts to higher temperatures (Figure 7). Furthermore, a broad low-intensity peak in the range of 300-600° is present which could be assigned to the reduction of NiO traces not detected by XRD and the partial reduction of Fe^{4+} to Fe^{3+} [50-54]. For $LaFe_{0.7-x}Ni_{0.3}Ru_xO_{3-\delta}$ samples, in the low-temperature region, three peaks of hydrogen consumption are observed: the low intensity peak at 260 - 270° and two overlapping peaks with the maximum at 315 - 370°C and 400 – 410 °C. The first peak and the peak at 315 - 370 °C could be assigned to reduction of Ni^{3+} to Ni^{2+} and Ru^{3+} to Ru^{2+} in perovskite [60, 87]. The peak at 400 - 410 °C corresponds to the reduction of NiO: the increase of its intensity at x=0.1 agrees with XRD data evidencing a higher NiO content in this sample (Figure 2). The low-intensity peaks in the 500 – 700 °C region could be assigned to the reduction of Fe^{4+} cations to Fe^{3+} within the perovskite lattice [51-54]. An unresolved intense high-temperature peak above 700°C is attributed to the partial destruction of perovskite structure with the formation of metallic phase and La_2O_3 confirmed by XRD data (Figure 2). In the case of $PrFe_{0.7-x}Ni_{0.3}Ru_xO_{3-\delta}$ series, the TPR profiles are similar (Figure 8), however, as compared with La-based perovskites, the low-temperature peaks are somewhat shifted to higher temperatures while the high-temperature peaks are shifted to lower temperatures. The higher temperature of the partial reduction of Ni^{3+} (Ru^{3+}) cations in Pr-based perovskites agrees with the data for Ru-free systems [75, 77]. The shift of the high-temperature peak to lower temperatures is determined by the lower lattice stability of Pr-containing perovskites [77].

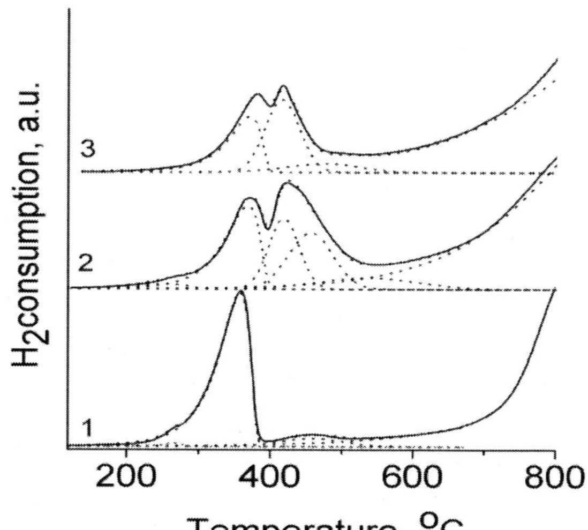

Figure 8. H_2 TPR profiles for $PrFe_{0.7-x}Ru_xNi_{0.3}O_{3-\delta}$: 1 – x=0; 2 – x=0.05; 3 - x=0.1. Adapted from refr. [55].

In general, when ruthenium is incorporated into perovskite, the peak corresponding to the partial reduction of Ni^{3+} (Ru^{3+}) is noticeably shifted to higher temperatures (from 315 to 370 °C and from 354 to 380 °C for $LaFe_{0.7-x}Ni_{0.3}Ru_xO_{3-\delta}$ and $PrFe_{0.7-x}Ni_{0.3}Ru_xO_{3-\delta}$, respectively (Figures 7, 8). A higher stability of both nickel and ruthenium in the perovskite lattice suggests strong inter-cations interaction which agrees with published data [60]. Some poisoning effect of Cl^- anions in the surface layer is possible as well.

Repeated reduction-reoxidation experiments (Figure 9) revealed some evolution of reactivity of perovskites apparently reflecting variation of their real structure/microstructure. In the case of perovskites without Ru reoxidized after 1st TPR run, the first peak at ~ 315 °C remains only as a shoulder of a new intense asymmetric peak with the main maximum at ~ 420 °C. Since the amount of hydrogen consumed in these low-temperature peaks corresponds to complete reduction of all Ni in perovskite to Ni^0 state, this clearly indicates the increase of sample reactivity after reoxidation (Figure 9a). Nearly identical H_2 TPR spectra for the second and the third runs demonstrate reproducibility of sample reactivity after reoxidation. Since reduction of both NiO and Ni-containing perovskites to metallic phase is a topochemical process including generation of Ni^0 nuclei followed by their growth [85], observed variation of perovskites reactivity after red-ox cycles can be explained by microheterogeneity of reoxidized perovskites caused by preferential location of Ni cations in the surface layer and within domain boundaries of defect La ferrite. For $LaFe_{0.6}Ni_{0.3}Ru_{0.1}O_{3-\delta}$ sample, TPR spectra change from run to run reflecting the increase of reactivity (Figure 9b). In the third run, practically all nickel and ruthenium cations contained in the sample are reduced to metal state in the low-temperature range. Since position of the main peak at ~ 400 °C is the same as for $LaFe_{0.7}Ni_{0.3}O_{3-\delta}$ sample, it clearly corresponds to the topochemical process of NiO_x species reduction. Peaks at 180 - 230 °C are due to reduction of RuO_x species formed as a result of ruthenium segregation from the perovskite during the first reduction [59-60], while peak at 340 - 390 °C can be assigned to reduction of mixed Ni-Ru oxide species [88]. The driving force for the increase of Ru-doped sample reactivity in red-ox cycles is clearly segregation of Ni and Ru in the surface layers as well as Cl removal from the surface.

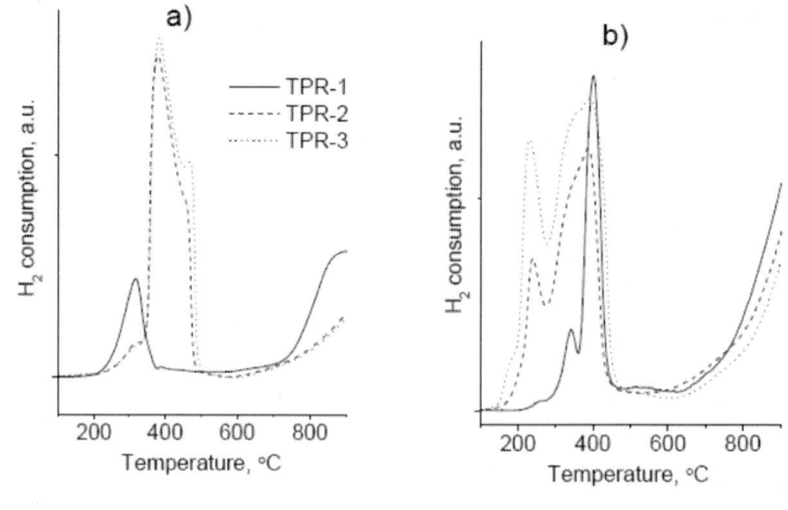

Figure 9. Repeated H_2 TPR after oxidation for $LaFe_{0.7}Ni_{0.3}O_{3-\delta}$ (a) and $LaFe_{0.6}Ru_{0.1}Ni_{0.3}O_{3-\delta}$ (b). Adapted from refr. [55].

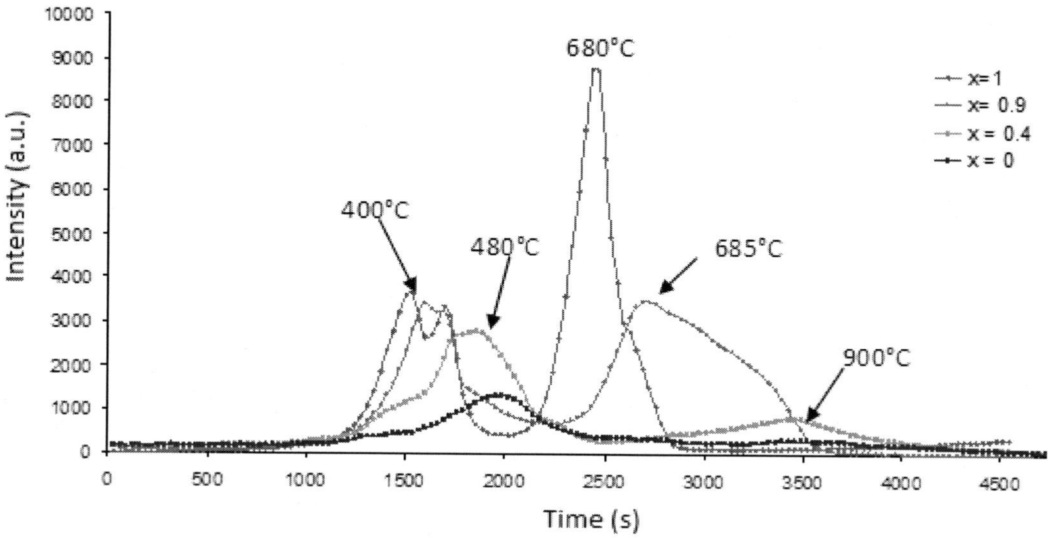

Figure 10. H_2 TPR profiles of $LaCo_xFe_{1-x}O_3$ catalysts.

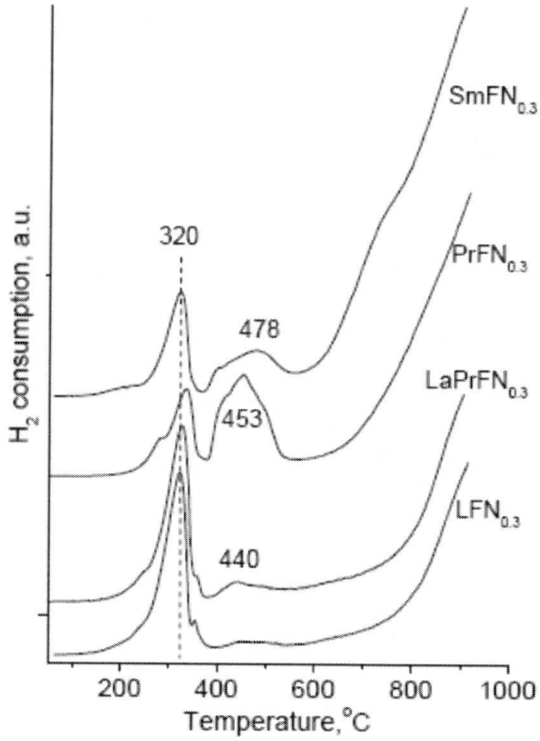

Figure 11. H_2-TPR profiles for the composites 90 wt.% $LnFe_{0.7}Ni_{0.3}O_{3-\delta}$ +10 wt.% GDC prepared via one-pot method and calcined at 900 °C.

The reducibility of $LaCo_xFe_{1-x}O_3$ (Figure 10) is strongly affected by the perovskite composition. At 900 °C the reduction of cobalt cations is almost total for $LaCoO_3$ (94% of reduction) whereas the percentage of reduction of Fe^{3+} is only 16% for $LaFeO_3$. Nevertheless, for the two types of cations the reduction occurs in two distinct steps: a first step at low temperature, between 350 and 550 °C depending on the Co/Fe contents, and a second step after 650 °C. The higher is the cobalt content, the lower are the reduction temperatures.

H_2–TPR spectra of composites 90 wt.% $LnFe_{0.7}Ni_{0.3}O_{3-\delta}$ +10 wt. % GDC prepared via one-pot method and calcined at 900 °C are presented in Figure 11.

For La-containing samples, the spectra profiles are similar to those for corresponding perovskites (Figure 7). The spectra of Pr- and Sm-containing composites show a broad peak at 400 - 500 °C which could be assigned to reduction of NiO admixture present in samples according to XRD data (Table 4).

H_2-TPR profiles for the composites (100-x) wt.% $PrFe_{0.7}Ni_{0.3}O_{3-\delta}$ +x wt.% GDC prepared via one-pot method and 50 wt.%$PrFe_{0.7}Ni_{0.3}O_{3-\delta}$ +50 wt.%GDC prepared by ultrasonic dispersion and calcined at 900 °C are shown in Figure 12.

For composites prepared via one-pot method, at increasing GDC content, the low-temperature peak assigned to the reduction of Ni^{3+} to Ni^{2+} shifts from 350 °C ($PrFe_{0.7}Ni_{0.3}O_{3-\delta}$) to 317°C (80 wt.% $PrFe_{0.7}Ni_{0.3}O_{3-\delta}$ +20 wt. % GDC) and, simultaneously, a shoulder at 280°C appears which could be related to the reduction of active oxygen of GDC. The broad peak at 450°C is assigned to reduction of NiO.

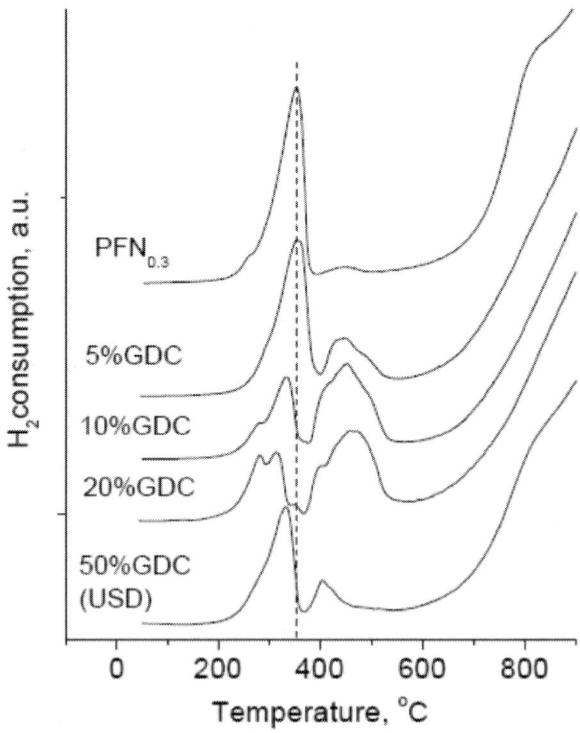

Figure 12. H_2-TPR profiles for the composites (100-x) wt.% $PrFe_{0.7}Ni_{0.3}O_{3-\delta}$ +x wt.% GDC prepared via one-pot method and 50 wt.% $PrFe_{0.7}Ni_{0.3}O_{3-\delta}$ +50 wt.% GDC prepared by ultrasonic dispersion and calcined at 900 °C.

4.2. Temperature-programmed Reduction by CH_4

Temperature-programmed reduction of perovskites by methane usually proceeds at higher temperatures as compared with the reduction by H_2 due to a strong C-H bond in CH_4 molecule. Typical profiles of CH_4 and product concentration variation during TPR for perovskite are presented in Figures 13, 14.

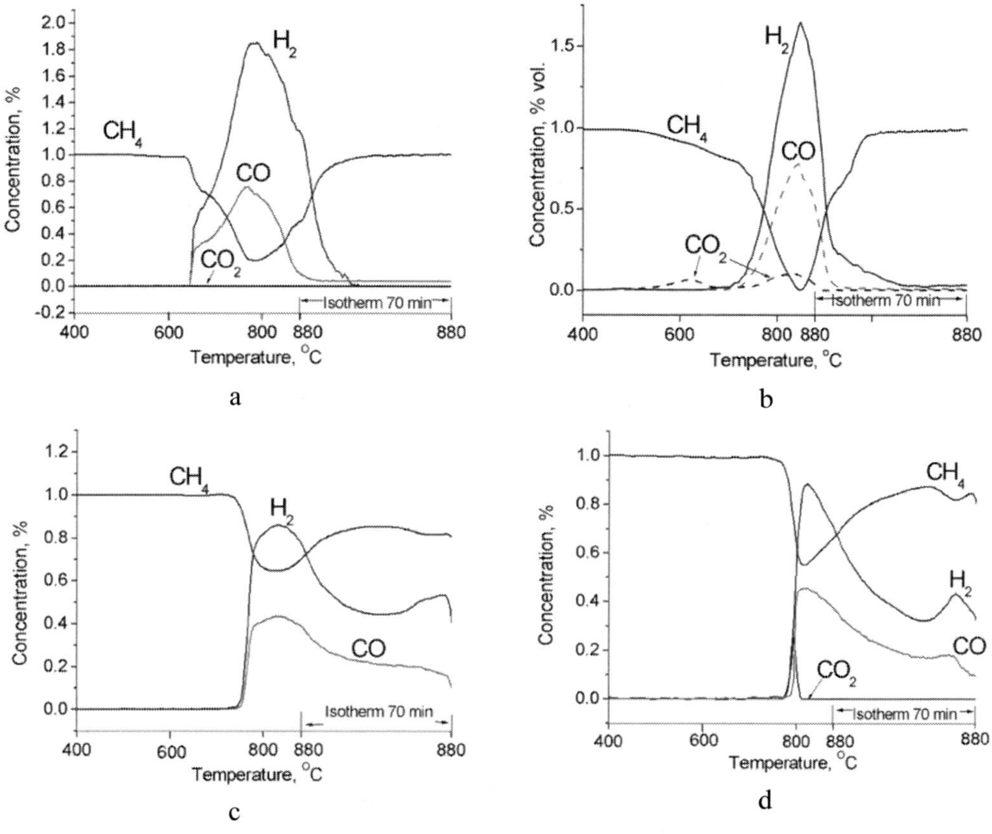

Figure 13. CH_4 TPR for $LnFe_{0.7-x}Ru_xNi_{0.3}O_{3-\delta}$ perovskites :
a) $LaFe_{0.7}RuNi_{0.3}O_3$; b) $PrFe_{0.7}RuNi_{0.3}O_3$; c) $LaFe_{0.65}Ru_{0.05}Ni_{0.3}O_3$; d) $PrFe_{0.65}Ru_{0.05}Ni_{0.3}O_3$. Adapted from refr. [55].

Reduction starts by formation of deep oxidation products (H_2O and CO_2) at 480 - 540 °C as a result of interaction of activated CH_x species with highly reactive surface oxygen. At 620–700 °C, CO and H_2 start to evolve due to interaction with strongly bound surface oxygen replenished by the lattice oxygen migration from the bulk of particles. These peaks along with a peak of CO_2 evolution are situated at 700 – 850 °C (Figures 13, 14). The complete reduction of perovskite proceeds with formation of Fe-Ni alloy and corresponding Ln oxide that is confirmed by XRD [54]. The temperatures of maximum rates of the product formation and their values for all LnFN perovskites are presented in Figures 15, 16. The relation among the rates of products formation evidently changes with the nature of Ln cation determining reactivity and mobility of the surface/bulk oxygen of perovskites. The profiles of CH_4 and products concentration during TPR for Ru-containing perovskites are presented in Figure 13 (c, d). The results show that incorporation of Ru in the perovskite strongly affects dynamics of its reduction. Thus, for $LaFe_{0.7}Ni_{0.3}O_{3-\delta}$ and $PrFe_{0.7}Ni_{0.3}O_{3-\delta}$ (Figure 13 a, b) reduction starts by formation of deep oxidation products (H_2O and/or CO_2) at 480 – 540 °C. CO and H_2 evolution begins at 620 - 700 °C. Reduction is completed during TPR run resulting in segregation of Fe-Ni alloy and corresponding Ln oxide within remaining perovskite (Figures 2-4). For all Ru-containing perovskites, the start of reduction product evolution shifts to 730 – 810 °C, and concentration of CO and H_2 passed through the maximum slowly declining with time (Figure 14 c, d). In agreement with H_2 TPR data, this evidences hampering of their reduction as compared with Ru-free perovskites.

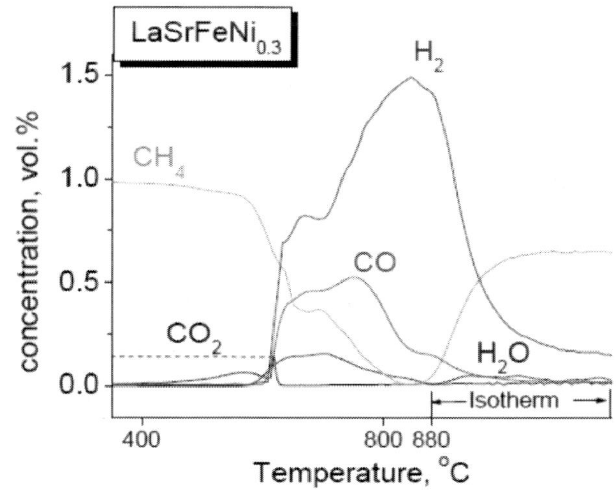

Figure 14. CH_4 TPR for perovskite $La_{0.8}Sr_{0.2}Fe_{0.7}Ni_{0.3}O_{3-\delta}$.

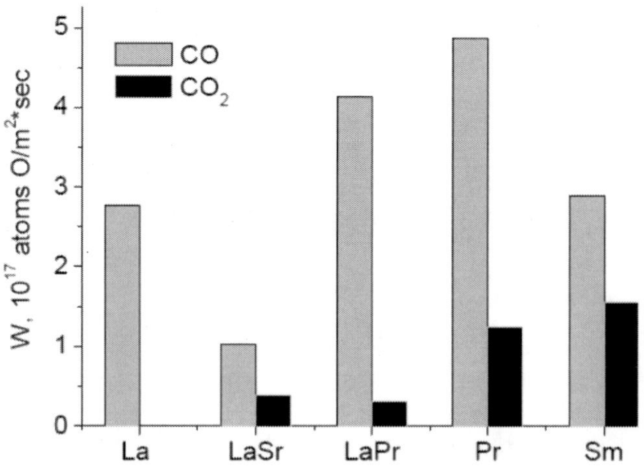

Figure 15. CH_4-TPR: maximum rates of CO and CO_2 formation for $LnFe_{0.7}Ni_{0.3}O_{3-\delta}$ perovskites. Adapted from refr. [54].

This could be caused by two reasons: inter-cation interaction stabilizing both nickel and ruthenium cations in the perovskite lattice [60] and the presence of oxychlorides which could block the surface of perovskite hampering the active metal formation required for methane activation. Further, the dynamics of perovskite reduction and composition of products are affected by the nature of Ln cation. Thus, for $PrFe_{0.65}Ru_{0.05}Ni_{0.3}O_{3-\delta}$, though the reduction products appear at higher temperature as compared with La-based perovskite (775 and 730 °C, respectively), the maximum rate of CO and H_2 formation is achieved at somewhat lower temperature (820 °C vs 830 °C) that could be due to a lower stability of $Pr(OH)Cl_2$ basic salt due to a lower basicity of Pr cation. In addition, in the case of Pr-based perovskite, along with CO and H_2, products of deep oxidation which are absent for La-containing perovskite are observed. This could be due to combustion of activated CH_x species by highly reactive oxygen of PrO_x [54, 56].

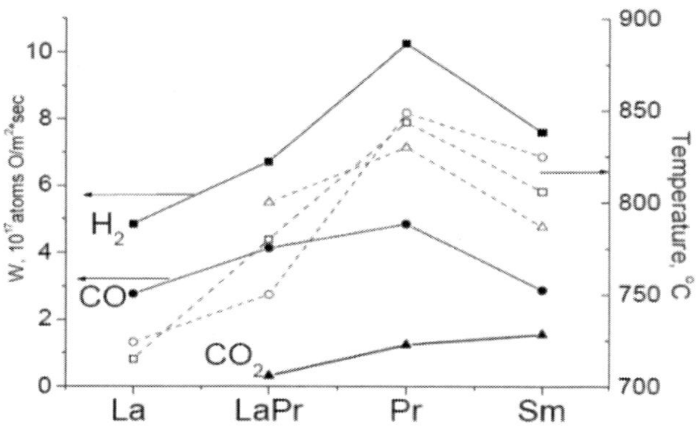

Figure 16. CH_4-TPR: maximum rates of H_2, CO and CO_2 formation and corresponding temperatures for $LnFe_{0.7}Ni_{0.3}O_{3-\delta}$ perovskites. Adapted from refr. [54].

CH_4-TPR spectra for composites 50 wt.% $PrFe_{0.7}Ni_{0.3}O_{3-\delta}$ +50 wt.%GDC prepared via one-pot method and by ultrasonic dispersion calcined at 900 °C are presented in Figure 17.

The one-pot composite containing NiO (Table 4) shows a higher H_2/CO ratio during TPR as compared with the composite prepared by ultrasonic dispersion, which is explained by CH_4 pyrolysis on Ni nanoparticles with coke accumulation of the GDC surface.

CH_4 TPR spectra for perovskite $La_{0.9}Ce_{0.1}Fe_{0.7}Ni_{0.3}O_{3-\delta}$ and composite 90 wt. % $LaFe_{0.7}Ni_{0.3}O_{3-\delta}$+10wt. % GDC prepared via one-pot method are shown in Figure 18.

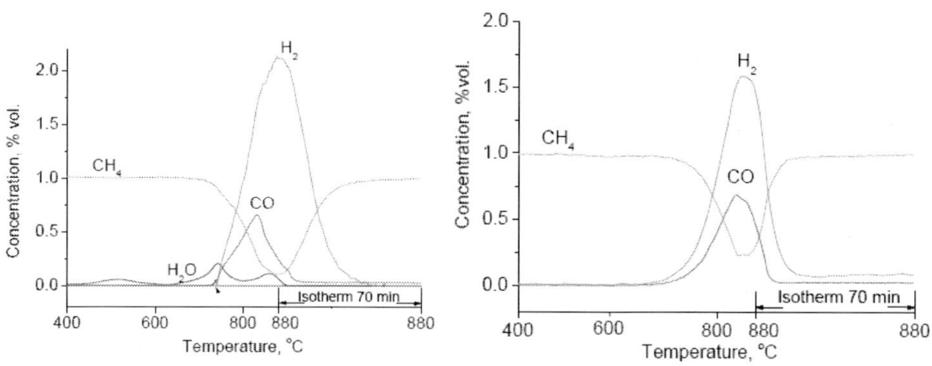

Figure 17. CH_4 TPR spectra for composites 50 wt. % $PrFe_{0.7}Ni_{0.3}O_{3-\delta}$+50 wt.% GDC prepared via one-pot method (left) and by ultrasonic dispersion (right) calcined at 900 °C.

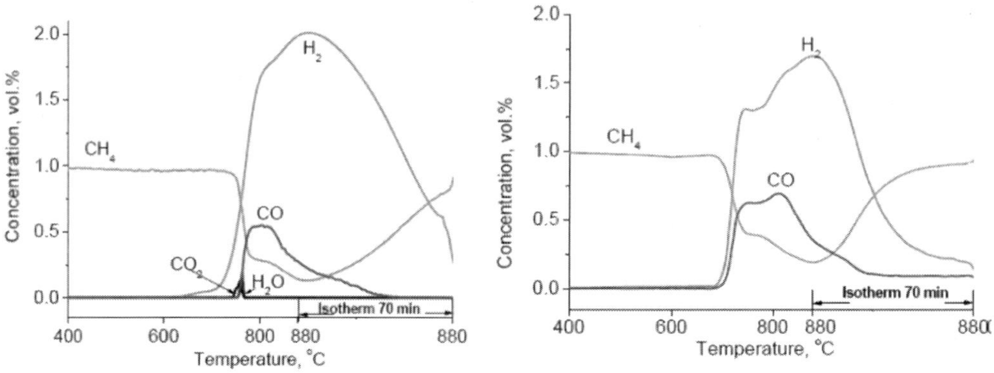

Figure 18. CH_4 TPR spectra for perovskite $La_{0.9}Ce_{0.1}Fe_{0.7}Ni_{0.3}O_{3-\delta}$ (left) and composite 90 wt.% $LaFe_{0.7}Ni_{0.3}O_{3-\delta}$ +10 wt.% GDC prepared via one-pot method (right) calcined at 900 °C.

The spectra show that interaction of methane with $La_{0.9}Ce_{0.1}Fe_{0.7}Ni_{0.3}O_{3-\delta}$ starts at ~630 °C with hydrogen evolution while CO and a small amount of CO_2 are formed at ~745 °C. In the case of composite, the formation of hydrogen and CO is observed practically at the same temperature (~685 °C). The formation of hydrogen at lower temperature over $La_{0.9}Ce_{0.1}Fe_{0.7}Ni_{0.3}O_{3-\delta}$ as compared with composite is due to easy reduction of surface NiO nanoparticles detected by XRD (Table 4) followed by CH_4 pyrolysis on surface Ni clusters.

Typical CH_4 TPR spectra for Ru/$La_{0.8}Pr_{0.2}Mn_{0.2}Cr_{0.8}O_3$/NiO/YSZ nanocomposites are shown in Figure 19. In this case, reduction starts at quite low temperatures (~400 °C) and is finished at ~ 600 °C, thus demonstrating a high reactivity of NiO nanoparticles incorporated into the surface layers of perovskite and promoted by Ru. Hydrogen evolution due to CH_4 decomposition continues even after samples reduction. Hence, coke thus formed is apparently deposited on the surface of perovskite particles while Ni-Ru alloy nanoparticles partially decorated by support oxidic species are free from coke being able to further activate CH_4 molecules [32].

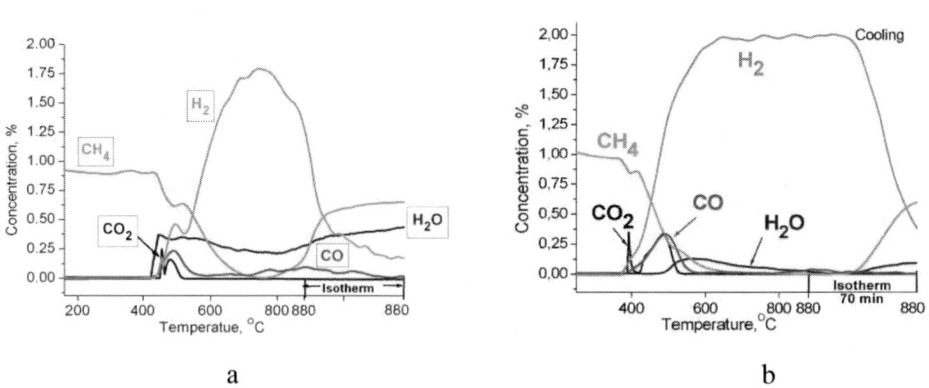

Figure 19. A typical CH_4 TPR data for the sample containing 1wt.%Ru, 80wt.% $La_{0.8}Pr_{0.2}Mn_{0.2}Cr_{0.8}O_3$, 10wt.% NiO and 10wt.% YSZ (a) and 1wt.%Ru, 50wt.% $La_{0.8}Pr_{0.2}Mn_{0.2}Cr_{0.8}O_3$, 30wt.% NiO and 20wt.% YSZ (b).

5. Methane Dry Reforming

5.1. Activation of Catalysts

Since in methane reforming reactions the rate -limiting stage –the rupture of C-H bond is well known to proceed on metal sites (Ni^0, Ru^0 etc), the activation of perovskite precursors must be done either via reductive pretreatment or directly in the reaction media [50-62, 89]. The peculiarities of the catalysts activation in the course of MDR and their activity have been firstly studied using the temperature-programmed reaction mode with the diluted feed (1% CH_4 + 1% CO_2 in He). Typical profiles of products concentration variation during temperature-programmed reaction of MDR are shown in Figures 20 and 21. For $LnFNi_{0.3}$ perovskites, the start of CH_4 and CO_2 consumption at 600 – 650 °C is accompanied by CO formation while hydrogen appears at higher temperatures. Such a sequence of products formation could be explained by occurrence of reverse water gas shift reaction (RWGS) $CO_2 + H_2 = CO + H_2O$ at temperature below 700 °C. Note that the starting temperatures of H_2 and CO formation depend on the nature of Ln cation in perovskite being identical and the lowest for $LaSrFNi_{0.3}$ and different for other perovskites.

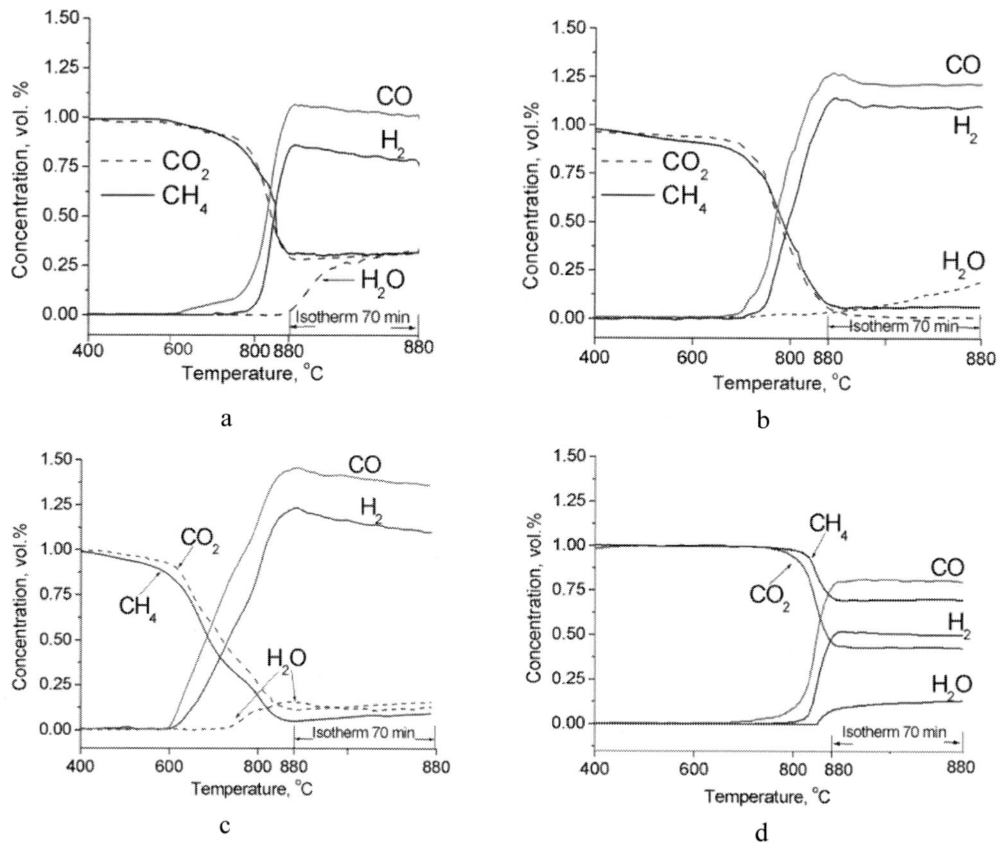

Figure 20. Typical curves of reagents and products concentration in the temperature –programmed reaction of CH_4 dry reforming for $LnFNi_{0.3}$ catalyst pretreated in O_2. Contact time 5 ms, feed 1% CH_4 +1%CO_2 in He; a-$PrFe_{0.7}Ni_{0.3}O_3$, b - $LaFe_{0.7}Ni_{0.3}O_3$, c - $La_{0.8}Sr_{0.2}Fe_{0.7}Ni_{0.3}O_3$, d -$La_{0.9}Ce_{0.1}Fe_{0.7}Ni_{0.3}O_3$.

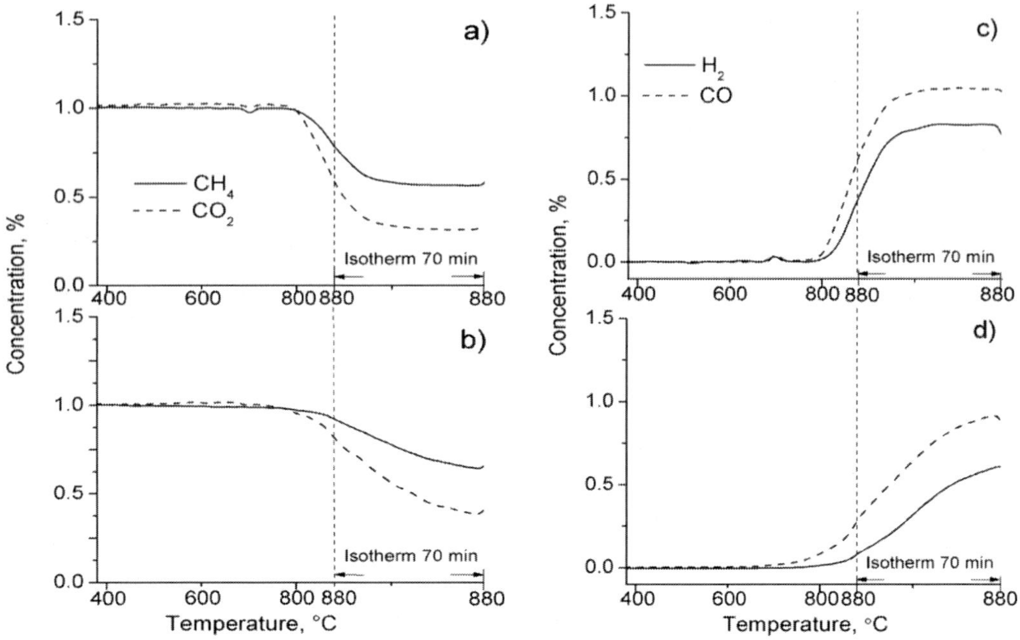

Figure 21. Typical curves of reagents (a, b) and products (c, d) concentration in the temperature-programmed reaction over $PrFe_{0.65}Ru_{0.05}Ni_{0.3}O_{3-\delta}$ (a, c) and $LaFe_{0.65}Ru_{0.05}Ni_{0.3}O_{3-\delta}$ (b, d). Feed 1% CH_4 + 1% CO_2 in He, contact time 0.001 s. Adapted from refr. [55].

For Ru-containing perovskites, the profiles of products concentration during temperature-programmed reaction of MDR presented in Figure 21 are, in general, similar to those for Ru-free samples. Higher temperatures of the products appearance agree with the CH_4-TPR data evidencing hampering of their reduction. Indeed, for Ru-free samples the reaction starts at 600 – 650°C, whereas a partial substitution of Fe by Ru increases reaction starting temperature up to 680-700 °C (Figures 22, 23).

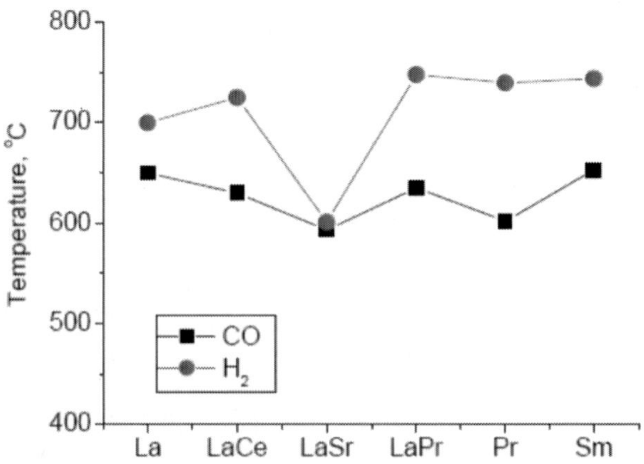

Figure 22. Starting temperature of products formation during the temperature-programmed reaction of CH_4 dry reforming for LnFN perovskites: ● H_2, ■ CO.

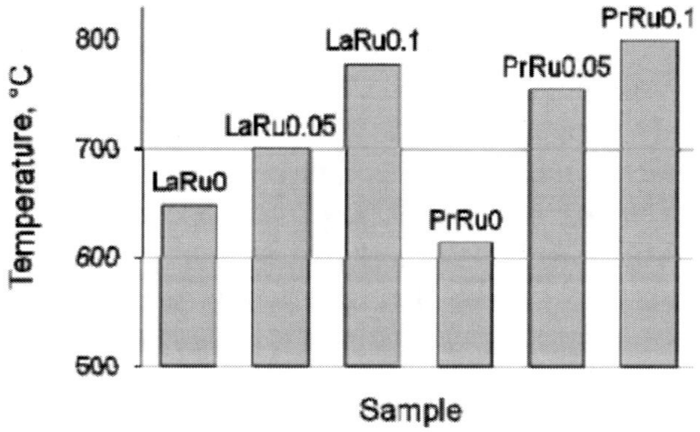

Figure 23. Temperature-programmed MDR: reaction starting temperature. Reaction conditions: 1% CH_4 + 1% CO_2 in He, contact time 0.001 s. Adapted from refr. [54].

In all cases, the consumption of CH_4 and CO_2 is followed by the formation of CO while hydrogen appears at higher temperatures (Figure 21). Such a sequence of products formation could be explained by a higher share of reverse water gas shift reaction (RWGS) in the overall process at lower temperatures. The reagents consumption and products formation depends on the Ln nature: for $PrFe_{0.65}Ru_{0.05}Ni_{0.3}O_{3-\delta}$, the steady-state concentrations are established at 880 °C in 30 - 40 min. whereas, in the case of $LaFe_{0.65}Ru_{0.05}Ni_{0.3}O_{3-\delta}$, the steady state is not attained even in 70 min.

Along with Ln nature, the activation process depends on the Ru content in perovskites. The variation of methane and CO_2 conversion during activation of $PrFe_{1-x}Ru_xNi_{0.3}O_{3-\delta}$ at 850°C in the reaction mixture 10%CH_4 + 10% CO_2 in He (Figure 24) evidences its retarding at increasing Ru content in perovskite from 5 to 10 mol%.

This is determined by a bigger amount of oxychloride in samples with a higher Ru concentration being revealed by XRD data (Figure 3). Hence, dynamics of catalyst activation during MDR reaction is clearly controlled by the same factors as their reduction with methane, namely, nickel and ruthenium inter-cation coupling in the perovskite lattice as well as an amount and stability of corresponding oxychloride.

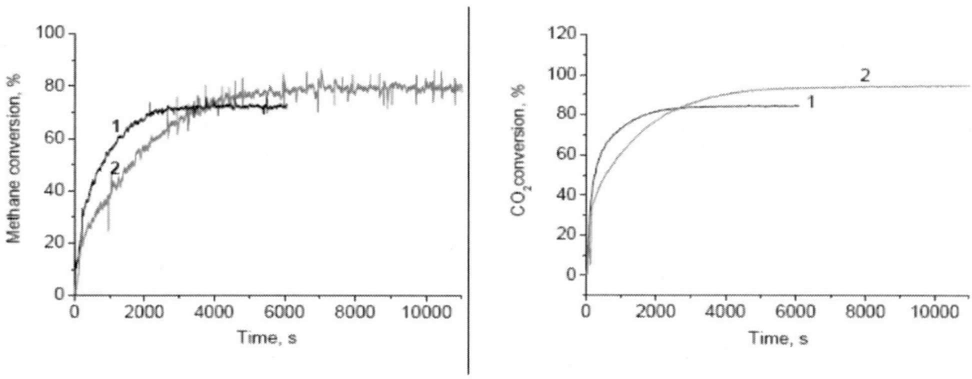

Figure 24. Time dependence of CH_4 and CO_2 conversion in MDR over oxidized $PrFe_{0.7-x}Ru_xNi_{0.3}O_{3-\delta}$ (1 - x=0.05, 2 – x = 0.1) at 850°C, feed 10%CH_4+10%CO_2, He balance, contact time 0.015 s. Adapted from refr. [55].

5.2. Steady State Activity

5.2.1. Catalysts Based on $LnFNi_{0.3}$ Perovskite Precursors

To study the effect of pretreatment on the catalyst activity, all initial $LnFNi_{0.3}$ perovskites were activated in the reaction mixture at 850 °C or in the H_2 flow at 600 °C (only PrFN was activated in H_2 at 800 °C as well) before the catalytic tests.

The temperature dependence of methane conversion for all perovskites activated and tested in the reaction mixture containing 10%CH_4+10%CO_2 are presented in Figure 25.

The values of methane conversions at 800 °C for perovskites activated in the reaction mixture and in the flow of H_2 at 600 °C are shown in Figure 26.

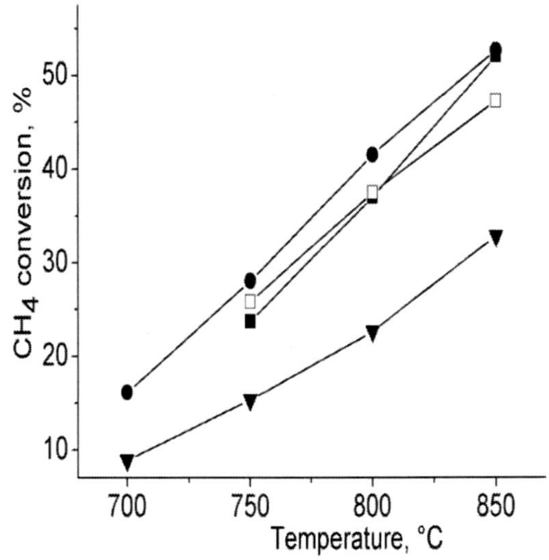

Figure 25. Temperature dependence of methane conversion in MDR over the $LnFe_{0.7}Ni_{0.3}O_{3-\delta}$ (Ln = La (▼), LaPr (■), Pr (●), Sm (□)) activated in the reaction mixture. Feed 10%CH_4+10%CO_2, He balance, contact time 0.01 s. Adapted from refr. [54].

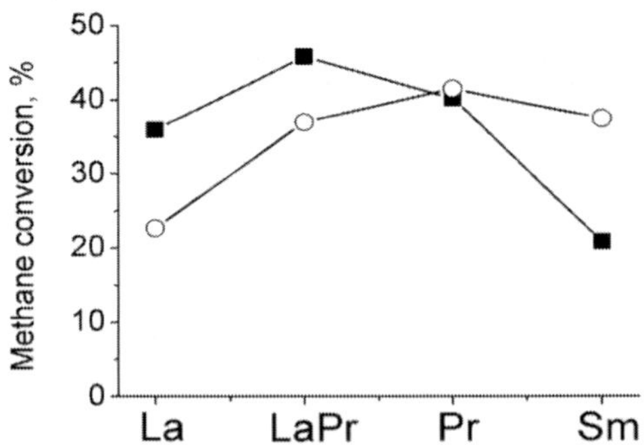

Figure 26. Methane conversion in CH_4 dry reforming over reduced (■) and oxidized (o) $LnFe_{0.7}Ni_{0.3}O_{3-\delta}$ (Ln = La, LaPr, Pr, Sm) catalysts. Adapted from refr. [54].

Among the samples activated in the reaction mixture, the conversion values decrease in the order PrFN > LaPrFN, SmFN >LaFN (Figures 25, 26). A low methane conversion over LaFN perovskite could be explained by its difficult reduction as shown by H_2 TPR (Figure 7). Indeed, pre-reduction of LaFN and LaPrFN at 600°C in H_2 leads to a higher methane conversion (Figure 26).

At the same time, the reduction of PrFN at 800 °C in H_2 leads to a lower activity (Figure 27) as compared to samples activated in the reaction mixture or reduced at 600°.

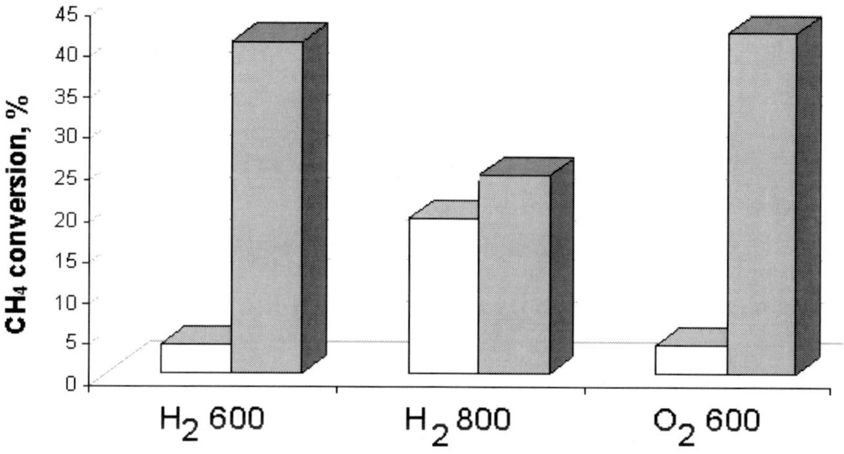

Figure 27. Initial (□) and steady-state (■) conversion of methane over PrFN oxidized at 600° in O_2 and reduced in H_2 at 600 and 800°C. Feed 10%CH_4+10%CO_2, He –balance, 800 °C. Adapted from refr. [54].

The comparison of methane conversion values for perovskites reduced in H_2 and activated in the reaction mixture shows that only in the case of PrFN they are close while for all other samples conversions differ noticeably. Such a difference in the activity appears to be determined by the composition of active Ni-Fe alloy particles formed in the oxide matrix depending on reducing media and the type of Ln cation in perovskite. Indeed, the XRD data show that, in the case of PrFN, Ni-Fe alloy with a higher Fe content is formed after reduction in H_2 at 800 °C (Table 3, Figure 4).

The conversion of reagents depends on the nickel content in perovskite. Thus, the highest activity is observed for perovskites with Ni content of 30 at.% (Figure 28) that correlates with its easier reduction according to H_2-TPR data (Figure 7).

Further, the data of TEM with EDX show that even minor amounts of free NiO present in the initial $LaFe_{0.6}Ni_{0.4}O_{3-\delta}$ catalyst lead to encapsulation of metal Ni particles with carbon (Figure 29) even in diluted reaction mixture 10%CH_4+10%CO_2, N_2 – balance, while in the case of single phase perovskites, for instance, $LaFe_{0.7}Ni_{0.3}O_{3-\delta}$, formed alloy clusters are not subjected to coking (Figure 30).

A higher activity of Sr-free perovskites is apparently caused by the negative effect of the surface $SrCO_3$ species forming along with $La_2O_2CO_3$ under contact with reaction feed. If the latter participates in the DR reaction and prevents Ni coking, more stable and nonreactive $SrCO_3$ can block the active Ni species.

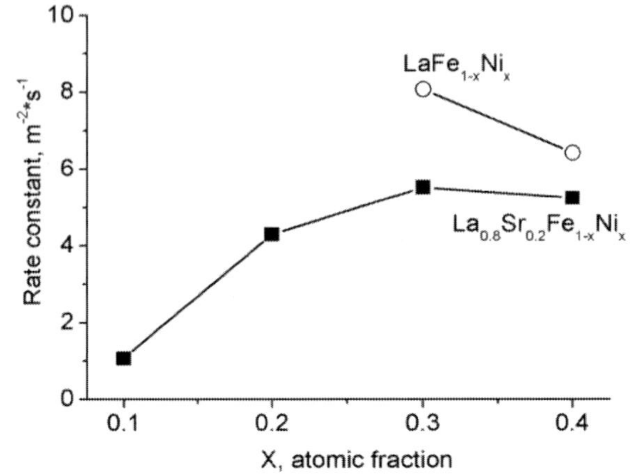

Figure 28. Dependence of the MDR specific constant on Ni content in perovskite precursors $La_{1-y}Sr_yFe_{1-x}Ni_xO_{3-\delta}$ (y = 0, 0.2; x = 0.1 - 0.4. Feed 1%CH_4+1%CO_2 in He, 800 °C.

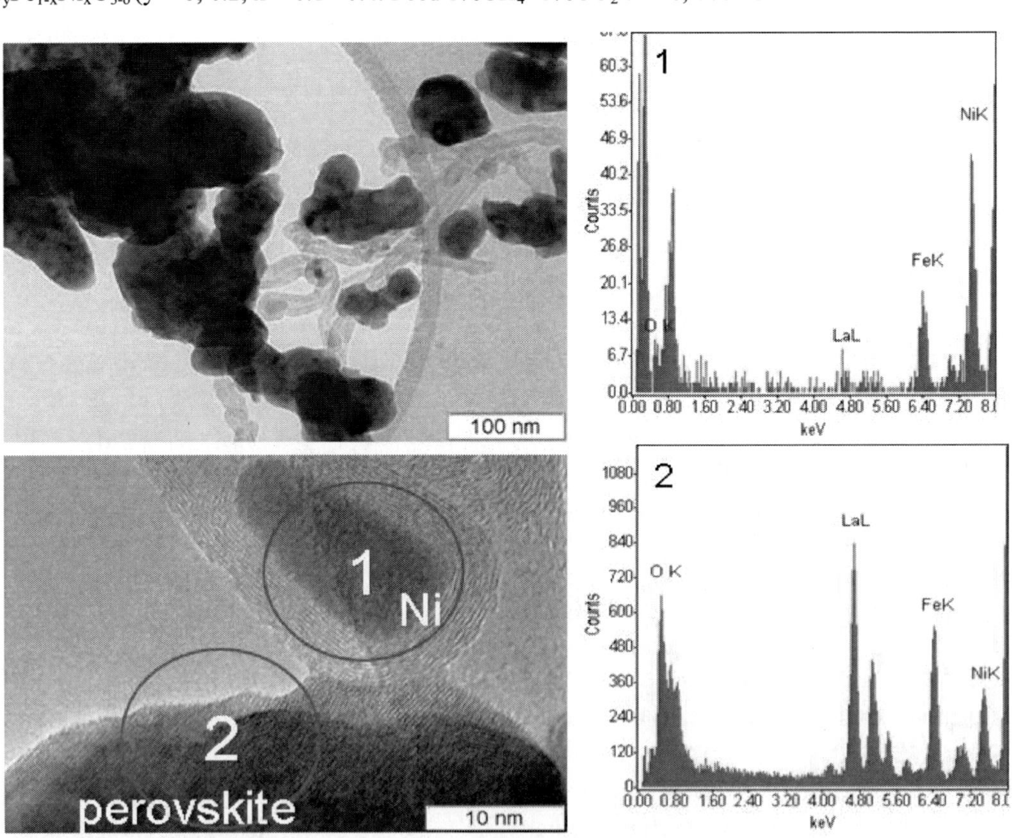

Figure 29. TEM image of the catalyst activated in reaction mixture (10%CH_4+10%CO_2 in He) after working in methane dry reforming. Precursor – $La_{0.8}Sr_{0.2}Fe_{0.6}Ni_{0.4}O_3$ perovskite.

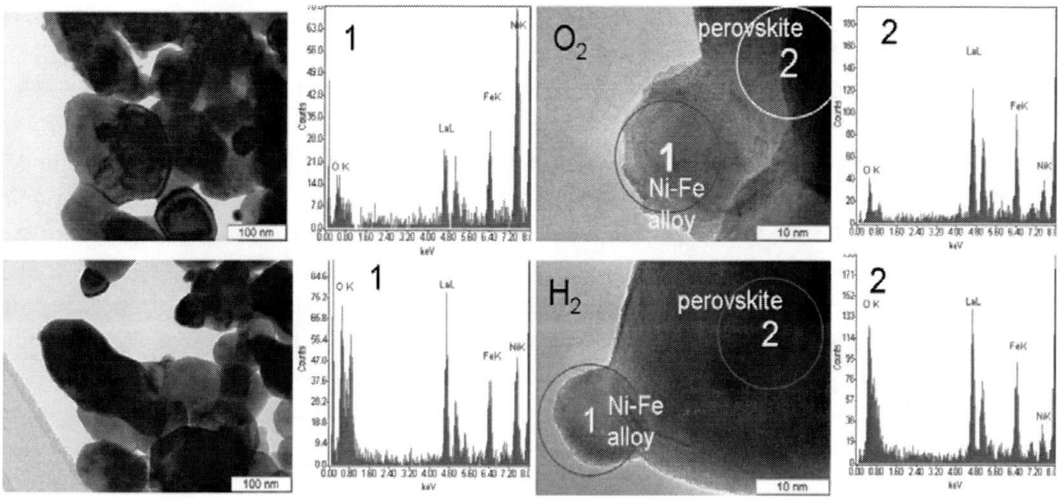

Figure 30. TEM image of the catalyst after working in methane dry reforming: the upper line – activated in the reaction mixture, the bottom line – pretreated in H_2 flow at 800 °C. Precursor - $LaFe_{0.7}Ni_{0.3}O_3$ perovskite.

5.2.2. Catalysts Based on $LnFe_{1-x}Ru_xNi_{0.3}O_{3-\delta}$ Perovskite Precursors

The typical temperature dependences of methane and CO_2 conversion and H_2/CO ratio along with equilibrium values are presented in Figure 31 for $PrFe_{0.65}Ru_{0.05}Ni_{0.3}O_{3-\delta}$ sample activated and tested in the reaction mixture containing 10%CH_4+10%CO_2 in He at contact time 0.015 s.

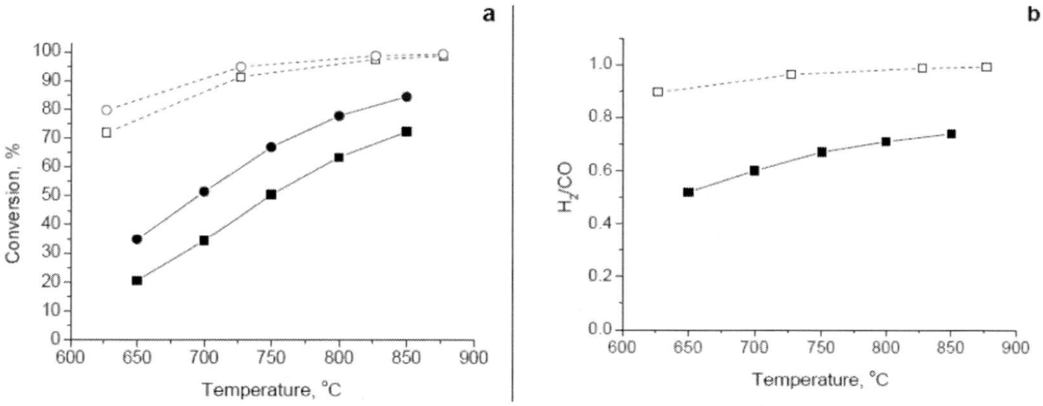

Figure 31. Temperature dependence of methane (square) and CO_2 (circle) conversion (a) and H_2/CO ratio (b). Open symbols - equilibrium values, solid symbols – the catalyst $PrFe_{0.65}Ni_{0.3}Ru_{0.05}O_{3-\delta}$ activated in the reaction mixture at 850 °C. Feed 10%CH_4+10%CO_2 in He, contact time 0.015 s. Adapted from refr. [55].

The reagents conversions increase with the temperature remaining below equilibrium values due to a short contact time, while approaching equilibrium at longer contact times [90, 91]. A higher conversion of CO_2 as compared to CH_4 in the whole temperature range is due to RWGS reaction occurring along with MDR, which results in H_2/CO ratios lower 1 [60, 92].

To analyze the effect of perovskite chemical composition and their pretreatment on the catalytic activity, the specific effective first-order rate constants of methane transformation were estimated using the model of a plug-flow reactor [93]. The values of specific rate constants of MDR at 800 °C for all catalysts activated in the reaction mixture and in the flow of H_2 at 800 °C are shown in Figure 32. These data show that, as in the case of Ru-free samples, the highest performance has been demonstrated by Pr-based catalysts, moreover, the activity of catalysts increases with Ru concentration independently of the activation mode. However, the activity of catalysts activated in the reaction mixture is slightly lower due to the presence of PrOCl detected by XRD (Figure 3). The pretreatment of catalysts in hydrogen leads to segregation of active metal alloy particles and partial decomposition of PrOCl blocking active centers that increases catalyst activity. Note that the values of specific rate constants for activated in the reaction mixture and reduced samples are close due to comparatively easy decomposition of PrOCl which is completely destroyed during long term testing in the real reaction mixture (Figure 3). In contrast to Pr-containing samples, for $LaFe_{1-x}Ru_xNi_{0.3}O_{3-\delta}$ catalysts, the activity order strongly depends on the activation mode. Thus, if the catalysts are activated in the reaction media, the next row of activity is observed: LFN > $LFNRu_{0.05}$ > $LFNRu_{0.1}$ (Figure 32). In this case, the decrease of activity at increasing Ru concentration is caused by inter-cation coupling stabilizing both nickel and ruthenium cations in the perovskite lattice [60] as well as by increasing content of stable LaOCl as shown by XRD data (Figure 2). The particles of LaOCl blocking the perovskite precursor surface hinder the formation of active metal centers. The results evidencing a poor activation of Ru-containing $LaFe_{1-x}Ru_xNi_{0.3}O_{3-\delta}$ samples in the reaction feed 10%CH_4 + 10% CO_2 in He agree with CH_4-TPR and MDR-TPR results (Figures 13, 21). The reversed activity order of catalysts is observed after the reductive activation in hydrogen. According to XRD data (Figure 2), the perovskite reduction in H_2 at 800 °C leads to segregation of metal alloy particles which are the active centers for methane activation, and activity of catalysts increases with Ru content.

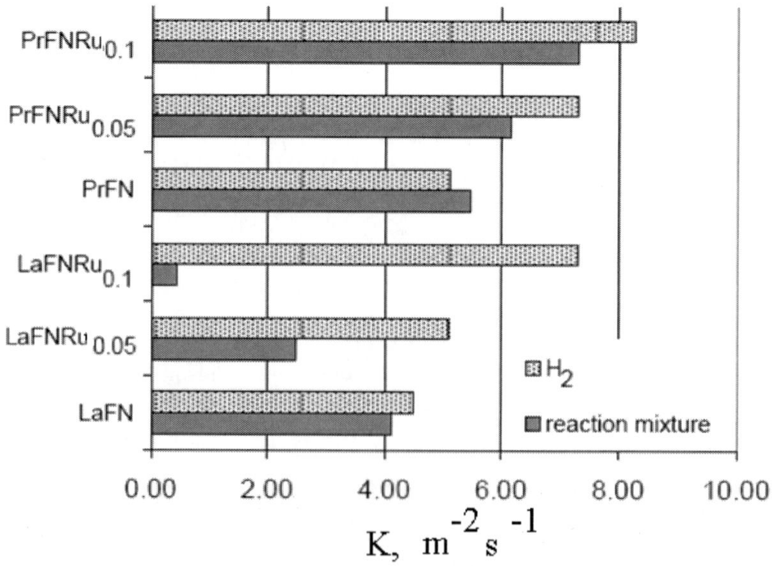

Figure 32. Specific rate constant of CH_4 dry reforming over perovskites reduced at 800 °C in H_2 or activated in the reaction mixture. Feed 10%CH_4+10%CO_2 in He, 800 °C, contact time 0.01 s. Adapted from refr. [55].

5.2.3. Catalysts Based on (100-x) wt. % LnFNi$_{0.3}$+x wt.% GDC Composite Precursors

The temperature dependence of specific rate constants and H_2/CO ratio for (100-x) wt.% LnFNi$_{0.3}$+x wt.% GDC composites is presented in Figures 33 - 35. For composites based on $PrFe_{0.7}Ni_{0.3}O_{3-\delta}$ prepared via one-pot method, activity decreases with increasing GDC content. The most active catalyst obtained from a composite containing 5% GDC shows much higher activity as compared with individual $PrFe_{0.7}Ni_{0.3}O_{3-\delta}$ perovskite (Figure 33). The data on MDR reaction for composites 50 wt.% $PrFe_{0.7}Ni_{0.3}O_{3-\delta}$ +50 wt. % GDC prepared by one-pot method and ultrasonic dispersion show that the first possesses a higher rate constant especially at high temperatures (Figure 34). For La-based samples containing Ce, the most effective catalyst was obtained from composite $LaFe_{0.7}Ni_{0.3}O_{3-\delta}$ + 10% GDC prepared via the one-pot method. Note that doping of perovskite by Ce decreases activity (Figure 35).

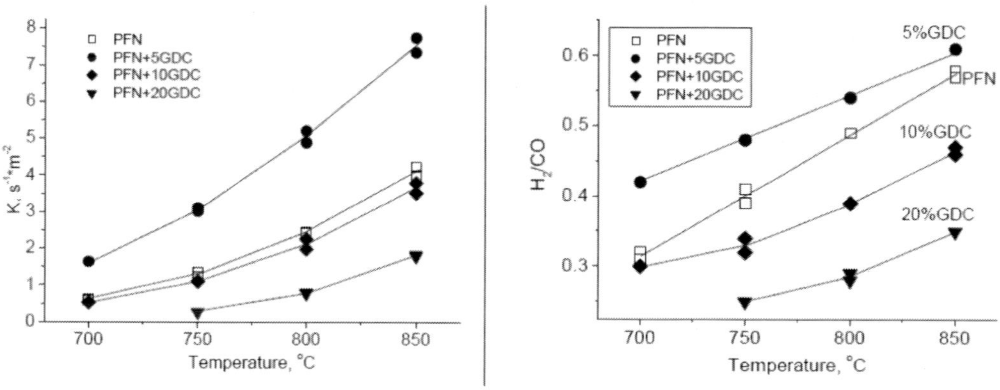

Figure 33. Temperature dependence of the specific rate constant and H_2/CO ratio for $PrFe_{0.7}Ni_{0.3}O_{3-\delta}$ and composites containing 5-20% GDC prepared via one-pot method. Samples were activated in the reaction mixture at 850°C. Feed 10%CH_4+10%CO_2 in He, contact time 0.015 s.

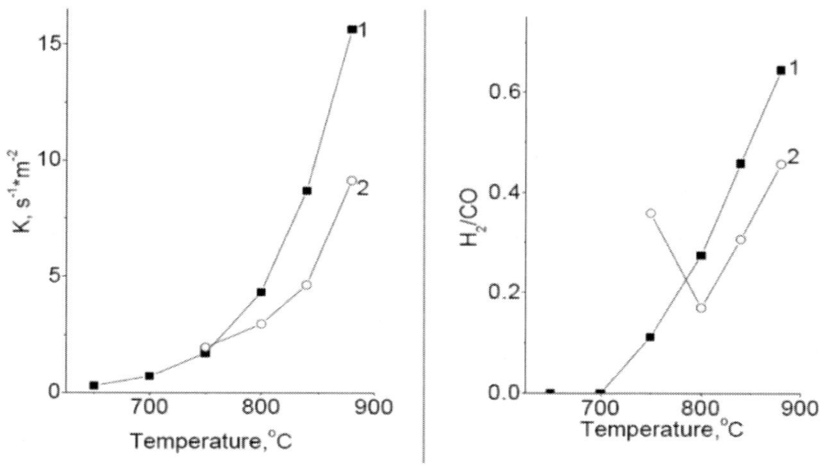

Figure 34. Temperature dependence of the specific rate constant and H_2/CO ratio for composite 50% wt $PrFe_{0.7}Ni_{0.3}O_{3-\delta}$ +50% wt. GDC prepared via one-pot method (1) and ultrasonic dispersion (2). Samples were activated in the reaction mixture at 850 °C; feed 1%CH_4+1%CO_2 in He, contact time 0.005 s

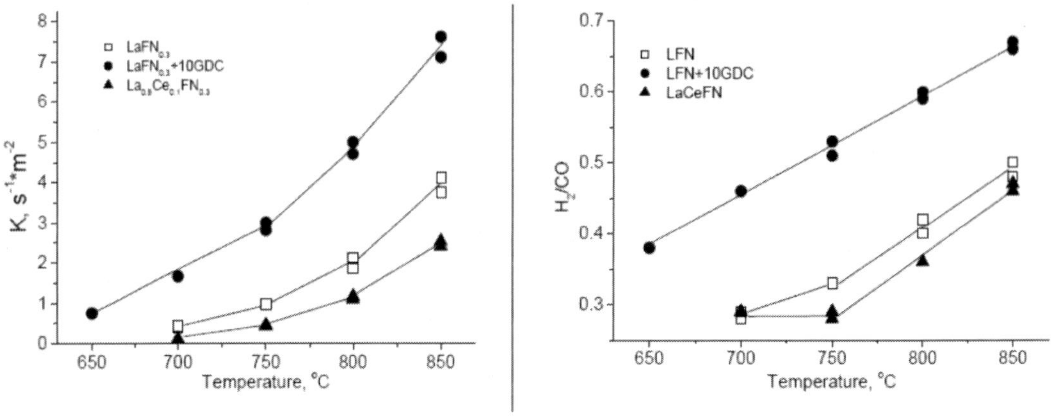

Figure 35. Temperature dependence of the specific rate constant and H_2/CO ratio for $La_{0.9}Ce_{0.1}Fe_{0.7}Ni_{0.3}O_{3-\delta}$, $LaFe_{0.7}Ni_{0.3}O_{3-\delta}$ and its composite containing 10% GDC prepared via one-pot method. Samples were activated in the reaction mixture at 850°C; feed 10%CH_4+10%CO_2 in He, contact time 0.015 s.

5.2.4. Catalysts Based on $La_{0.8}Pr_{0.2}Mn_{0.2}Cr_{0.8}O_3$+NiO+YSZ Nanocomposite Precursor

Typical temperature dependence of the specific rate constant for 2.8%Ru/10%Ni/80%LaMnCrPr/10%YSZ nanocomposite is given in Figure 36.

As follows from comparison of this catalysts performance with that of other perovskite-containing systems (Figures 32-35), mixed manganite-chromite/YSZ nanocomposite as support provides specific activity of Ru+Ni alloy nanoparticles even somewhat higher than that for other perovskite-like supports. A high CH_4 conversion was also observed in concentrated feeds containing natural gas with up to 5% admixture of C_2-C_4 hydrocarbons with different CO_2/natural gas ratios (Figures 37, 38).

Cyclic variation of contact time for this catalyst in concentrated feed demonstrated a good reversibility of performance and, hence, its required stability (Figure 39).

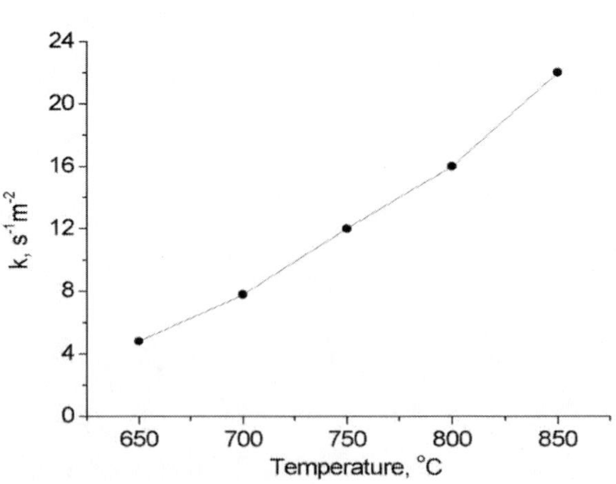

Figure 36. Temperature dependence of specific rate constant for Ru/Ni/LaMnCrPr/YSZ. Feed 10%CH_4 +10%CO_2 in He, contact time 15 ms.

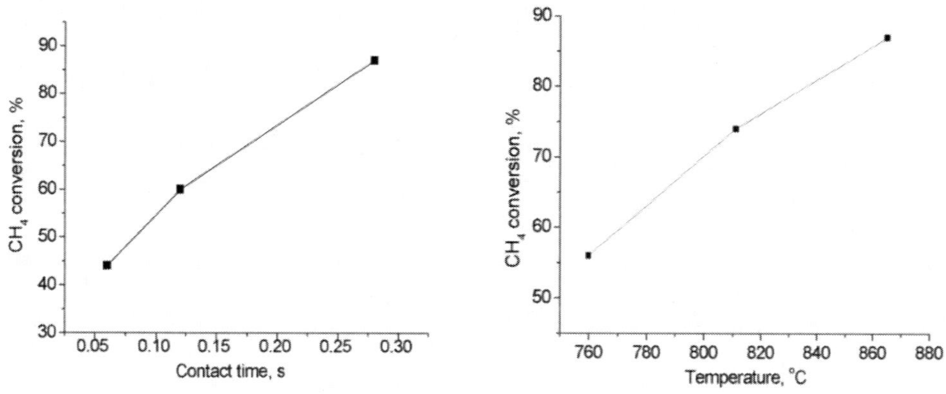

a b

Figure 37. Methane conversion vs. contact time (a) or temperature (b) in CH_4 DR on 2.8%Ru/Ni/LaMnCrPr/YSZ nanocomposite. Feed composition 47.5% natural gas + 47.5%CO_2 + 5% N_2 , temperature 800 °C (a), contact time 0.12 s (b).

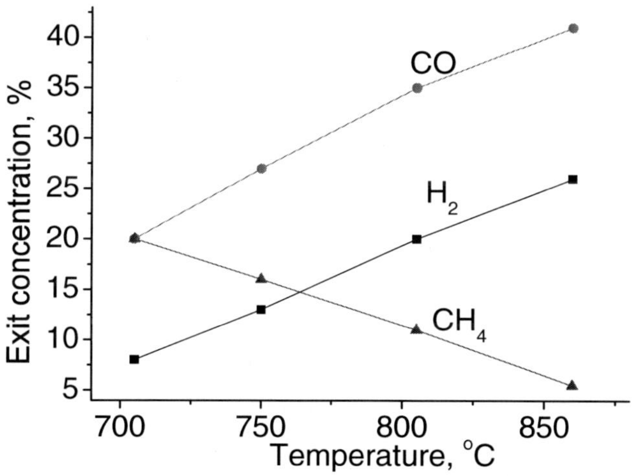

Figure 38. Temperature dependence of exit concentrations in the process of natural gas dry reforming on 2.8%Ru/LaPrMnCr/10%YSZ/10NiO at contact time 0.09 s, feed 53%CO_2+38% NG+N_2.

5.2.5. Long-term Testing and Post-reaction Analysis of Perovskite Catalysts

The time-on-stream dependence of methane conversion and H_2/CO ratio during long-term stability testing of catalysts in the real feed is presented in Figure 40.

For Pr-based samples with and without Ru, a high methane conversion close to equilibrium value is observed in the first 30 minutes after 2 hours pretreatment in the same feed. In the case of $LaFe_{0.65}Ru_{0.05}Ni_{0.3}O_3$, a high methane conversion is achieved only after 1 - 1.5 hour due to slow reduction of this sample (Figure 13, 21). Ru-doped catalysts demonstrate a better performance stability, which can be explained by prevention of coking due to formation of Ni-Fe-Ru alloys. Moreover, formation of such alloys could also help to suppress coarsening of Ni clusters. Similar to Ru+Ni/LaPrCrMn/YSZ catalyst, a higher level of activity was obtained here for Pr-containing catalysts as well, which suggests importance of these redox cations for the efficient activation of CO_2 as well as providing higher oxygen mobility in the oxide support [54, 55].

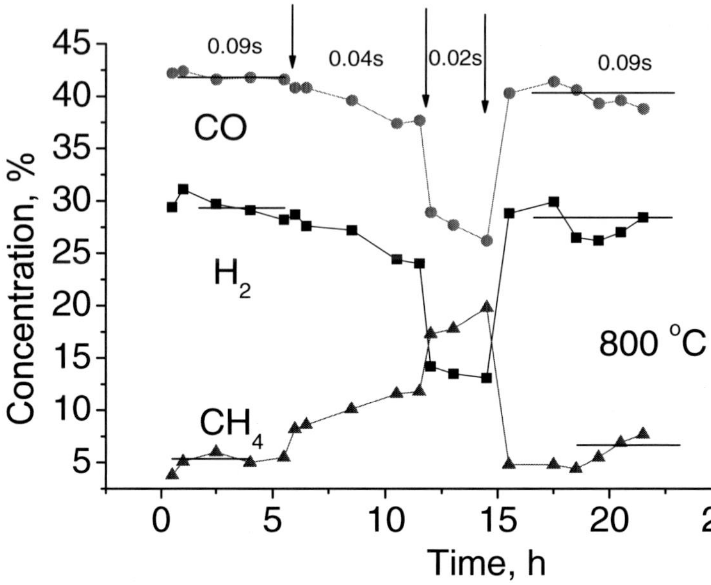

Figure 39. Variation of exit concentrations with time-on-stream in the process of natural gas dry reforming on fresh 2.8%Ru/LaPrMnCr/10%YSZ/10NiO catalyst heated to 800 °C in the stream of N_2 followed by switch to 53%CO_2+38% NG+N_2 stream.

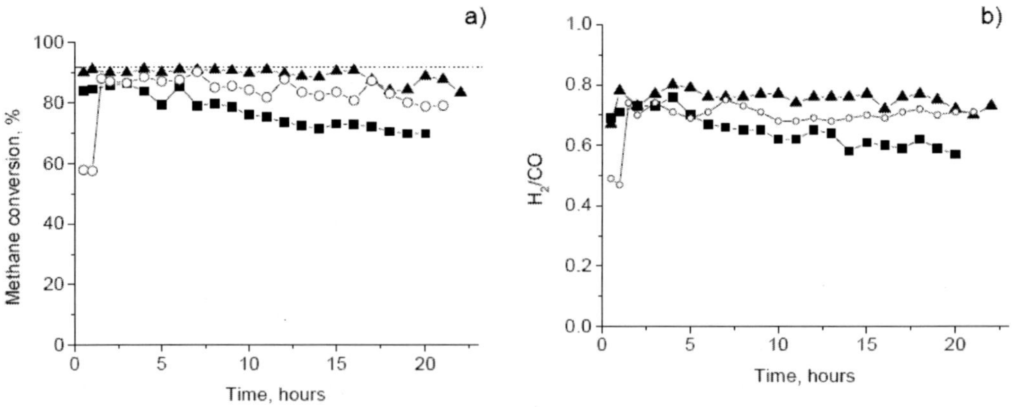

Figure 40. Methane conversion and H_2/CO ratio in long-term tests of $PrFe_{0.7}Ni_{0.3}O_3$ (■), $PrFe_{0.65}Ru_{0.05}Ni_{0.3}O_3$ (▲) and $LaFe_{0.65}Ru_{0.05}Ni_{0.3}O_3$ (**O**). Dotted line – equilibrium value. Reaction conditions: the feed 45% CO_2 + 45 % natural gas + 10% N_2 at 800°C, pretreatment in reaction media 2 h at 850°C, contact time 0.24 s. Adapted from refr. [55].

The amount of carbon deposited on Ru-containing catalysts during long-term testing has been determined using TPO. The results showed no CO_2 evolution up to 880 °C evidencing the absence of coke formation over the catalysts during the reaction. The absence of any type of coke on the used catalysts has been confirmed by TEM study of $PrFe_{1-x}Ru_xNi_{0.3}O_3$ samples (Figure 41).

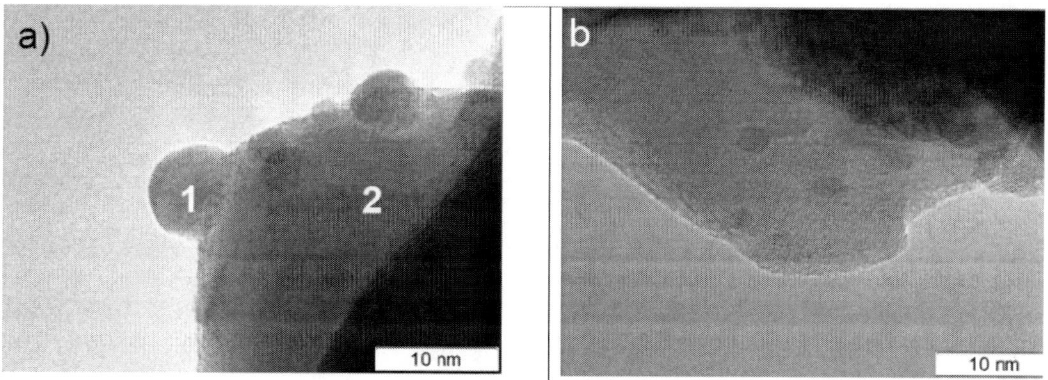

a) atomic %

	Ru	Pr	Fe	Ni
1	46.48	40.03	3.24	10.24
2	7.38	83.17	6.65	2.8

b) atomic %

Ru	Pr	Fe	Ni
3.59	54.21	33.5	8.71

Figure 41. TEM with EDX of used samples: a) $PrFe_{0.7}Ru_{0.1}Ni_{0.3}O_3$, b) $PrFe_{0.65}Ru_{0.05}Ni_{0.3}O_3$. Adapted from refr. [55].

In agreement with XRD, TEM data show that $PrFe_{0.6}Ru_{0.1}Ni_{0.3}O_3$ catalyst is comprised of Ni-Ru [JCPDF 06-663] nanoparticles (marked by 1 in Figure 41a, lattice spacing 2.08 Å) strongly interacting with the surface of Pr ferrite $PrFeO_3$ [JCPDF 47-0065] domains (marked by 2 in Figure 41a, (021) lattice spacing 2.66 Å). In the case of $PrFe_{0.65}Ru_{0.05}Ni_{0.3}O_3$, small alloy clusters are observed in a good epitaxy with the oxide particle containing both Pr and Fe (Figure 41b), thus corresponding to perovskite phase ((020) lattice spacing 2.79 Å). This certainly provides strong metal – support interaction required for preventing sintering and coking. Indeed, no carbon deposits or fibers were detected by TEM in catalysts discharged after reaction.

The XPS study of $PrFe_{0.7-x}Ru_xNi_{0.3}O_{3-\delta}$ perovskites shows that the surface composition of catalysts does not vary appreciably after contact with reaction feed (Table 5).

Initial perovskite contains oxidized forms of ruthenium (III, IV) according to well-defined $Ru3d_{5/2}$ peak observed at 282.6 eV (Figure 42).

A weak component at 280.5 eV could be assigned to very small (<2 nm) clusters characterized by noticeable relaxation processes in the photoemission that leads to shift of Ru3d peaks. The samples tested in MDR have mainly metal Ru in the surface as indicated by well-defined $Ru3d_{5/2}$ peak observed at 280.0 and 280.3 eV for PrRu0.1 and PrRu0.05 samples, respectively. Mathematical analysis with Doniach-Sunjic and Gauss-Lorentz function revealed the presence of a low intensity component at 282.6 eV, which can be assigned to Ru (III, IV) state. Since samples were transferred through air after discharging from the reactor, this feature can be explained by their partial oxidation due to this manipulation.

The oxidation state of Fe is identified by $Fe2p_{3/2}$ core-level peak as corresponding to Fe (III). It remained almost unchanged after treatment in the reaction media followed by transfer through the air. The state of Ni in the initial perovskite is characterized mainly by $Ni2p_{3/2}$ peaks at 854.1 and 856.1 corresponding to Ni(II) and Ni(III) states, respectively (Figure 43).

Table 5. The bulk and surface composition of initial and used $PrFe_{0.7-x}Ru_xNi_{0.3}O_{3-\delta}$

x	Bulk composition		Surface composition	
	Ni/Pr	Ru/Pr	Ni/Pr	Ru/Pr
0.1 (initial)	0.30	0.10	0.12	0.041
0.1 (used)	0.30	0.10	0.11	0.044
0.05 (used)	0.30	0.05	0.15	0.031

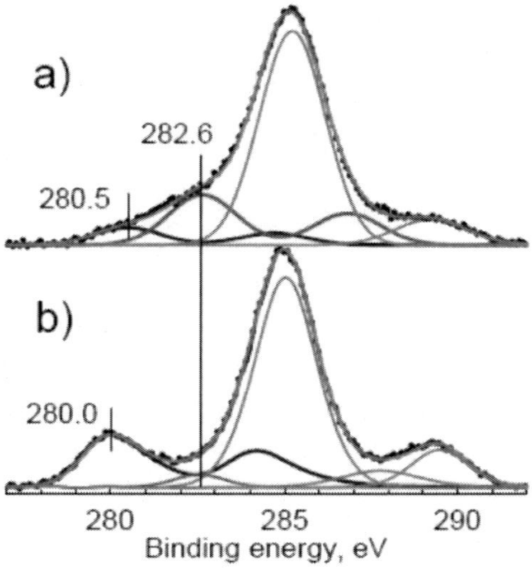

Figure 42. The profile of Ru 3d core-level spectra of $PrFe_{0.6}Ru_{0.1}Ni_{0.3}O_{3-\delta}$ perovskites: initial (a) and used (b).

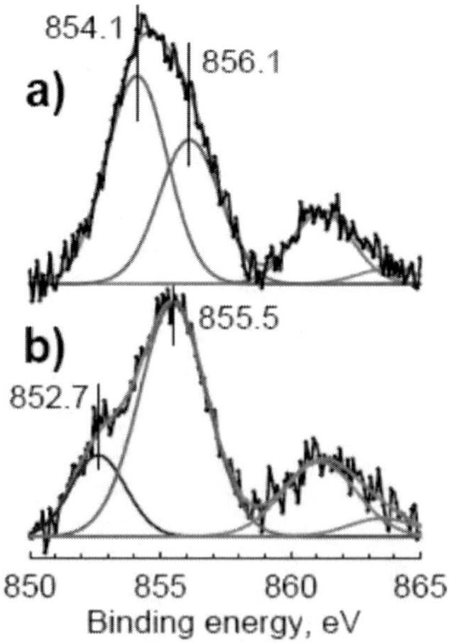

Figure 43. The profile of Ni $2p_{3/2}$ core-level spectra of $PrFe_{0.6}Ru_{0.1}Ni_{0.3}O_{3-\delta}$ perovskites: initial (a) and used (b). Adapted from refr. [55].

The additional component at 852.5-852.7 eV, corresponding to Ni^0 state, appears in spectra of samples after MDR reaction. With a due regard for the presence of mainly metal Ru as well as XRD and TEM data, this Ni metal state can correspond to Ni-Ru alloy. In addition, the main part of Ni is present in Ni (II) and Ni (III) states as indicated by components at 854.5 and 856.3 eV, respectively. This feature can be again caused by the partial reoxidation of Ni metal under contact of discharged catalyst with air, pure Ni^0 particles (generated by reduction of NiO admixture) being oxidized much easier as compared with Ni-Ru alloy nanoparticles.

It is known that activity and coking stability of Ni-containing catalysts are controlled by the ensemble effect when, for reforming reaction, the ensemble of Ni atoms is smaller than that required for nucleation of carbon whiskers [94]. Therefore, a higher dilution of metal Ni with Fe favors a higher activity and stability of the catalyst formed from PrFN as compared with LaFN and LaPrFN. At the same time, a higher activity of catalysts activated in the reaction mixture could be conditioned by the optimal composition of Ni-Fe alloy. Furthermore, if activation of methane is determined by the metallic phase, activation of CO_2 occurs with the participation of the oxide matrix and could be governed by its red-ox properties. The data of oxygen thermodesorption show the maximal rate of oxygen desorption at high temperature for PrFN as compared with other perovskites (Table 6).

Table 6. Data of O_2 TPD: desorption of bulk oxygen

Sample	W_{max} 10^{-14}, O_2 molecules/$m^2\cdot s$	T_{max}, °C
$LP_{0.1}FN$	6	889
PFN	9	879
SFN	6	892

Probably, praseodymium oxides possessing a high red-ox activity [56] may react with CO_2 and with carbon residues thus facilitating reforming reaction and preventing catalyst coking. However, it is not sufficient to provide a high and stable performance of Ni-Fe alloy particles as clearly demonstrated by much lower activity of PrO_x-supported catalysts as compared with those derived from perovskite precursors (Table 7).

Table 7. Specific surface area of praseodymium oxides and supported Ni-Fe catalysts

Sample	SSA, m^2/g
Pr_xO_y(n)	2.3
Pr_xO_y(P)	3.7
3% Ni-Fe/Pr_xO_y(n)	5.2
5% Ni-Fe/Pr_xO_y(P)	4
7% Ni-Fe/Pr_xO_y(P)	4

Hence, perovskite phase remaining even after prolonged contact of catalysts with concentrated feed (CH_4:CO_2=1:1) at high temperatures (Figures 44, 45) appears to be the most important factor providing a high and stable performance. This can be explained by a high reactivity and mobility of the surface/lattice oxygen of these ferrite-type perovskites disordered due to reductive segregation of Ni (Figure 44). This helps to provide an efficient route of CO_2 activation on the surface of perovskites and a fast transfer of oxygen atoms or oxygen–containing species (hydroxyls, hydroxocarbonates , etc) to the interface with segregated Ni-Fe alloy particles where they consume activated CH_x fragments – coke precursors. Though composition of these alloy particles varies broadly (Figure 45), neither coke deposits nor fibers were detected in discharged catalysts. This implies that a high lattice oxygen mobility and surface reactivity of nanostructured $PrO_x/LnFeO_y$ supports generated by reductive decomposition of complex perovskite precursors in the reaction media play more important role in preventing coking of supported Ni-Fe alloy particles than their exact chemical composition (Fe content).

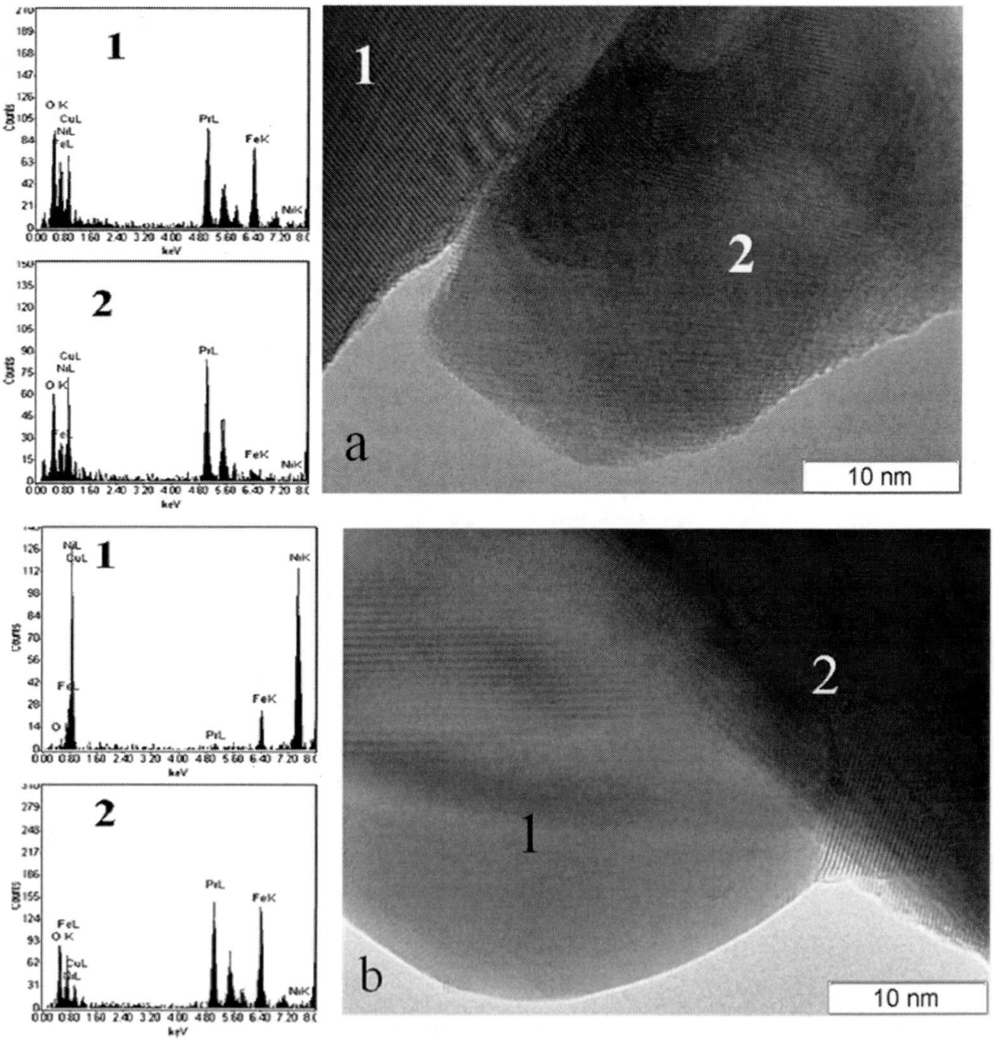

Figure 44. High-resolution TEM image of used $PrFN_{0.3}$ catalyst with EDX: a) 1 – perovskite, 2 – Pr_xO_y; b) 1 – Ni-Fe alloy, 2 – perovskite. Adapted from refr. [54].

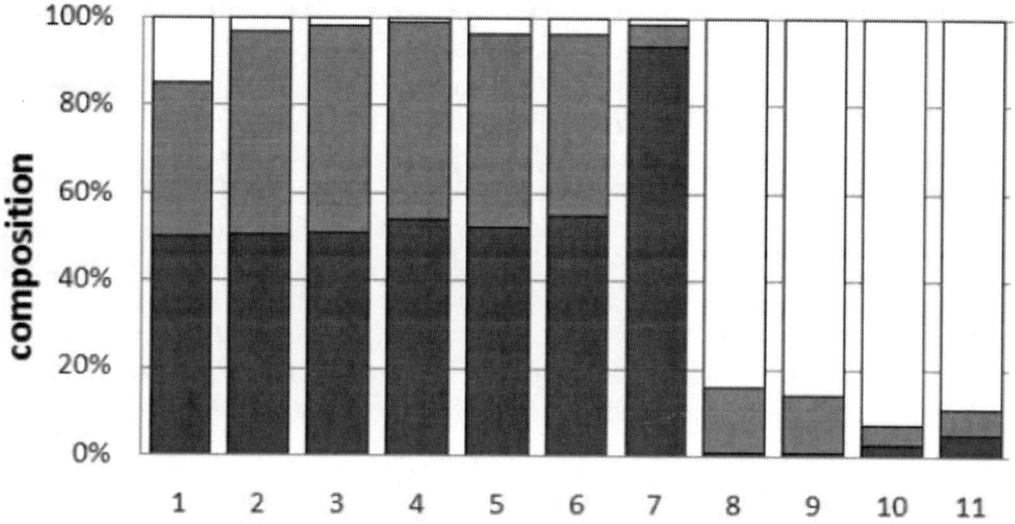

Figure 45. EDX elemental distribution for $PrFe_{0.7}Ni_{0.3}O_{3-\delta}$-based catalyst after testing in the reaction mixture CH_4:CO_2=1:1. 1 – integral; 2-6 – perovskite particles; 7 – praseodimium oxide; 8-11 – Ni-Fe-alloy particles. Ni □, Fe ■, Pr ■. Adapted from refr. [54].

Table 8 presents methane conversion, H_2/CO ratio and amount of carbon deposits obtained at different GHSV and temperatures during MDR using an undiluted feed with CH_4/CO_2=1 for Pr(La)$Fe_{0.65}Ru_{0.05}Ni_{0.3}$ O_3 catalysts and various catalysts reported in the literature [17-20, 45, 56, 59-60, 94].

Table 8. Methane conversion, H_2/CO ratio and amount of carbon deposits during MDR using an undiluted feed with CH_4/CO_2 = 1 for Pr(La)$Fe_{0.65}Ru_{0.05}Ni_{0.3}$ O_3 catalysts and various catalysts reported in the literature

Sample	Methane conversion, %	H_2/CO ratio	Carbon Deposits, wt.%	GHSVL/g*h	T, °C	Ref
6.6%Ni/Al_2O_3	68	-	67	22	700	[17]
4.5%Ni/SiO_2	50	-	1.2	21	750	[18]
10%Ni/3% CeO_2/3%La_2O_3/γ-Al_2O_3	77	1.09	17.5	12	800	[19]
10%Ni/8%Y_2O_3-ZrO_2	70	0.7	coke	36	800	[20]
1.5%Ru/CeO_2-ZrO_2	80	0.97	0.06	12	800	[30]
La_2NiO_4/ZSM-5	63	-	20	48	700	[95]
$La_{0.9}Pr_{0.1}NiO_3$	48	0.81	0.0	60	700	[56]
$LaNi_{0.9}Ru_{0.1}O_3$	85	0.62	20.3	72	750	[60]
$La_{3.5}Ru_4O_3$	76	0.62	0.938	72	750	[60]
$LaRu_{0.8}Ni_{0.2}O_3$	79	0.94	98.2*	48	700	[59]
$PrFe_{0.65}Ru_{0.05}Ni_{0.3}$ O_3	90	0.76	0.0	22	800	This
$LaFe_{0.65}Ru_{0.05}Ni_{0.3}$ O_3	87	0.72	0.0	22	800	work

* Carbon balance.

Note that the different conditions of catalysts testing make difficult the comparison of the catalysts activity, however, some conclusions could be drawn. The data show that performance of only Ni-based catalysts depends on the support nature, thus, the amount of coke changes from 1.2% (4.5%Ni/SiO_2) [17] to 67% (6.6%Ni/Al_2O_3) [18] while it is 0.06 % in the case of 1.5%Ru/CeO_2-ZrO_2 [30].

For catalysts based on pure nickel supported on different substrates the sintering is a main problem which could be avoided by using alloys of nickel with other metals (e.g. Fe, Co, Ru) [3-7, 15-16, 27-29]. Highly dispersed active metal alloy species of various composition fixed in oxide matrix are formed under MDR reaction conditions as a result of perovskite reduction [36-45, 50-62]. $LaNi_{1-x}Ru_x$ (x= 0 - 0.2) [60], $Ln_{1-x}Ca_xRu_{0.8}Ni_{0.2}O_3$ (Ln= La, Pr, Sm) [57, 59] have been used as precursors for bimetallic Ni-Ru catalysts containing metal particles distributed in a matrix of corresponding oxides (e.g. CaO-La_2O_3) which are formed after reductive treatment.

Though such systems are sufficiently active and stable in MDR, the formation of filamentous coke is observed in the case of $LaNi_{1-x}Ru_x$ catalysts with high Ni concentration [60]. As $Ln_{1-x}Ca_xRu_{0.8}Ni_{0.2}O_3$ catalyst is concerned, high concentration of Ru makes it rather expensive.

The use of $LnFe_{1-x}Ni_{0.3}Ru_xO_{3-\delta}$ (Ln=La, Pr, x=0÷0.1) precursors with low nickel and ruthenium concentrations provides a high dispersion of Ni-Fe-Ru alloy species epitaxially bound with the oxide matrix and, hence, a high activity and stability to coking and sintering. Furthermore, if activation of methane is determined by the metallic phase, activation of CO_2 occurs with the participation of the oxide matrix and could be controlled by its red-ox properties [54, 85].

In contrast to catalysts studied in [59-61], when perovskite phase completely decomposes in reaction conditions, nickelate-ferrite phase used as precursor in this work is only partially destroyed even after prolonged contact of catalysts with concentrated feed at high temperatures. Stable perovskite phase along with alloy clusters appear to be the most important factors providing a high and stable performance.

This can be explained by a high reactivity and mobility of the surface/lattice oxygen of these ferrite-type perovskites disordered due to reductive segregation of Ni and Ru [54, 85]. This helps to provide an efficient route of CO_2 activation on the surface of perovskites and fast transfer of oxygen atoms or oxygen–containing species (hydroxyls, hydroxocarbonates, etc) to the interface with segregated Ni-Fe-Ru alloy particles where they consume activated CH_x fragments – coke precursors.

6. Steam Reforming of Ethanol

6.1. C_2H_5OH TPR

Typical curves of CO, CO_2 and H_2 evolution during perovskites and nanocomposite temperature-programmed reduction by C_2H_5OH are presented in Figures 46-48.

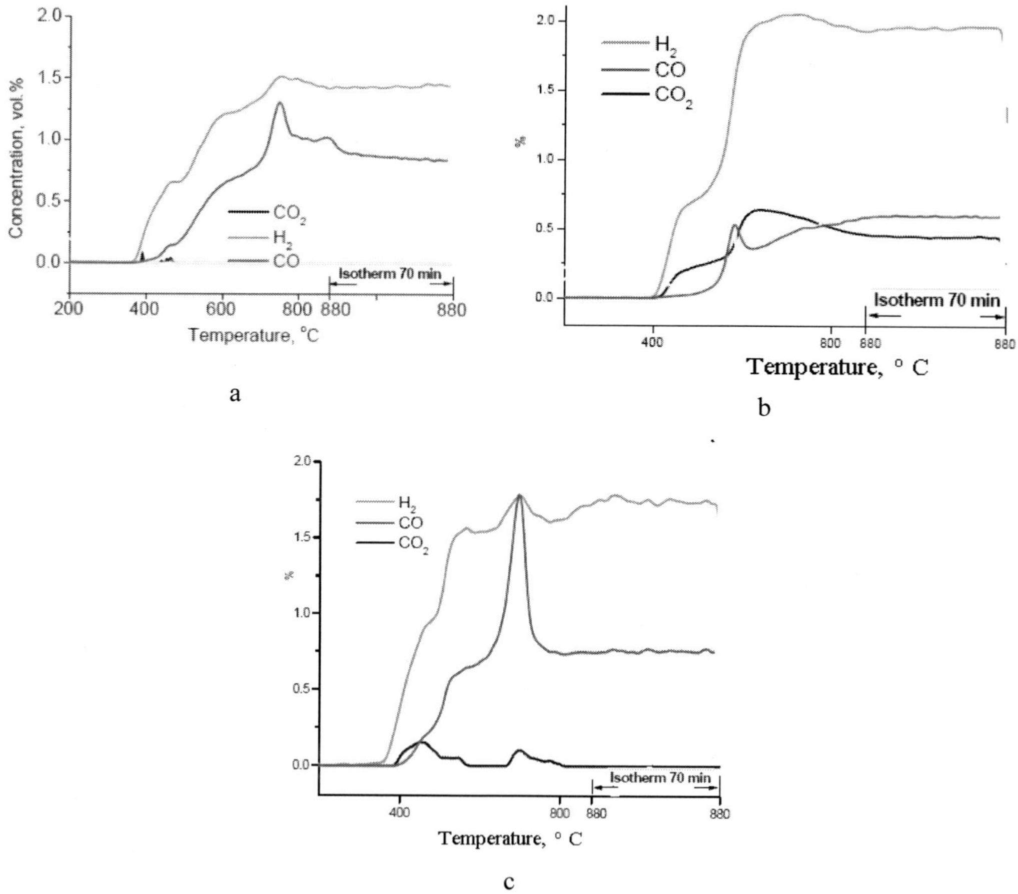

Figure 46. CO, CO_2 and H_2 evolving during temperature programmed reduction of perovskites $LnFe_{0.7}Ni_{0.3}O_{3-\delta}$ with C_2H_5OH: a – La, b – Pr, c - Sm. Temperature ramp 5°/min, 1% C_2H_5OH in He, feed rate 5 l/h.

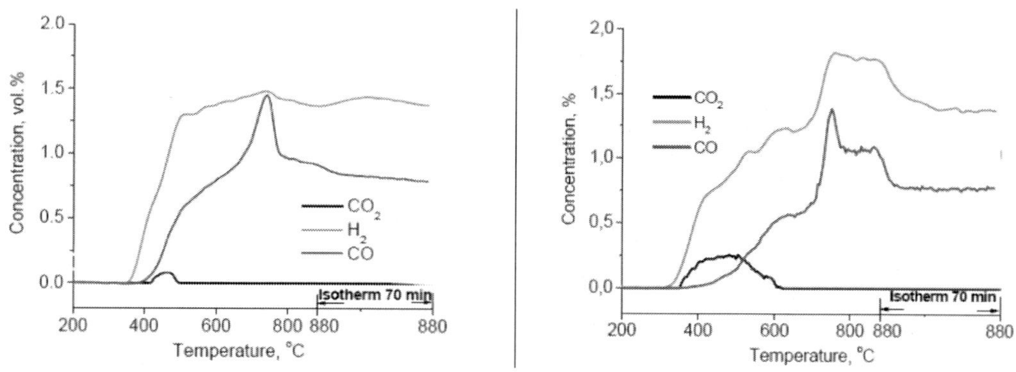

Figure 47. CO, CO_2 and H_2 evolving during temperature programmed reduction of $PrFe_{0.65}Ru_{0.05}Ni_{0.3}O_{3-\delta}$ (left) and $LaFe_{0.4}Mn_{0.3\,5}Ni_{0.3}O_{3-\delta}$ (right) by C_2H_5OH. Temperature ramp 5°/min, 1% C_2H_5OH in He, feed rate 5 l/h.

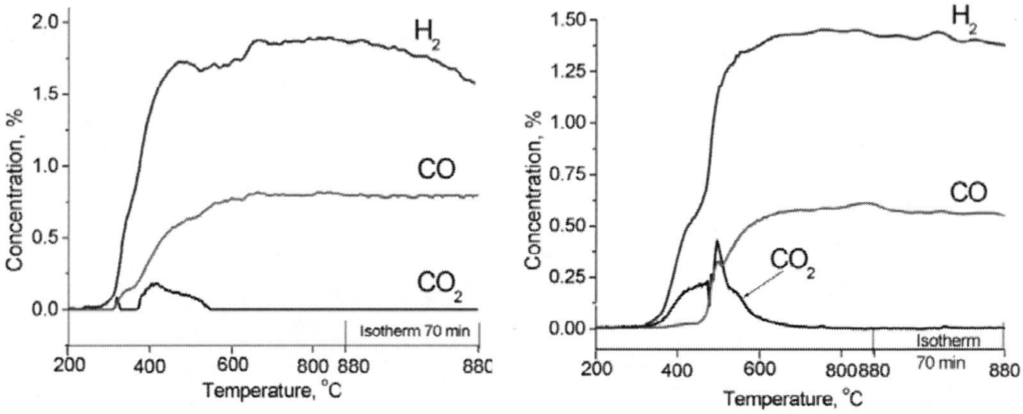

a b

Figure 48. Typical curves of C_2H_5OH TPR for Ru+Ni/LaPrMnCrO (a) and 80%LaPrMnCrO+10% NIO + 10% YSZ (b) samples. Temperature ramp 5°/min, 1% C_2H_5OH in He, feed rate 5 l/h.

Along with main products, admixtures of water and CH_x not shown in Figures were observed. Unreacted ethanol and liquid products were frozen in the trap which precludes strict estimation of ethanol conversion and carbon balance. However, the data on dynamics of main products evolution contain useful information about reactivity and ethanol transformation routes for different perovskites.

In general, the interaction of ethanol with all LnFN perovskites is similar: at 320-400°C hydrogen appears followed by other products. For perovskites $LnFe_{0.7}Ni_{0.3}O_3$, the starting temperature of H_2 evolution changes with the nature of Ln cation: a lower temperature is observed in the case of $SmFe_{0.7}Ni_{0.3}O_3$.

Simultaneously, CO and CO_2 profiles differ noticeably: for $LaFe_{0.7}Ni_{0.3}O_3$, CO_2 is practically absent while its concentration is rather high in the case of $PrFe_{0.7}Ni_{0.3}O_3$ caused by a high red-ox activity of praseodymium oxides [56, 95, 96]. Partial substitution of Fe with Ru or Mn shifts the start of H_2 evolution and, therefore, ethanol decomposition, towards low temperatures (Figure 48). Similarly, reduction of Ru+Ni-supported LaPrMnCr perovskite and Ru+Ni/LaPrMnCr /YSZ nanocomposite starts at the lowest (~ 250 °C) temperatures (Figure 48).

For all catalysts, evolution of H_2 and CO continues in the isothermal mode at 880 °C when active metal alloy particles are formed from partially reduced perovskites as indicated above [54, 55]. This evidences continuous decomposition of ethanol [97-99] and though, without doubt, formation of surface carbonaceous species simultaneously occurs, the performance of samples during isothermal run is practically stable (Figures 46-48) showing a loose nature of coke deposits.

At the steady state, the value of H_2/CO ratio in products changes from 1.6-1.76 for La-containing perovskites up to 3.25 for $PrFe_{0.7}Ni_{0.3}O_3$ (Figure 49) showing the influence of Ln-cation nature on the selectivity of ethanol decomposition.

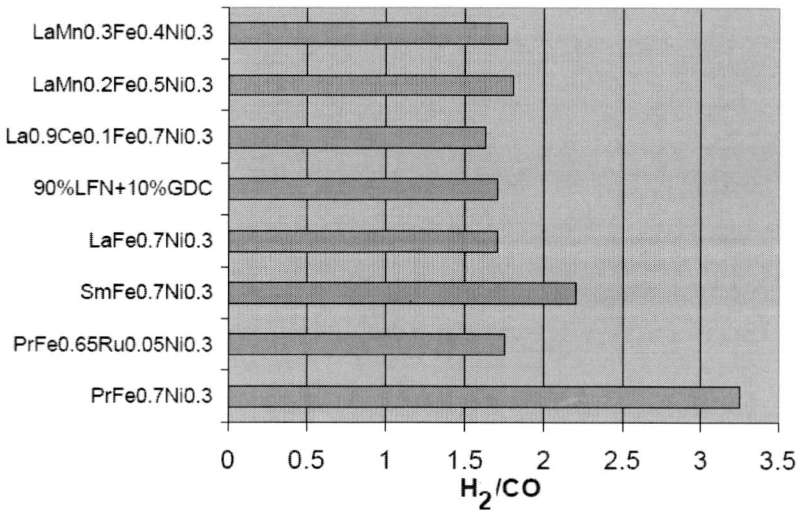

Figure 49. Steady-state H_2/CO ratio during temperature programmed reduction of perovskites by C_2H_5OH.

The data of temperature-programmed reoxidation by H_2O (H_2O-TPO) of deposited carbonaceous species for different catalysts are presented in Figure 50.

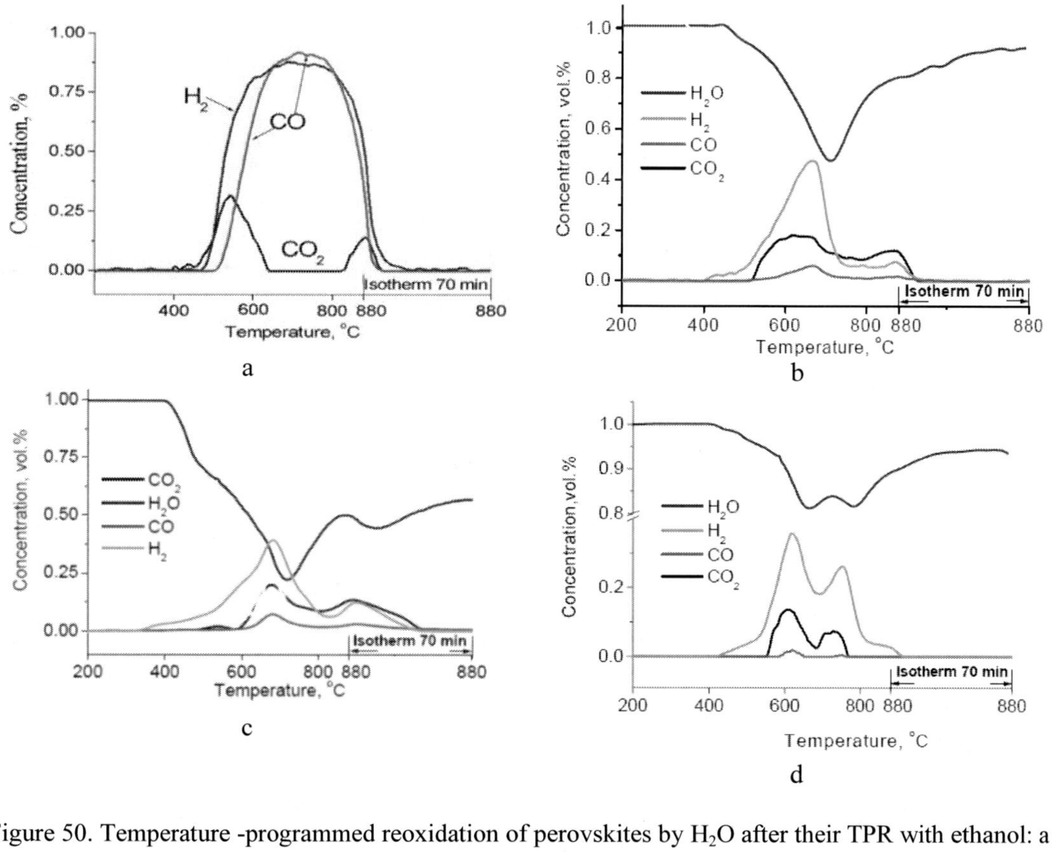

Figure 50. Temperature -programmed reoxidation of perovskites by H_2O after their TPR with ethanol: a – Ru+Ni/LaPrMnCr, b - $LaFe_{0.7}Ni_{0.3}O_{3-\delta}$; c - $La_{0.9}Ce_{0.1}Fe0_{.7}Ni_{0.3}O_{3-\delta}$; d - $LaFe_{0.4}Mn_{0.3}Ni_{0.3}O_{3-\delta}$

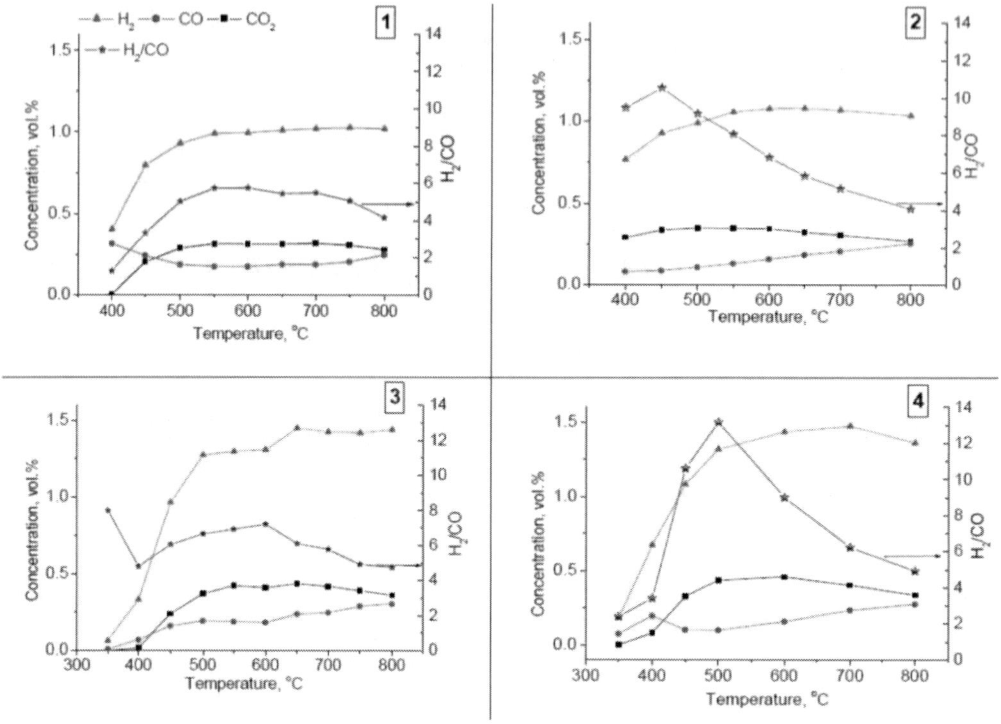

Figure 51. Temperature dependence of H_2, CO, CO_2 concentrations and H_2/CO ratio in EtOH steam reforming on $PrFe_{0.7}Ni_{0.3}O_{3-\delta}$ (1), $PrFe_{0.65}Ru_{0.05}Ni_{0.3}O_{3-\delta}$ (2), $LaFe_{0.5}Mn_{02}Ni_{0.3}O_{3-\delta}$ (3) and $SmFe_{0.7}Ni_{0.3}O_{3-\delta}$. (4). Feed 0.5 % C_2H_5OH + 2 % H_2O in He, contact time 36 ms.

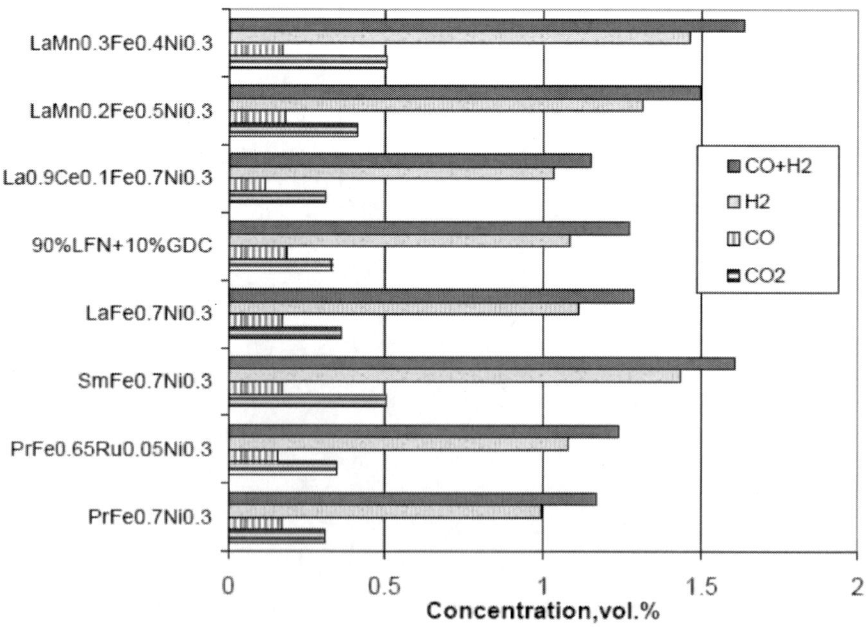

Figure 52. Comparison of H_2, CO, CO_2 concentrations and CO+H_2 sum at 600 °C in ethanol steam reforming on perovskites. Feed 0.5 % C_2H_5OH + 2 .% H_2O in He, contact time 36 ms.

The profiles of H_2O and product concentration as well as a starting temperature of their evolution depend on the perovskite composition. For ferrites, the evolution of H_2 due to oxidation of reduced perovskite (both surface layers and bulk of particles) by H_2O yielding hydrogen starts at rather low temperatures (340 – 400 °C). After achieving some level of surface coverage by reactive oxygen species, they begin to oxidize the surface coke which is reflected in CO_2 and CO formation. In some cases, positions of peaks of H_2, CO and CO_2 evolution are correlated, thus suggesting occurrence of WGS reaction due to interaction of CO produced by coke oxidation with water yielding H_2 and CO_2. For Ni+Ru-suppported LaPrMnCr H_2, CO and CO_2 appear simultaneously (Figure 50), which suggests a higher rate of the surface reoxidation by water as well as a high reactivity of surface oxygen forms bound with Pr, Mn and Cr cations. For all catalysts, the process of coke oxidation is completed at temperatures < 800 °C, thus demonstrating a high reactivity of surface coke species.

6.2. Ethanol Steam Reforming in Diluted Feed: Screening Experiments

To estimate the scope of catalytic activity variation for perovskite-based catalysts, the screening experiments in diluted reaction mixture (0.5vol.% C_2H_5OH + 2.0vol.% H_2O in He) were carried out. In these experiments, concentrations of H_2 and CO were chosen as parameters characterizing catalytic performance. Catalysts were reoxidized by 1% H_2O in He after C_2H_5OH TPR (vide supra) and brought into the steady-state under contact with reaction feed 0.5.% C_2H_5OH + 2.0% H_2O in He at 800 °C, then temperature was step-wise decreased. The typical temperature dependence of product concentrations is presented in Figure 51.

For all perovskites, a high hydrogen concentration is already attained at 500-600 °C (Figure 52) thus their catalytic performance in ESR is sufficiently effective.

Performance of Ru+Ni/LaPrMnCrO catalyst is rather close to that of the best perovskites with a higher Ni content containing Mn cations (Figure 53).

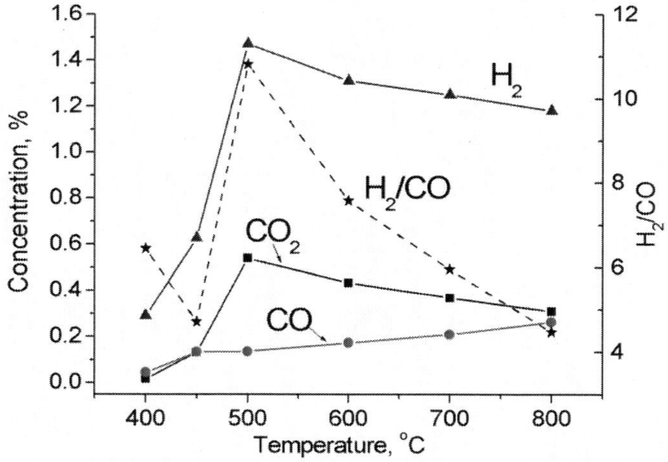

Figure 53. Temperature dependence of products concentration and H_2/CO ratio for Ru+Ni/LaPrMnCrO catalyst in ethanol steam reforming. Feed 0.5 % C_2H_5OH + 2.0 % H_2O in He, contact time 36 ms.

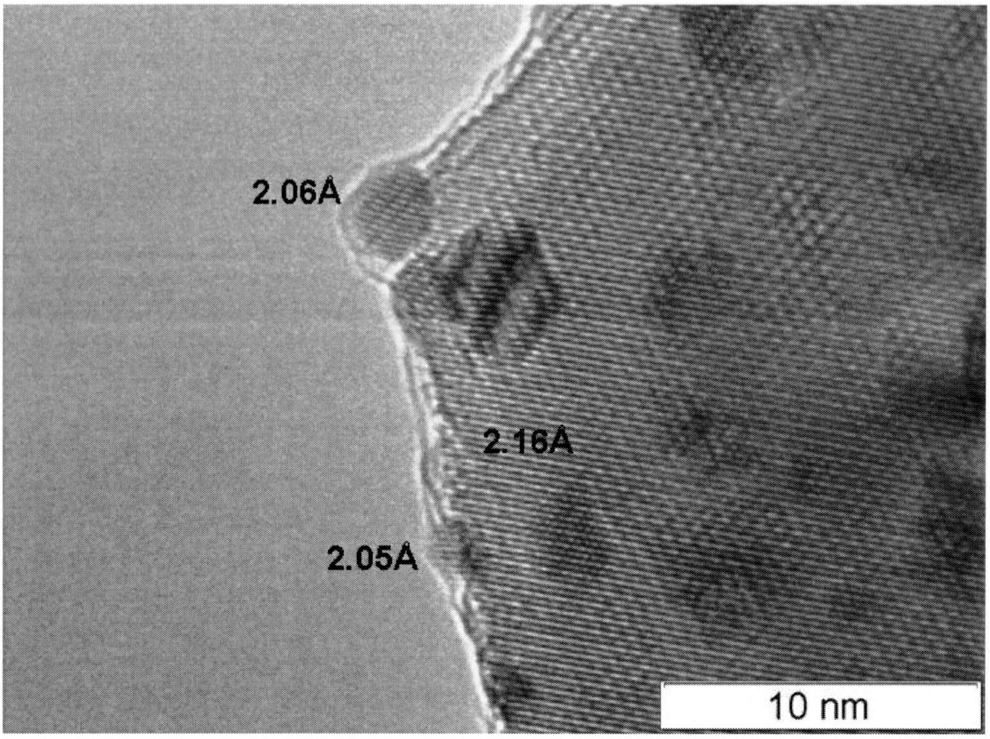

Figure 54. High resolution TEM image of Ru+Ni/LaPrMnCrO catalyst particle after reaction of EtOH steam reforming in diluted feed 0.5 % C_2H_5OH + 2.0% H_2O in He. Lattice spacing corresponding to Ru-Ni alloy indicated.

A high specific activity of this catalyst is apparently ensured by high concentration of small Ru-Ni alloy nanoparticles/clusters strongly interacting with LaPrMnCr perovskite support due to a good epitaxy (Figure 54).

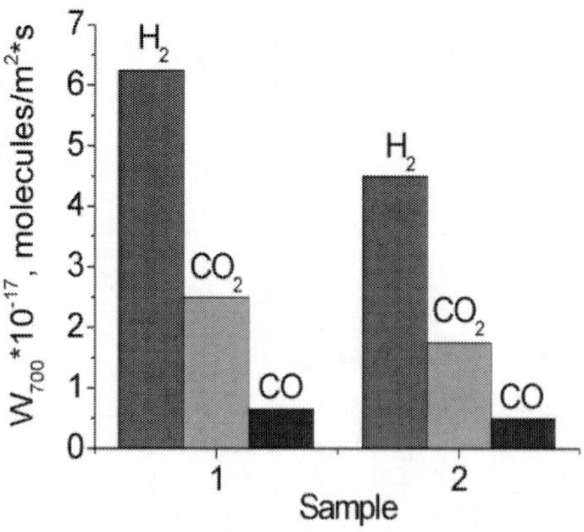

Figure 55. Rates of products formation in ethanol steam reforming at 700°C on nanocomposites. Contact time 36ms, feed 0.5% EtOH + 2.4%H_2O in He. 1-50%LaMnCrPr+30%NiO+20%YSZ, 2-80%LaMnCrPr +10%NiO+10%YSZ.

For nanocomposite 80%LaMnCrPr+10%NiO+10%YSZ even without Ru a similar concentrations of products (1.1%H_2, 0.5%CO_2, 0.15% CO) were achieved in the same experiments at 600 °C. The increase of Ni content in nanocomposite allows increasing further the rates of products formation (Figure 55).

6.3. Ethanol Steam Reforming in Realistic Feeds

The activity of some catalysts has been studied in the reaction mixture with composition: EtOH:H_2O:N_2= 1:4:5. The ethanol conversion over studied perovskites is high (75 - 100%) already at 700 °C (contact time 70 ms) and substantially exceeds the values in a blank experiments (Figure 56).

The typical temperature dependence of products concentration is shown in Figure 57.

In all temperature range, the main products are hydrogen and CO. The presence of acetaldehyde (~1 vol.%) at 650°C and the only trace concentration of ethylene evidence that the main route of ESR reaction over studied perovskites is the dehydrogenation of ethanol [35]. At increasing temperature the concentration of CH_4 increases due to ethanol cracking [35, 65-67]. In blank experiments (Figure 58) selectivity of ethanol transformation is different, main products being CO, CH_4 and C_2H_4. Hence, such processes as ethanol cracking and dehydration apparently dominate for gas-phase reactions catalyzed by quartz reactor walls.

The temperature dependence of hydrogen yield and selectivity of other products are presented in Figures 59 and 60.

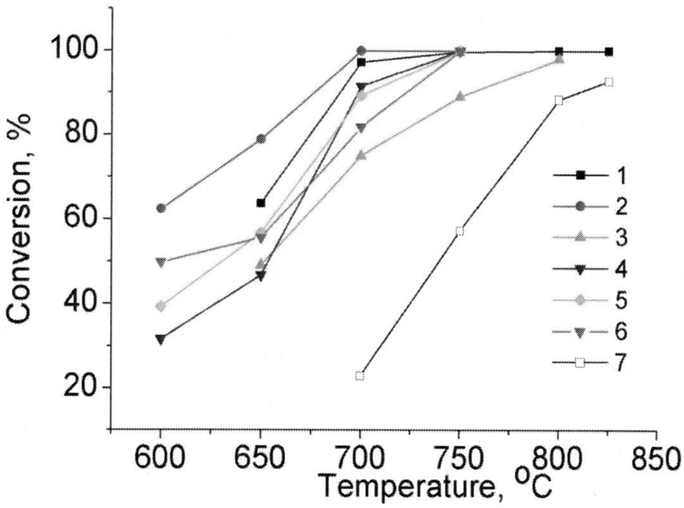

Figure 56. Temperature dependence of ethanol conversion in ESR. Reaction mixture EtOH:H_2O:N_2= 1:4:5, contact time 70 ms. 1- $PrFe_{0.7}Ni_{0.3}O_{3-\delta}$, 2 - $PrFe_{0.65}Ru_{0.05}Ni_{0.3}O_{3-\delta}$, 3 – $LaFe_{0.7}Ni_{0.3}O_{3-\delta}$,4 – $La_{0.7}Pr_{0.3}Fe_{0.7}Ni_{0.3}O_{3-\delta}$, 5 - $LaFe_{0.5}Mn_{0.2}Ni_{0.3}O_{3-\delta}$, 6 - $LaFe_{0.4}Mn_{0.3}Ni_{0.3}O_{3-\delta}$, 7 – blank experiment. Perovskites were activated at 850°C in the reaction mixture.

The yield of hydrogen is higher for Pr-containing perovskites as compared with $LaFe_{0.7}Ni_{0.3}O_{3-\delta}$ and only slightly increases when Ru is introduced in perovskite (Figure 59).

At 650 – 750 °C, CO selectivity decreases in the row: $PrFe_{0.7}Ni_{0.3}O_{3-\delta}$ >$PrFe_{0.65}Ru_{0.05}Ni_{0.3}O_{3-\delta}$> $LaFe_{0.7}Ni_{0.3}O_{3-\delta}$, being constant at increasing temperature for only Ni-containing perovskites while for $PrFe_{0.65}Ru_{0.05}Ni_{0.3}O_{3-\delta}$ its decreases. The increasing of CH_4 selectivity with the temperature depends on the perovskite composition: it is lowest in the case of $PrFe_{0.7}Ni_{0.3}O_{3-\delta}$ (Figure 59).

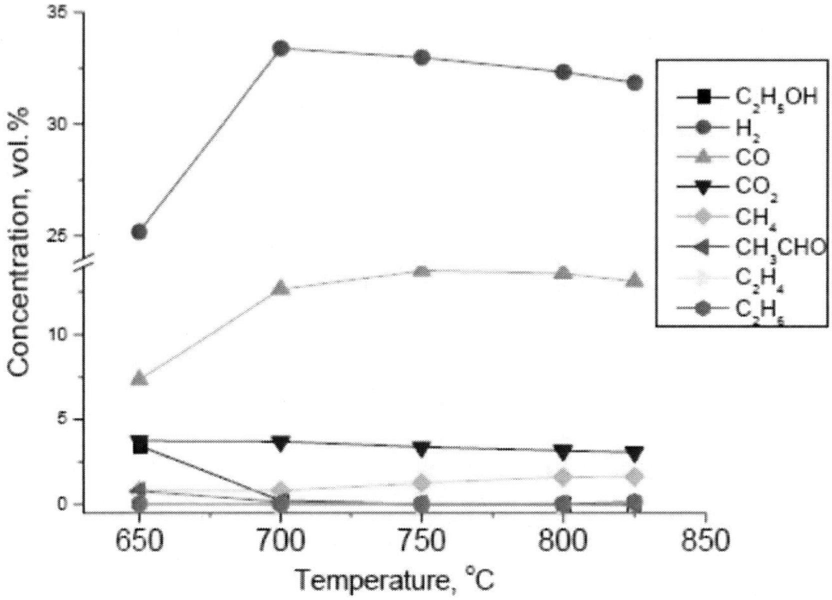

Figure 57. Temperature dependence of the exit concentrations in ESR on $PrFe_{0.7}Ni_{0.3}O_{3-\delta}$ catalyst activated at 850°C in reaction mixture EtOH:H_2O:N_2= 1:4:5, contact time 70 ms.

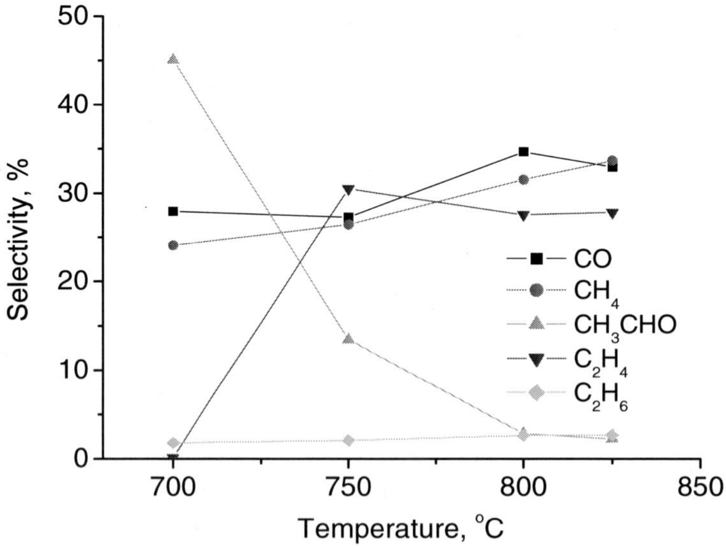

Figure 58. Selectivity of products formation in blank experiments

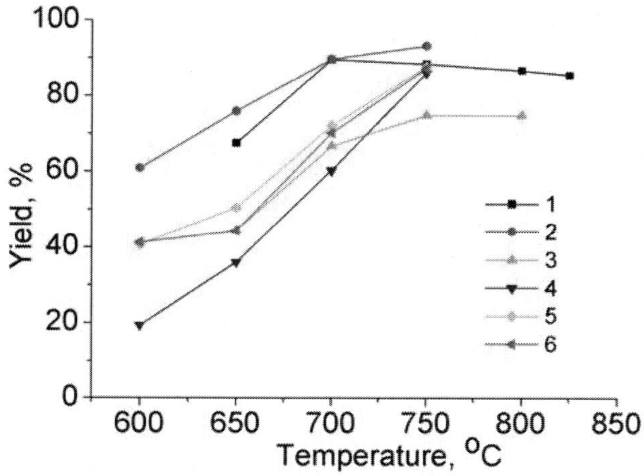

Figure 59. Temperature dependence of H_2 yield in ESR. 1- $PrFe_{0.7}Ni_{0.3}O_{3-\delta}$, 2 - $PrFe_{0.65}Ru_{0.05}Ni_{0.3}O_{3-\delta}$, 3 – $LaFe_{0.7}Ni_{0.3}O_{3-\delta}$, 4 – $La_{0.7}Pr_{0.3}Fe_{0.7}Ni_{0.3}O_{3-\delta}$, 5 - $LaFe_{0.5}Mn_{0.2}Ni_{0.3}O_{3-\delta}$, 6 - $LaFe_{0.4}Mn_{0.3}Ni_{0.3}O_{3-\delta}$ activated at 850°C and tested in a reaction mixture EtOH:H_2O:N_2= 1:4:5, contact time 70 ms.

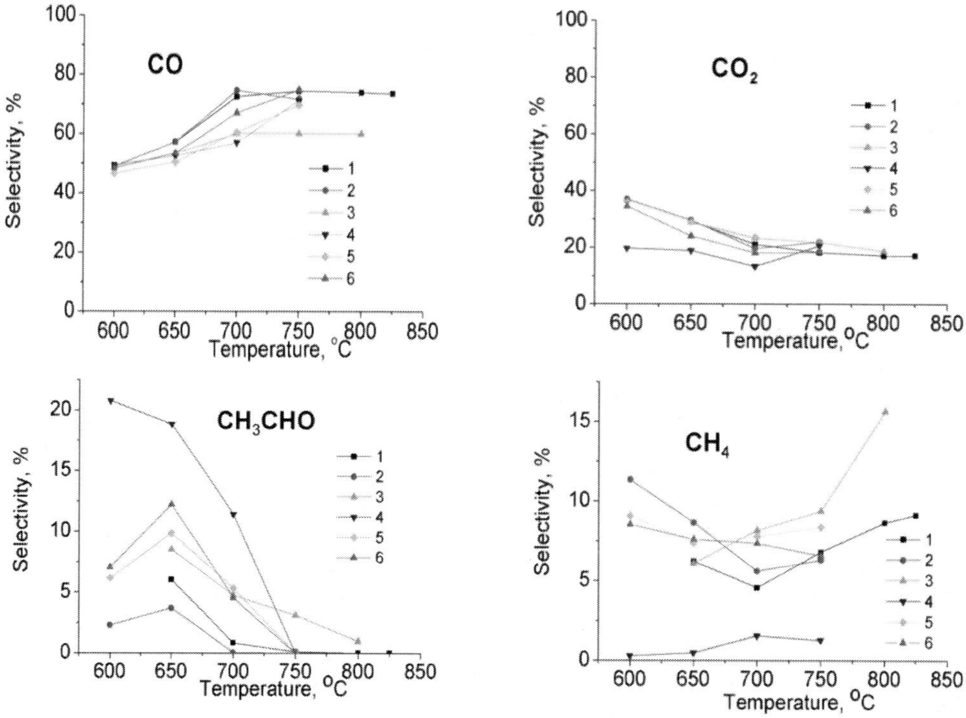

Figure 60. Temperature dependence of products selectivity in ESR. 1- $PrFe_{0.7}Ni_{0.3}O_{3-\delta}$, 2 - $PrFe_{0.65}Ru_{0.05}Ni_{0.3}O_{3-\delta}$, 3 – $LaFe_{0.7}Ni_{0.3}O_{3-\delta}$, 4 – $La_{0.7}Pr_{0.3}Fe_{0.7}Ni_{0.3}O_{3-\delta}$, 5 - $LaFe_{0.5}Mn_{0.2}Ni_{0.3}O_{3-\delta}$, 6 - $LaFe_{0.4}Mn_{0.3}Ni_{0.3}O_{3-\delta}$ activated at 850°C and tested in a reaction mixture EtOH:H_2O:N_2= 1:4:5, contact time 70 ms.

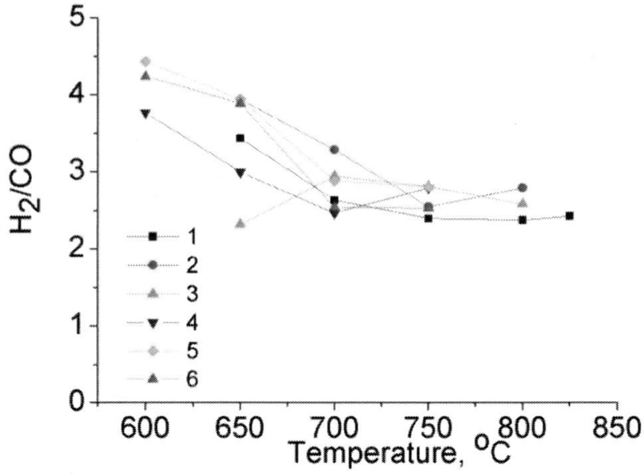

Figure 61. Temperature dependence of H_2/CO ratio in ESR. 1- $PrFe_{0.7}Ni_{0.3}O_{3-\delta}$, 2 - $PrFe_{0.65}Ru_{0.05}Ni_{0.3}O_{3-\delta}$, 3 – $LaFe_{0.7}Ni_{0.3}O_{3-\delta}$, 4 – $La_{0.7}Pr_{0.3}Fe_{0.7}Ni_{0.3}O_{3-\delta}$, 5 - $LaFe_{0.5}Mn_{0.2}Ni_{0.3}O_{3-\delta}$, 6 - $LaFe_{0.4}Mn_{0.3}Ni_{0.3}O_{3-\delta}$. Reaction mixture $EtOH:H_2O:N_2$= 1:4:5, contact time 70 ms.

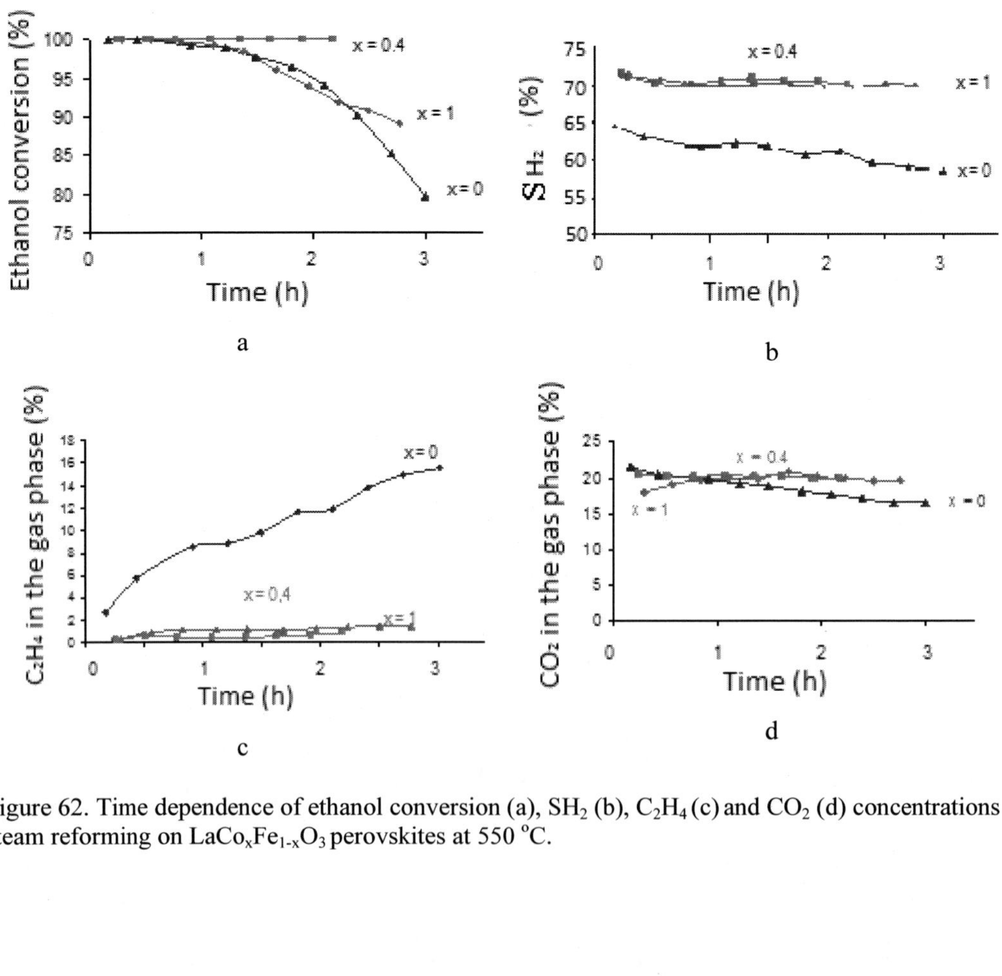

Figure 62. Time dependence of ethanol conversion (a), SH_2 (b), C_2H_4 (c) and CO_2 (d) concentrations in steam reforming on $LaCo_xFe_{1-x}O_3$ perovskites at 550 °C.

The values of H_2/CO ratio (~2.5) in the products are close for all perovskites in the temperature range of 700 - 750°C being different at lower and higher temperatures (Figure 61).

Dynamics of the ethanol conversion variation in the course of its steam reforming for $LaFeO_3$ (x=0), $LaCo_{0.4}Fe_{0.6}O_3$ (x=0.4) and $LaCoO_3$ (x=1) at 550°C is given in Figure 62a.

The total conversion is obtained for $LaCo_{0.4}Fe_{0.6}O_3$ mixed perovskite over the whole period of test. For $LaFeO_3$ and $LaCoO_3$, the ethanol conversion rapidly declines after the first hour of test. Hydrogen selectivity (Figure 62b) remains at a stable and high level (around 71%) for the Co-containing catalysts. For $LaFeO_3$ it slowly decreases from 65 to 59%, along with the decrease of conversion. For this system, the decreasing H_2 selectivity is clearly explained by the increase of ethylene formation with time on stream, from 2 to 16% (Figure 62c). In the case of $LaCo_{0.4}Fe_{0.6}O_3$ and $LaCoO_3$, the ethylene selectivity is always lower than 1%. Let's note that the tendency of the catalysts to form CH_4 follows the same trend. The initial CH_4 selectivity is of 7.6, 1.2 and 1.1 % for x=0, 0.4 and 1, respectively. As shown in Figure 6d, the formation of CO_2 is less affected by the composition of perovskites.

The intermediate composition $LaCo_{0.4}Fe_{0.6}O_3$ is clearly the best formulation for a good activity in ethanol steam reforming, a good selectivity to hydrogen and a stable catalytic behaviour.

The high H_2 selectivity is clearly related to the Co content in perovskites, which is required for efficient steam reforming reaction. For lanthanum ferrite only formation of methane and ethylene takes place to the detriment of H_2.

The ethanol conversion is optimal for $LaCo_{0.4}Fe_{0.6}O_3$. In that case, the intermediate reducibility is optimal to stabilize a part of Co^{2+} in the perovskite, in addition to the generation of Co° metallic phase under the reaction conditions. The $Co°/Co^{2+}$ interaction is crucial for a stable catalytic behaviour in steam reforming reactions [101-102]. In addition, similar to the case of mixed nickelate-ferrite perovskites, a high and stable performance of this composition can be provided by stabilization of small Co^{o} clusters on the surface of defect La ferrite with a high mobility and reactivity of surface/lattice oxygen species (vide supra). Since performance of La-Co-Fe-O perovskites was tested at comparable contact times (~ 140 ms vs. 70 ms for LnNiFe perovskites), complete conversion of ethanol at 550 °C suggests substantially higher specific activity of Co clusters on perovskite support as compared with Ni (Ni-Ru) clusters (cf. Figure 56, maximum EtOH conversion ~60 % at 600 °C), which certainly requires further studies.

7. Testing of Structured Catalysts in Oxygenates Reforming

Testing of structured/monolithic catalysts in concentrated feeds in steam/autothermal reforming of oxygenates in pilot-scale reactors allows to assess the effect of the heat and mass transfer on performance and stability. Since heat- conducting substrates based on metal alloys (foils, gauzes, etc) and SiC are apparently attractive for such application, they were used for preparation of structured catalysts as described in details in previous publications [73, 74, 85, 86]. Here only some examples are given to demonstrate efficiency of perovskite-based active components in structured monolithic catalysts of oxygenates reforming.

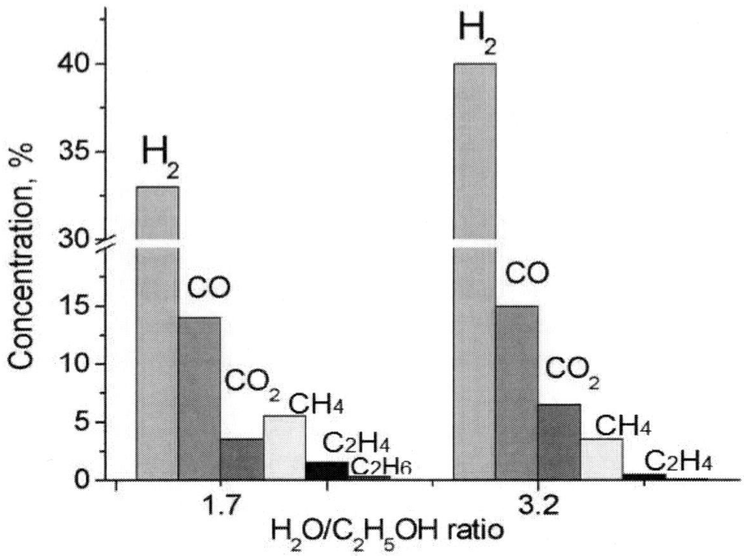

Figure 63. Concentration of products in the effluent stream of ethanol oxysteam reforming on stack comprised of 12 Ni-Al-foam plates and 11 sheets of Fecralloy gauzes loaded with 2%Ru/(80%$La_{0.8}Pr_{0.2}Mn_{0.2}Cr_{0.8}O_3$ + 10%NiO + 10%YSZ) (volume 34x34x34 mm^3). Feed composition: 5% O_2+ H_2O +25% EtOH + N_2; T inlet 700 °C, contact time 0.3 s.

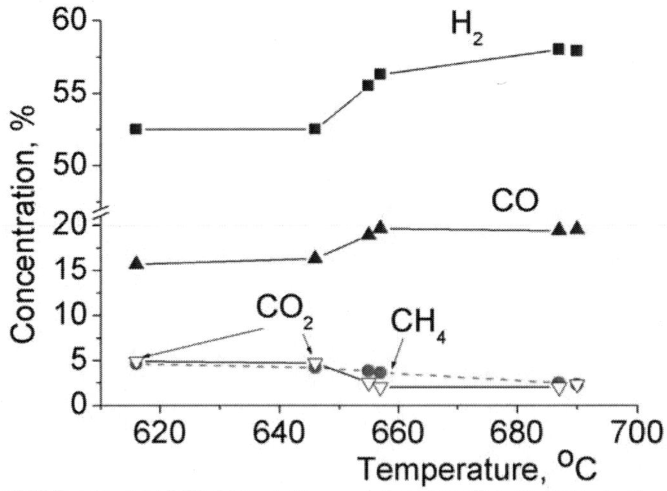

Figure 64. Temperature dependence of products concentration for LaNiPt/PrCeZrO catalyst supported on SiC monolith in ethanol steam reforming. Feed composition 30% ethanol + 52%H_2O + 18% N_2, contact time 0.83 sec.

Figure 63 demonstrates performance characteristics in ethanol autothermal reforming of nanocomposite active component Ru+Ni/80%$La_{0.8}Pr_{0.2}Mn_{0.2}Cr_{0.8}O_3$+10%YSZ loaded in amount corresponding to ~ 10 wt.% on stack comprised of 12 compressed Ni-Al foam platelets alternating with Fechraloy gauzes [73].

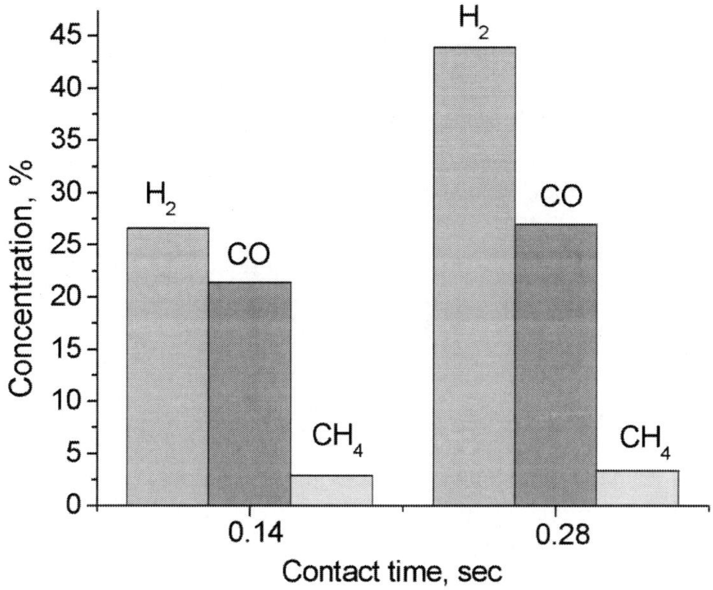

Figure 65. Effect of contact on concentrations of products of glycerol steam reforming on stack comprised of 12 Ni-Al-foam plates and 11 sheets of Fecralloy gauzes loaded with 2%Ru/(80%$La_{0.8}Pr_{0.2}Mn_{0.2}Cr_{0.8}O_3$ + 10%NiO + 10%YSZ) active component (volume 34x34x34 mm^3). Feed 27% Glycerol + 47% H_2O, N_2 balance, inlet temperature 550-600 °C.

As follows from these results, in the intermediate temperature range a high yield of syngas in the oxysteam reforming of ethanol is obtained even at a small excess of steam. Along with syngas, usual by-products – CH_4, C_2H_4, C_2H_6 were observed due to concurrent dehydration, hydrogenation and cracking reactions (vide supra). The stack performance was stable at least for 20 h, which implies that even ethylene –well known coke precursor is not only produced but also consumed on the perovskite surface due to interaction with active oxygen-containing species. Indeed, at a higher steam excess concentrations of by-products strongly decline while concentration of syngas increases (Figure 63).

For another type of heat-conducting substrate - SiC monolith with porous walls and square channels nanocomposite perovskite-fluorite active component also provides a high and stable performance in steam reforming of ethanol in the intermediate temperature range (Figure 64), though at somewhat longer contact times.

At last, in steam reforming of another very popular type of biofuel-glycerol, nanocomposite active component Ru+Ni/80%$La_{0.8}Pr_{0.2}Mn_{0.2}Cr_{0.8}O_3$ +10% YSZ on heat-conducting substrate –stack of Ni-Al foam plates and FeChraloy gauzes also demonstrated a high and stable performance at a small steam excess (Figure 65).

CONCLUSION

The most efficient approach to design of efficient and stable to coking nanocomposite catalysts of natural gas and oxygenates reforming into syngas consists in synthesis of complex perovskite precursors containing transition metal cations with broadly varying redox properties such as easily reduced to metal Ni, Co, Ru cations and much less prone to reduction Cr, Fe and Mn cations. Under reaction conditions, these perovskites are transformed into nanocomposites comprised of metal/ alloy nanoparticles and LnO_x patches epitaxially bound with disordered Ln–(Fe, Mn, Cr)–O perovskite particles. This provides developed interface between metal components activating fuel molecules and complex oxides involved in activation of mild oxidants (H_2O and CO_2). Fast transfer of oxygen species from the oxide sites to the metal/oxide interface where they transform activated CH_xO_y molecules into syngas ensures operation of the bifunctional mechanism of fuels reforming. High catalytic activity and coking stability is provided by tuning the oxygen mobility and reactivity in complex oxide supports (via changing the type and content of A (La, Pr, Sm) and B (Fe, Cr, Mn) cations in perovskites and/or by adding fluorite-like oxides with a high oxygen mobility) as well as by optimizing the size and composition of metal/alloy nanoparticles. Strong metal –oxide support interaction provides also stability of metal nanoparticles to sintering in real operation conditions.

Best compositions of nanocomposite active components supported as porous layers on heat-conducting substrates demonstrated a high and stable performance in dry refroming of real natural gas and steam/oxysteam reforming of ethanol and gycerol in concentrated feeds.

Acknowledgments

The authors gratefully acknowledge support from FP 7 OCMOL Project, Integration Project 8 of SB RAS-NAN Belarus, Project 50 of Presidium RAS, Russian Fund of Basic Research Projects RFBR-CNRS 09-03-93112 and 12-03-93115, Federal Program “Scientific and Educational Cadres of Russia”.

References

[1] Graves C.; Ebbesen S. D.; Mogensen M. and Lackner, K. S. *Renewable and Sustainable Energy Reviews*, 2011, *15*, 1 - 23.
[2] Vlachos, D.G.; Caratzoulas, S. *Chemical Engineering Science*, 2010, *65*, 18-29.
[3] Bradford, M. C. J. and Vannice, M. A. *Catalysis Reveiws*, 1999, *41(1),* 1-42.
[4] Spivey, J. J. and Inui, T. *Catalysis*, *The Royal Society of Chemistry*, 2002, *16,* 133-151.
[5] Hu, Y. H. and Ruckenstein, E. *Advances in Catalysis*, 2004, *48*, 297-345.
[6] York, A. P. E.; Xiao, T.; Green M. L. H. and Claridge, J. *Catalysis Reviews*, 2007, *49*, 511-560.
[7] Ma, J.; Sun, N.; Zhang, X.; Zhao, N.; Xiao, F.; Wei, W. and Sun, Y. *Catalysis Today*, 2009, *148*, 221-231.
[8] Adesina, A. A. *Current Opinion in Chemical Engineering,* 2012, *1*, 272–280.
[9] Demirbas, A. *Progress in Energy and Combustion Science,* 2007, *33*, 1–18.
[10] Sarkar, S.; Kumar, A. *Bioresource Technology,* 2010, *101*, 7350-7361.
[11] Gupta, K. K.; Rehman, A.; Sarviy, R. M. *Renewable and Sustainable Energy Reviews* 2010, *14*, 2946 - 2955.
[12] Serrano-Ruiz, J. C.; Dumesic, J. A. *Energy Environ. Sci.* 2011, *4,* 83-99.

[13] Yung, M. M.; Jablonski, W. S. and Magrini-Bair, K. A. *Energy & Fuels,* 2009, *23,* 1874 –1887.

[14] Kırtay, E. *Energy Conversion and Management*, 2011, *52*, 1778 – 1789.

[15] Gonzalez-Delacruz, V. M.; Ternero, F.; Pereniguez, R.; Caballero, A. and Holgado, J. P. *Applied Catalysis A: General*, 2010, *384*, 1 - 9.

[16] Daza, C. E.; Kiennemann, A.; Moreno, S. and Molina, R. *Applied Catalysis A: General*, 2009, *364*, 65 - 74.

[17] Juan-Juan, J.; Roman-Martınez, M. C.; Illan-Gomez, M. J. *Applied Catalysis A: General*, 2006*, 301*, 9 - 15.

[18] Effendi, A.; Hellgard, K.; Zhang, Z.-G.; Yoshida, T. *Catal. Comm.*, 2003, 4, 203-207.

[19] Yang, R.; Xing, C.; Lv, C.; Shi, L.; Tsubaki, N. *Applied Catalysis A: General,* 2010, *385*, 92 - 100.

[20] Bellido, J.D.A.; Assaf, E.M. *Applied Catalysis A: General*, 2009, *352*, 179 - 187.

[21] Biswas, P.; Kunzru, D. *Hem. Eng. J.*, 2008, *136*, 41 – 49.

[22] Haga, F.; Nakajima, T.; Miya, H.; Mishima, S. *Catal. Lett.* 1997, *48*, 223 - 227.

[23] Rodrigues, C. P.; da Silva, Victor T.; Schmal, *M. Cat. Com.*, 2009, *10*, 1697 – 1701.

[24] Resinia, C.; Concepcio, M.; Delgado, H.; Presto, S.; Alemany, L.; Riani, J.; Marazza, P.; Ramis, R.; Busca, G.; Int. G. *Int. J. Hydrogen Energy*, 2008, 33, 3728 – 3735.

[25] Velu, S.; Satoh, N.; Gopinath, C. S.; Suzuki, K. *Catal. Lett.* 2002, *82* (1), 145 - 152.

[26] Yaseneva, P.; Pavlova, S.; Sadykov, V.; Alikina, G.; Lukashevich, A.; Rogov, V.; Belochapkine, S.; Ross. J.; *Cat. Today*, 2008, *137*, 23 – 28.

[27] Souza, M. M. V. M.; Aranda, D. A. G.; and Schmal, M. *Journal of Catalysis*, 2001, *204,* 498 - 511.

[28] Safariamin, M.; Tidahy, L. H.; Abi-Aad, E.; Siffert, S. and Aboukais, A. *Comptes Rendus Chimie*, 2009, *12*, 748 - 753.

[29] Carrara, C.; Roa, A.; Cornaglia, L.; Lombardo, E.; A. Mateos-Pedrero, C. and Ruiz, P. *Catalysis Today*, 2008, *133–135*, 344 – 350.

[30] Chen, J.; Yao, C.; Zhao, Y.; Jia, P. *Int. J. Hydr. Energy*, 2010*, 35*, 1630 – 1642.

[31] Domine, M. E.; Iojoiu, E. E.; Davidian, T.; Guilhaume, N.; Mirodatos, C. *Catalysis Today,* 2008, *133–135*, 565 – 573.

[32] Sadykov, V.; Mezentseva, N.; Alikina, G.; Bunina, R.; Rogov, V.; Krieger, T.; Belochapkine, S.; Ross, J. *Catalysis Today*, 2009, *145*, 127 – 137.

[33] Busca, G.; Montanari, T.; Resini, C.; Ramis, G.; Costantino, U. *Catalysis Today*, 2009. *143*, 2 - 8.

[34] Takanabe, K.; Aika, K.; Seshan, L.; Lefferts, *J. Catal*. 2004, *227,* 101 – 108.

[35] Vaidya, P.D.; Rodrigues, A.E. *Chem. Eng. J.*, 2006, *117*, 39 – 49.

[36] Hayakawa, T.; Suzuki, S.and Nakamura, J. *Applied Catalysis A: General*, 1999, *183*, 273 - 285.

[37] Mawdsley, J. R.; Krause, T. R. *Applied Catalysis A: General*, 2008, *334*, 311–320.

[38] Batiot-Dupeyrat, C.; Gallego, G. A. S F.; Mondragon, J.; Barrault, J-M. *Catalysis Today,* 2005, *107–108*, 474–480.

[39] Pereniguez, R.; Gonzalez-DelaCruz, V. M.; Holgado, J. P.; Caballero, A. *Applied Catalysis B: Environmental*, 2010, *93*, 346 – 353.

[40] Moradi, G.R.; Rahmanzade, M.; Sharifni, S. *Chemical Engineering Journal*, 2010, *162*, 787–791.

[41] Nam, J. Woo; Chae, H.; Lee, S. Ho; Jung, H. and Lee, K.-Y. *Studies in Surface Science and Catalysis*, 1998, *119*, 843 - 848.

[42] Valderrama, G.; Goldwasser, M. R.; Urbina de Navarro, C.; Tatibouet, J. M.; Barrault, J.; Batiot-Dupeyrat, C.; Martınez, F. *Catalysis Today*, 2005, *107 – 108*, 785 – 791.

[43] Rynkowski, J.; Samulkiewicz, P.; Ladavos, A. K.; Pomonis, P. J. *Applied Catalysis A: General*, 2004, *263*, 1 – 9.

[44] Chettapongsaphan, C.; Charojrochkul, S.; Assabumrungrat, S.; Laosiripojana, N. *Applied Catalysis A: General*, 2010, *386*, 194 – 200.

[45] Valderrama, Gustavo; Kiennemann, Alain; Goldwasser, Mireya R. *J. Power Sources*, 2010, *195*, 1765 – 1771.

[46] Mota N.; Álvarez-Galván, M.C.; Al-Zahrani, S.M.; Navarro, R.M.; Fierro, J.L.G. Int. J. Hydr. Energy, 2012, *37,* 7056–7066

[47] Gallego, G. S.; Batiot-Dupeyrat, C.; Barrault, J.; Florez, E. and Mondragon, F. *Applied Catalysis A: General*, 2008, *334*, 251 - 258.

[48] Valderrama, G.; Kienneman, A. and Goldwasser, M. R. *Catalysis Today*, 2008, *133*, 142 - 148.

[49] Goldwasser, M. R.; Rivas, M. E.; Lugo, M. L.; Pietri, E.; Pe´rez-Zurita, J.; Cubeiro, M. L.; Griboval-Constant, A.; Leclercq, G. *Catalysis Today*, 2005, *107–108*, 106 – 113.

[50] Provendier, H.; Petit, C.; Estournes C. and Kiennemann, A. *Studies in Surface Science and Catalysis*, 1998, *119*, 741-746.

[51] Provendier, H.; Petit, C.; Estournes C. and Kiennemann, A. *Chemistry*, 2001, *4,* 57-66.

[52] Provendier, H.; Petit, C.; Estournes C.; Libs, S.; Kiennemann, A. *Applied Catalysis A: General*, 1999, *180*, 163 - 173.

[53] Falcon, H.; Baranda, J.; Campos-Martin, J. M.; Pena, M. A. and Fierro, J. L. G. *Studies in Surface Science and Catalysis*, 2000, *130*, 2195 - 2200.

[54] Kapokova, L.; Pavlova, S.; Bunina, R.; Alikina, G.; Krieger, T.; Ishchenko, A.; Rogov V. and Sadykov, V. *Catalysis Today*, 2011, *164,* 227 - 233.

[55] Pavlova, S.; Kapokova, L.; Bunina, R.; Alikina, G.; Sazonova, N.; Krieger, T.; Ishchenko, A.; Rogov, V.; Gulyaev, R.; Sadykov V. and Mirodatos, C. *Catal. Sci. Technol*., 2012, *2,* 2099 – 2108.

[56] Gallego, G. S.; Marín, J. G.; Batiot-Dupeyrat, C.; Barrault, J. and Mondragon, F. *Applied Catalysis A: General*, 2009, *369*, 97 - 103.

[57] Rivas, M. E.; Fierro, J. L. G.; Goldwasser, M. R.; Pietri, E.; Peŕ rez-Zurita, M. J.; Griboval-Constant, A.; Lecler, G.; *Applied Catalysis A: General*, 2008, *344*, 10 – 19.

[58] Goldwasser, M. R.; Rivas, M. E.; Pietri, E.; Perez-Zurita, M. J.; Cubeiro, M. L.; Gingembre, L.; Leclercq, L.; Leclercq, G. *Applied Catalysis A: General*, 2003, *255,* 45 - 57.

[59] Goldwasser, M. R.; Rivas, M. E.; Pietri, E.; Perez-Zurita, M. J.; Cubeiro, M. L.; Grivobal-Constant, A.; Leclercq, G. *Journal of Molecular Catalysis A: Chemical*, 2005, *228*, 325 - 331.

[60] Araujo, G. C.; Lima, S. M.; Assaf, J. M.; Pena, M. A.; Garcia Fierro, J. L. and M. D. C. Rangel, *Catysis Today*, 2008, *133-135*, 129-135.

[61] Mota, N.; Alvarez-Galvána, M.C. ; Navarro, R.M. ; Al-Zahrani, S.M.; Goguet, A. ; Daly, H.; Zhang,W.; Trunschke, A.; Schlögl, R.; Fierro, J.L.G. *Applied Catalysis B: Environmental,* 2012, *113–114,* 271–280.

[62] Labhsetwar, N. K.; Watanabe, A.; Mitsuhashi, T.; Haneda, H. *Journal of Molecular Catalysis A: Chemical*, 2004, *223,* 217 - 223.

[63] Chen, H.; Yu, H.; Peng, F.; Yang, G.; Wang, H.; Yang, J.; Tang, Y. *Chemical Engineering Journal,* 2010, *160*, 333 – 339.

[64] Liu, J. Y.; Lee, C. C.; Wang, C. H.; Yeh, C. T.; Wang, C. B.; *Int. J. Hydr. Energy*, 2010, *35*, 4069 – 4075.

[65] de Lima, S. M.; da Silva, A. M.; da Costa, L. O. O.; Assaf, J. M.; Jacobs, G.; Davis, B. H.; Mattos, L. V.; Noronha, F. B. *Applied Catalysis A,* 2010, *377,* 181 – 190.

[66] Chen, S. Q.; Wang, H.; Liu, Y.*; Int. J. Hydr. Energy*, 2009, *34*, 7995 - 8005.

[67] Chen, S. Q.; Li, Y. D.; Liu, Y.; Bai, X. *Int. J. Hydr. Energy*, 2011, *36*, 5849 – 5856.

[68] de Lima, S. M.; da Silva, A. M.; da Costa, L. O. O.; Assaf, J. M.; Mattos, L. V.; Sarkari, R.; Venugopale, A.; Noronha, F. B. *Appl. Catal. B: Env*., 2012, *121– 122*,1– 9.

[69] Urasaki, K.; Tokunaga, K.; Sekine, Y.; Matsukata, M.; Kikuchi, E. *Catalysis Communications*, 2008, *9*, 600 – 604.

[70] Pechini, M. P. *US Patent 3*, 1967, *330*, 697.

[71] Kuznetsova, T. G.; Sadykov, V. A.; Moroz, E. M.; Trukhan, S. N.; Paukshtis, E. A.; Kolomiichuk, V. N.; Burgina, E. B.; Zaikovskii, V. I.; Fedotov, M. A.; Lunin V. V.; and Kemnitz, E. *Studies in Surface Science and Catalysis*, 2002, *143*, 659 - 667.

[72] Bedel, L.; Roger, A. C.; Estournes, C.; Kienneman, A. *Catal. Today* , 2003, *85*, 207 - 218.

[73] Sadykov, Vladislav; Sobyanin, Vladimir; Mezentseva, Natalia; Alikina, Galina; Vostrikov, Zakhar; Fedorova, Yulia; Pelipenko, Vladimir; Usoltsev, Vladimir; Tikhov, Sergey; Salanov, Aleksei; Bobrova, Lyudmila; Beloshapkin, Sergey; Ross, Julian R. H.; Smorygo, Oleg; Ulyanitskii, Vladimir; Rudnev, Vladimir Fuel, 2010, 89, 1230 – 1240.

[74] Sadykov, V.; Bobrova, L.; Pavlova, S.; Simagina, V.; Makarshin, L.; Parmon, V.; Ross, J. R. H.; Van Veen, A. C.; Mirodatos, C. Syngas Generation from Hydrocarbons and Oxygenates with Structured Catalysts. Series Energy Science, Engineering and Technology, *Nova Science Publishers, Inc*, New York, 2012, 140p, ISBN: 978-1-60876-323-8.

[75] Ciambelli, P.; Cimino, S.; De Rossi, S.; Lisi, L.; Minelli, G.; Porta P. and Russo, G. *Applied Catalysis B: Envoiropmental*, 2001, *29*, 239 - 250.

[76] Kumar, R.; Choudhary, R. J.; Ikram, M.; Shukla D. K.; Mollah, S.; Thakur P.; Chae, K. H.; Angadi, B. and Choi, W. K. *J. Appl. Physics,* 2007, *102*, 073707.

[77] Porta, P.; Cimino, S.; De Rossi, S.; Faticanti, M.; Minelli, G. and Pettiti, I. *Materials Chemistry and Physics*, 2001, *71*, 165 - 173.

[78] Shannon, R. D. *Acta Crystallogr., Sect.* A, 1976, *32*, 751.

[79] Falcón, H.; Goéta, A.E.; Punte, G. and Carbonio, R.E. *J. Solid State Chem.* 1997, *133*, 379 - 385.

[80] Chiba, R.; Yoshimura, F.; Sakurai, Y. *Solid State Ionics*, 1999, *124*, 281 – 288.

[81] Kumar, R.; Choudhary, R. J.; Khan, M. W;. Srivastava, J. P.; Bao, C. W.; Tsai, H. M.; Chiou, J. W.; Asokan, K.; Pong. W. F. *J. Appl. Physics*, 2005, *97*, 093526.

[82] Bontempi, E.; Garzella, C. ;Valetti, S.; Depero. L. E. *Journal of the European Ceramic Society*, 2003, *23 (12)*, 2135-2142.

[83] Tsipis, E.V.; Kiselev, E.A.; Kolotygin, V.A.; Waerenborgh, J.C;. Cherepanov, V.A.; Kharton, V.V. *Solid State Ionics*, 2008,*179*, 2170 - 2180.

[84] Electronic sources http://abulafia.mt.ic.ac.uk/shannon/ptable.php.

[85] Sadykov, V. A.; Pavlova, S. N.; Kharlamova, T. S.; Muzykantov, V. S.; Uvarov, N. F.; Okhlupin, Yu. S.; Ishchenko, A. V.; Bobin, A. S.; Mezentseva, N. V.; Alikina, G. M.; Lukashevich, A. I.; Krieger, T. A.; Larina, T. V.; Bulgakov, N. N.; Tapilin, V. M.; Belyaev, V. D.; Sadovskaya, E. M.; Boronin, A. I.; Sobyanin, V. A.; Bobrenok, O. F.; Smirnova, A. L.; Smorygo O. L.; and Kilner, J. A. *Perovskites: Structure, Properties and Uses, Nova Science Publishers, Inc*, 2010, 67-178.

[86] Sadykov, V.; Mezentseva, N.; Alikina, G.; Bunina, R.; Pelipenko, V.; Lukashevich, A.; Tikhov, S.; Usoltsev, V.; Vostrikov, Z.; Bobrenok, O.; Smirnova, A.; Ross, J.; Smorygo, O.; Rietveld, B. *Catal. Today*, 2009, *146*, 132 – 140.

[87] 83. Hurst, N. W.; Gentry, S. J.; Jones, A. and McNicol, B. D. *Catalysis Reviews*, 1982, *24 (2)*, 233 - 309.

[88] Crisafulli, C.; Scire, S.; Minico, S. and Solarino, L. *Applied Catalysis A: General*, 2002, *225*, 1 - 9.

[89] Juan-Juan, J.; Roman-Martinez, M. C. and Illan-Gomez, M. J. *Applied Catalysis A: General*, 2009, *355*, 27 - 32.

[90] Liu, D.; Lau, R.; Borgna, A.; Yang, Y. *Appl. Catal. A*, 2009, *358*, 110 – 118.

[91] Chen, J.; Yao, C.; Zhao, Y.; Jia, P. *Int. J. Hydr. Energy* 2010, *35*, 1630 - 1642.

[92] Swaan, H. M.; Kroll, V. C. H.; Martin, G. A.; Mirodatos, C. *Catal. Today*, 1994, *21*, 571 - 578.

[93] Sazonova, N. N.; Sadykov, V. A.; Bobin, A. S.; Pokrovskaya, S. A.; Gubanova, E. L.; Mirodatos, C. *React. Kinet. Catal. Lett.*, 2009, *98*, 35 - 41.

[94] Rostrup Nielsen, J. R. *Stud. Surf. Sci. Catal.* 1991, *68*, 85 - 101.

[95] Zhang, W. D.; Liu, B. S.; Zhu, C.; Tian, Y. L. *App. Catal. A*, 2005, *292*, 138 - 143.

[96] Barroso, M. N.; Galetti, A. E.; Abello, M. C. *Appl. Catal, A: General*, 2011, *394*, 124 – 131.

[97] Haryanto, A.; Fernando, S.; Murali, N.; Adhikari, S. *Energy & Fuels*, 2005, *19*, 2098 - 2106.

[98] Vaidya, P.D.; Rodrigues, A.E. *Chem. Eng. J.*, 2006, 117, 39 - 49.

[99] Domine, M. E.; Iojoiu, E. E.; Davidian, T.; Guilhaume, N.; Mirodatos, C. *Catalysis Today*, 2008, *133–135*, 565 – 573.

[100] de Lima, S. O.; da Cruz, I.; Jacobs, G.; Davis, B.H.; Mattos, L.V.; Noronha F. B., *J. Catal.*, 2008, *257, 2,* 356 – 368.

[101] Bedel L.; Roger A.C.; Rehspringer J.L.; Kiennemann A. *Stud. Surf. Sci. Catal.*, 2004, *147*, 319-324.

[102] Araque M.; Vargas J. C.; Zimmermann Y.; Roger A. C. *Int. J. Hydrogen Energy* 2011, *36*, 1491 - 1502.

In: Perovskite
Editors: Jinghua Zhang and Huan Li

ISBN: 978-1-62417-800-9

Chapter 2

BI CONTAINING MULTIFERROIC PEROVSKITE OXIDE THIN FILMS

***Rainer Schmidt*[1,*], *Eric Langenberg*[2,3], *Jofre Ventura*[2] and *Manuel Varela*[2]**

[1]Universidad Complutense de Madrid, GFMC,
Dpto. Física Aplicada III, Facultad
de Ciencias Físicas, Madrid, Spain
[2]Universitat de Barcelona, Dpto. Física Aplicada i Óptica,
Diagonal Sud, Facultats
de Física i Química, Barcelona, Spain
[3]Universidad de Zaragoza, Instituto de Nanociencia de Aragón,
Mariano Esquillor, Zaragoza, Spain

ABSTRACT

In this work multiferroic thin films of Bi containing perovskite oxides are discussed, where the driving force for ferroelectricity are the Bi^{3+} lone-pair electrons. First, a brief introduction of Bi containing multiferroic perovskite oxides will be presented to describe the mechanisms for establishing magnetic and ferroelectric order in these compounds and some recent developments in this field of research are reported.

The second section addresses experimental aspects of epitaxial thin film growth of $BiMnO_3$, $(Bi_{0.9}La_{0.1})_2NiMnO_6$ and $BiFeO_3$ thin films by pulsed laser deposition. The third section is dedicated to the physical properties of such films in terms of the structural characterization and the magnetic and ferroelectric properties and their correlations in form of magnetoelectric coupling (MEC) effects.

* Corresponding author. Email: rainerxschmidt@googlemail.com.

1. INROUCTION

1.1. Pathways to the Coexistence of Magnetism and Ferroelectricity in Perovskites

Multiferroics are materials where at least two ferroic orders coexist in the same phase. The definition of the term ferroic aims to characterise materials in which domains are formed with a corresponding net macroscopic magnitude that can be hysteretically switched by applying an external field, i.e. ferroelectric, ferromagnetic, ferroelastic and ferrotoroidic orders (primary ferroics) [1, 2, 3]. Nonetheless, antiferroic orders consisting in antiferromagnetism, ferrimagnetism and antiferroelectricity are widely accepted to be also included in the extended definition of the term multiferroic. Among all kinds of multiferroics, those showing both ferroelectric and magnetic order draw special attention in terms of their magnetoelectric properties, which is the focus of this work. The particular interest in such multiferroics is driven by industrial aspirations to design charge storage devices, where information could be written electrically onto a data bit and retrieved magnetically. For the sake of simplicity the term "multiferroics" is used here to refer to ferroelectric and (anti-) ferromagnetic multiferroics.

For multiferroic charge storage devices it is required that the magnetization (polarization) can be modified by applying an electric (magnetic) field and vice versa, which implies that ferroelectric and magnetic orders need to be strongly coupled to each other (Figure 1). The so-called magnetoelectric coupling (MEC) effect has therefore great technological potential for novel magnetoelectric applications and spintronics [4-8]. Yet it is worth mentioning that the sole coexistence of the two ferroic orders is not a sufficient condition for MEC to occur, which is a more restrictive material property.

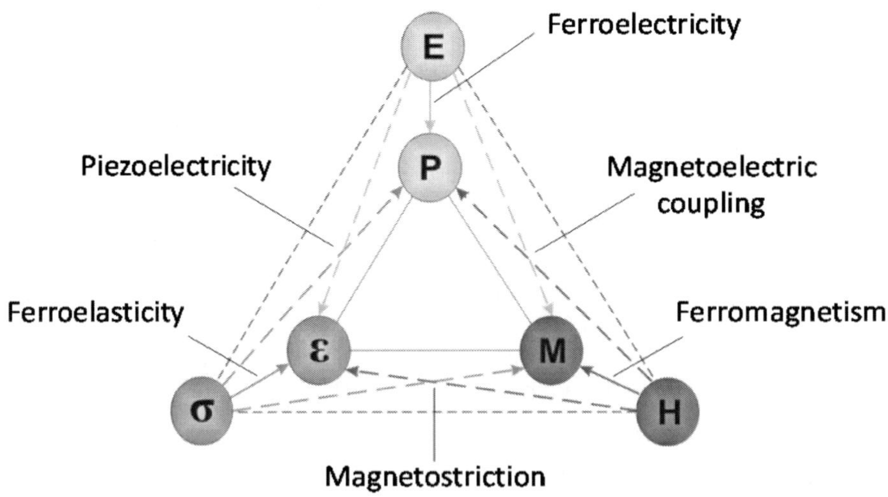

Figure 1. Summary of all the functionalities that multiferroic materials may offer. E and P stand for electric field and electric polarization, H and M stand for magnetic field and magnetization, and σ and ε stand for stress and deformation, respectively. Adapted from Reference [3].

Despite the large number of ferroelectric or magnetic materials existing in nature, the combination of ferroelectric and ferromagnetic orders in one intrinsic material seems to be quite elusive [9-11]. Whereas magnetic materials can be either insulating or conductive, ferroelectric materials can solely be insulating, otherwise an applied electric field would induce an electric current rather than switching ferroelectric domains. Second, from crystal symmetry considerations (see Figure 2) ferroelectricity breaks the spatial inversion symmetry, which requires non-centrosymmetric crystal structures for ferroelectric order to occur. Conversely, magnetism breaks time-reversal symmetry [2]. Combination of the two symmetry restrictions leaves only 13 point groups (among 233) that would enable ferroelectricity and magnetism to coexist [9-11].

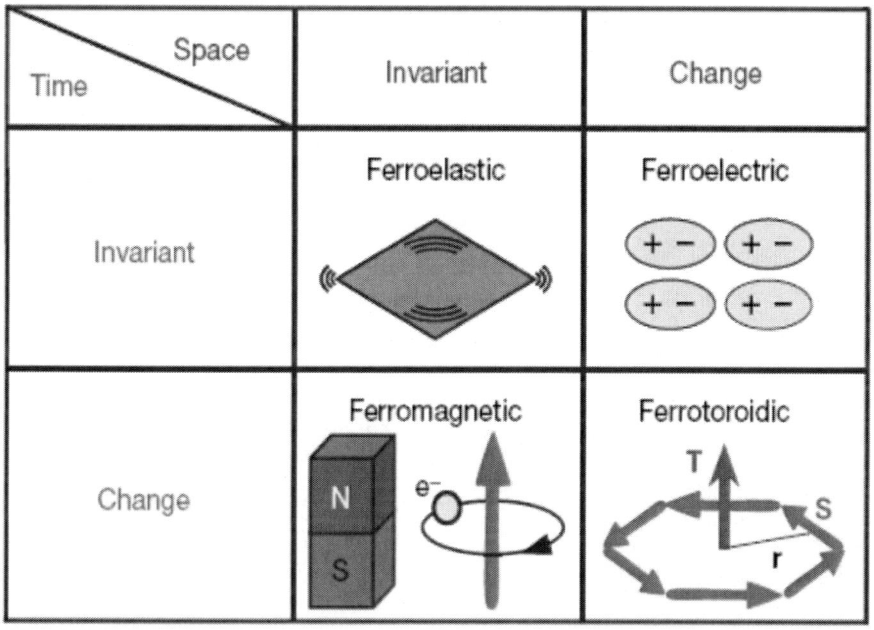

Figure 2. Ferroic orders under inversion symmetry of time and space [2].

Since most ferroelectrics and a large number of magnetic materials are transition metal oxides with the ABO_3 perovskite structure (e.g. $BaTiO_3$ and (La, Sr)MnO_3), much of the attention was reasonably drawn to this crystal structure type for the search for multiferroic materials. Both, in ferroelectric and magnetic perovskites it is usually the B-cation which drives the ferroelectric displacement and the magnetic order, respectively. In the case of ferroelectrics, the B-cation displaces from the centre of the surrounding oxygen octaheadra, breaking the centrosymmetry and creating an electric dipole. Yet this condition requires the B-cation to have a d^0 electron configuration, i.e. empty *d* orbitals like in Ti^{4+}, which minimizes the Coulombian electrostatic repulsion of the surrounding oxygen anions [9]. Magnetism in transition metal perovskites arises from superexchange interactions (or double exchange for conductive magnetic oxides) between the B-cations mediated by the adjacent oxygen anions [9, 12]. This condition requires the B-cation to possess a magnetic moment, which is only possible when the *d* orbitals of the B-cation are partially occupied. Thus, a mutual exclusion between ferroelectricity and magnetism in perovskites exists to some extent [9].

To date, most attempts to design multiferroic materials have been based on looking for new sources for ferroelectricity, while maintaining the same recipes for magnetism. Based on such considerations, multiferroics have been classified in two types as described by Khomskii [13]: In Type-I multiferroics, ferroelectricity and magnetism rely on two independent mechanisms, whereas in Type-II the ferroelectricity arises from the magnetic order, i.e. ferroelectricity exists only in a magnetically ordered state. The latter is expected to produce large magnetoelectric coupling as ferroelectric order is intrinsically related to the magnetic one, but such ferroic properties tend to order at rather low temperatures. This coupled type of ferroelectric and magnetic order often involves commensurate and incommensurate spiral spin structures, which induce ferroelectricity. This occurs in several rare-earth manganites [14-16]. Contrarily, Type-I multiferroics tend to show much higher ferroic transition temperatures at the expense of a small MEC.

1.1.1. Ferroelectricity in the Presence of Bi^{3+} Lone-pair Electrons

The work presented here focuses on one particular subgroup of Type-I multiferroics, where ferroelectricity is induced by means of lone-pair electrons. In order to achieve both ferroic orders in ABO_3 perovskite oxides it is obvious that one of the cations may be utilized for inducing ferroelectric order, while the other may induce magnetism. One possibility is the exploitation of a d^0 transition metal cation located on the *B*-site and a magnetic cation on the A-site, for example in $EuTiO_3$ [17]. The second possibility is the use of stereochemical active cations like Bi^{3+} or Pb^{2+} on the perovskite A-site inducing ferroelectricity due to lone-pair electrons, and a magnetic cation located on the B-site. The latter approach applies to all systems studies here, $BiMnO_3$, $(Bi_{0.9}La_{0.1})_2NiMnO_6$ and $BiFeO_3$. The electronic configuration of Bi^{3+} (and also Pb^{2+}) is $[Xe]4f^{14}5d^{10}6s^26p^0$ with empty 6*p*-states and the two 6*s* outer lone-pair electrons do not participate in chemical bonds.

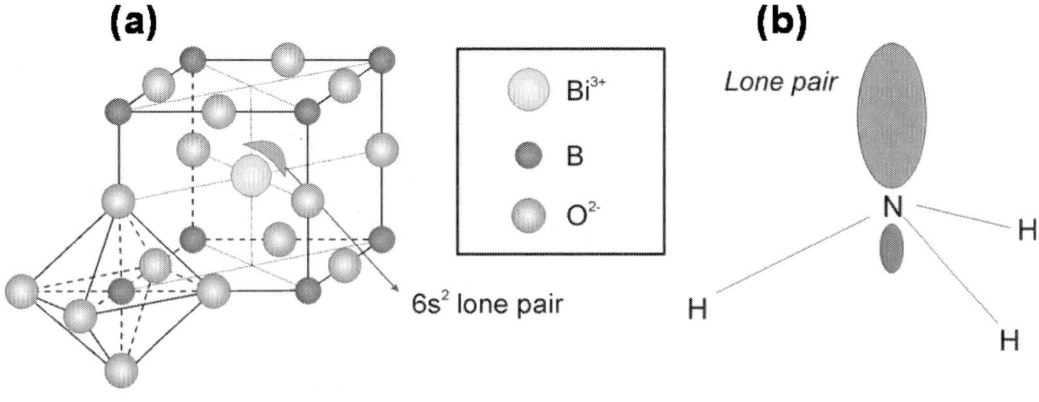

Figure 3. (a) Schematic representation of the lobe-like distribution of the $6s^2$ lone-pair electrons in Bi-based perovskite structures, $BiBO_3$, breaking the spatial inversion symmetry. (b) Lone-pair $2s^2$ in the ammonia molecule.

In the absence of interactions, the lone-pairs are nearly spherically distributed, but when surrounded by oxygen anions they shift away from the centrosymmetric position due to the Coulombian electrostatic repulsion forming a localized lobe-like distribution, very much alike the ammonia molecule (Figure 3) [18]. Thus, the lone-pairs form an electric dipole, which breaks the spatial inversion symmetry. This mechanism is the driving force for ferroelectricity in all Bi-based multiferroic perovskites.

1.1.2. Magnetic Ordering in Bi Containing Perovskite Oxides

Bi^{3+} is always situated on the perovskite A-site, thus allowing the location of a magnetic transition metal oxide cation on the B-site with a (partially) occupied outer electron shell. Since B-site cation chains are interrupted by oxygen anions, long range magnetism arises from magnetic superexchange interactions mediated by the adjacent oxygens: B – O – B.

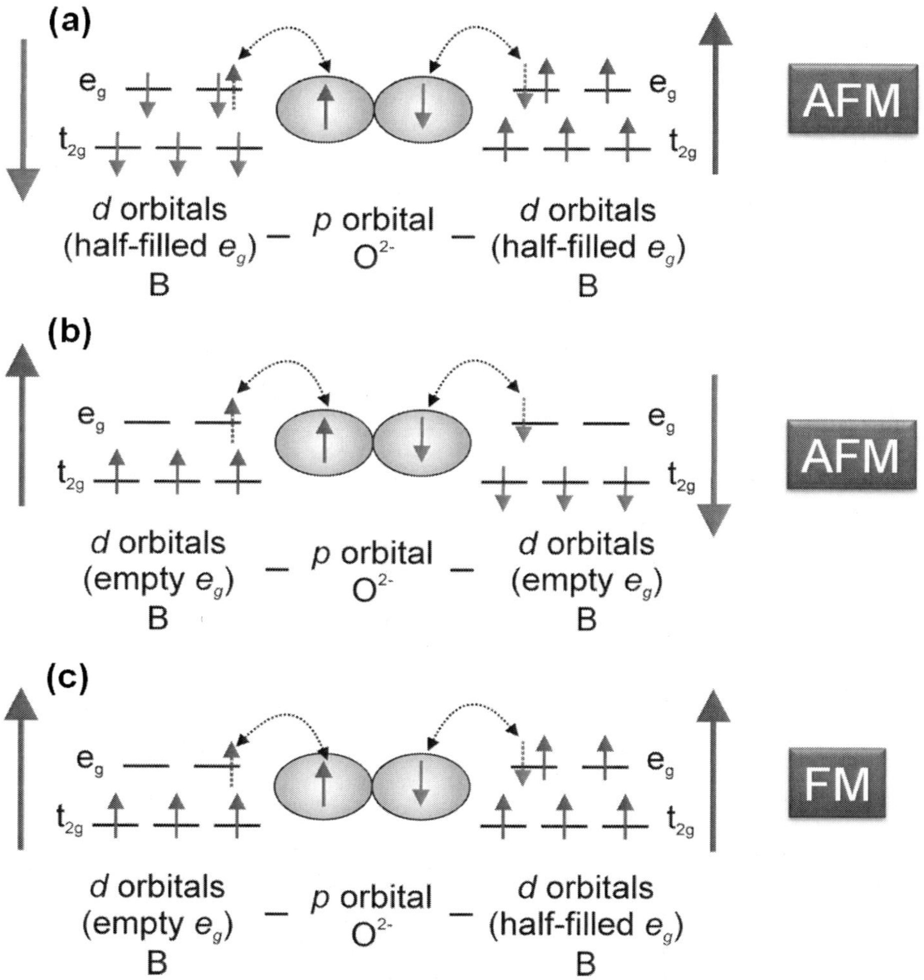

Figure 4. Schematic representation of superxchange interaction in ABO_3 perovskite with an ideal 180° B – O – B bonding-angle. Depending on the occupation of the e_g orbitals, the type of magnetic interaction is antiferromagnetic (a, b) or ferromagnetic (c).

The surrounding oxygen octahedra of the B-site splits *d*-levels into two energy states: two high-energetic e_g and three low-energy t_{2g} orbitals [19]. Whether the magnetic interaction is ferromagnetic or antiferromagnetic depends on the filling of the e_g orbitals according to the Goodenough-Kanamori's (GK) rules. Bearing in mind *(i)* the Pauli's exclusion principle stating that two electrons in the same orbital must possess antiparallel spins, and *(ii)* the Hund's rule stating that electrons in degenerated *d* orbitals minimize their energy in a parallel spin alignment, Goodenough-Kanamori's rules can be summarized as follows:

a) Either B (empty e_g orbitals) – O – B (empty e_g orbitals) or B (half-filled e_g orbitals) – O – B (half-filled e_g orbitals) give rise to antiferromagnetic interactions (Figure 4a, b).
b) B (empty e_g orbitals) – O – B (half-filled e_g orbitals) give rise to ferromagnetic interaction (Figure 4c).

Note that these GK rules are deduced in the simplest scenario of an ideal perovskite where the B – O – B bond angle is 180º. Distorted perovskites, rotation of the oxygen octahedral and different B – O – B bond angles may give rise to different magnetic interactions [12].

As the perovskite B-site is quite often occupied by only one kind of cation with the same oxidation state and the same e_g orbital filling, antiferromagnetic interactions are much more common than ferromagnetic ones. Therefore, most multiferroic $BiBO_3$ are antiferromagnetic.

In order to induce ferromagnetism, two different magnetic transition metal cations at the B-site in a double perovskites oxide $Bi_2BB'O_6$, can be employed to provide ferromagnetic exchange according to the GK rules, if the different cations are long-range ordered in an alternating fashion. The appropriate choice of B - B' cations should therefore be aimed at such B-site cation long-range order to be likely and the following electron configuration: B (empty e_g orbitals) – O – B' (half-filled e_g orbitals).

Still, if the antiferromagnetic interactions dominate in a specific B-site ordered double-perovskite, long range ferrimagnetism is likely to occur due to the fact that every transition metal oxide possesses different magnetic moment (different *d* orbital filling), where the magnetic moments are not compensated. For this reason, Bi-based double perovskite oxides tend to be either ferromagnetic or ferrimagnetic, showing net magnetization. However, it is worth noting that due to the different electronic configurations of the B - B' cations, double-perovskite oxides tend to be less insulating than their counterpart 'single' perovskite oxides.

1.2. Bi-Based Single Perovskites, $BiBO_3$

Almost all $BiBO_3$ single perovskite multiferroics are antiferromagnetic, with the unique exception of $BiMnO_3$. Therefore, $BiMnO_3$ may be one of the most prominent multiferroics, together with $BiFeO_3$, which is the only known single phase compound which is ferroelectric and antiferromagnetic at room temperature. These two most important Bi containing single perovskite multiferroics will be reviewed in detail here. Furthermore, several more recently discovered Bi-based single perovskite oxides will be mentioned briefly.

1.2.1. BiMnO₃

$BiMnO_3$ is commonly synthesized in bulk form at≈ 600 - 700ºC under high pressure of ≈ 3 GPa- 6 GPa to stabilize the phase [20-24]. Single-phase $BiMnO_3$ can also be obtained in thin film form by pulsed laser deposition, where the high-pressure requirement for phase stabilization is replaced by the epitaxial strain imposed by the substrate [25-28]. Nonetheless, the high volatility of Bi together with the multiphase tendency of the Bi – Mn – O ternary system entails difficulty to obtain single-phase stabilization even in thin films [29-32].

$BiMnO_3$ is expected to crystallize in an orthorombic structure similar to $LaMnO_3$ since La^{3+} and Bi^{3+} have similar cationic radii [33]. However, this is not the case at low temperatures, where $BiMnO_3$ crystallizes in a highly distorted non-centrosymmetric monoclinic *C2* structure [20] due to the stereochemical activity of Bi^{3+} lone-pairs [18]. On increasing temperature $BiMnO_3$ undergoes two phase transitions at ≈ 450 K and ≈ 770 K [22]. Between ≈ 450 -770 K the structure can still be indexed as non-centrosymmetric monoclinic *C2*, whereas above ≈ 770 K the structure is centrosymmetric *Pbnm* which no longer allows spontaneous polarization (Figure 5).

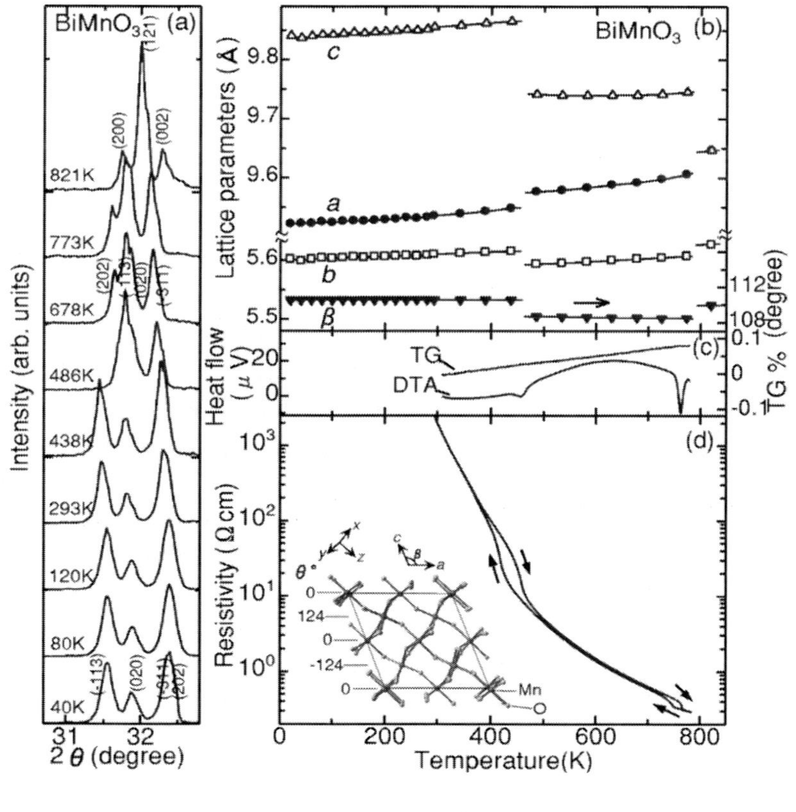

Figure 5. (a) Powder X-ray diffraction of $BiMnO_3$ at different temperatures. (b) Temperature dependence of lattice parameters. Above ≈ 770 K the structure can be indexed as centro-symmetric Pbnm. (c) Thermogravimetry and differential thermal analysis showing two phase transitions. (d) Temperature dependence of resistivity signalling the two phase transitions. Reproduced from Reference [22], with permission from the American Physical Society (APS).

Thus, $BiMnO_3$ is considered to be ferroelectric up to ≈ 770 K well above room temperature, although neither conclusive ferroelectric hysteresis loops nor ferroelectric domain switching current measurements have been reported due to high leakage. In fact, some revised analysis of the crystal structure [34, 35] and first principles calculations [36] have questioned the non-centrosymmetric structure of $BiMnO_3$. However, the four-resistive states reported for La-doped $BiMnO_3$ multiferroic tunnel junctions [4] can only be understood in the frame of a ferromagnetic and ferroelectric material, which strongly indicates ferroelectric behaviour, at least in thin film form. In addition, nonlinear optical measurements on $BiMnO_3$ thin films have revealed changes in the polar symmetry of the second harmonic generation signal by applying electric fields, consistent with changes in the ferroelectric domain structure [37]. Nevertheless, a clear-cut evidence for switching ferroelectric domains by the application of electric fields is still missing, despite strong indications for ferroelectricity.

The ferromagnetic order in $BiMnO_3$ sets in below ≈ 105 K [22]. Despite the similar orbital ordering compared to antiferromagnetic $LaMnO_3$, two out of three Mn – O – Mn orbital configurations in $BiMnO_3$ favour ferromagnetic interactions, which results in an overall long-range ferromagnetism, overcoming the antiferromagnetic interactions [22]. It should be noted that the Mn – O – Mn bond angles in the highly distorted monoclinic structure of $BiMnO_3$ are significantly smaller than the ideal 180º (≈ 140º - 160º) [22]. In bulk the saturated magnetization is found to be ≈ 3.6 Bohr magneton (μ_B) per formula unit (f.u.), [22] whereas lower values were reported in thin films (≈ $2.2\mu_B$/f.u.) [27].

1.2.2. $BiFeO_3$

No other multiferroic material has been more extensively studied than $BiFeO_3$, because its multiferroic properties are robustly maintained well above room temperature. The crystal structure of $BiFeO_3$ is rhombohedral $R3c$ [38, 39], which is a polar group and thus permits the establishment of ferroelectric order. The ferroelectric transition temperature was proved to be rather high, T_{FE} = 1103 K, above which a centrosymmetric paraelectric phase with space group $R\bar{3}c$ can be indexed [40]. The polar axis is found along the [111] direction and a large saturated polarization, ≈ 80 – 100 $\mu C/cm^2$, is reported as expected for a high-Curie temperature ferroelectric [41].

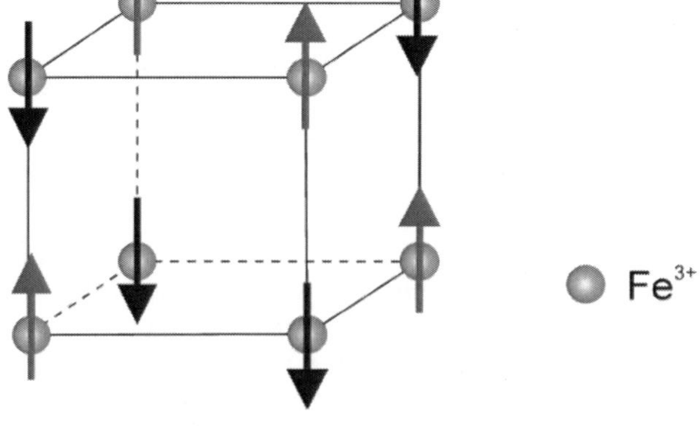

Figure 6. The magnetic structure of $BiFeO_3$: G-type antiferromagnetism.

$BiFeO_3$ exhibits antiferromagnetism with G-type arrangement (Figure 6) and a Néel temperature of ≈ 620 K [42].

Additionally, superimposed on the antiferromagnetic order, an incommensurate cycloidal spiral arrangement of the spins with a long period of 620 Å was reported [43]. Moreover, the symmetry of the $BiFeO_3$ structure allows small canting of the spins to occur (Figure 7), which would give rise to a net magnetization in terms of weak ferrimagnetism of the Dzyaloshinski-Moriya type [44].

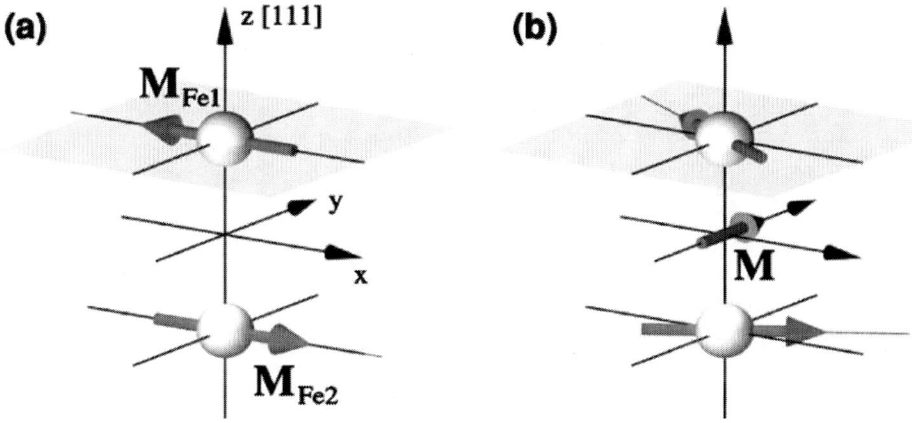

Figure 7. Weak ferromagnetism in $BiFeO_3$. Reproduced from Reference [44], with permission from the American Physical Society (APS).

This weak ferrimagnetism requires the spiral spin structure to be suppressed though, which can be attained in thin films due to the epitaxial constraints or enhanced anisotropy [44]. The expected magnetization would be ≈ 0.1 μ_B/f.u., which was confirmed experimentally [45]. Despite early claims of higher magnetisation [41], the true value is now quite well established to be close to the theoretical 0.1 μ_B/f.u. [45].

Most importantly, by combining photoelectron emission microscopy and piezoforce microscopy it was shown that the antiferromagnetic and ferroelectric domains in $BiFeO_3$ are strongly coupled [46], which was confirmed by neutron scattering experiments on single crystals [47]. The magnetoelectric coupling was argued to be diminished in thin films because of the absence of the spiral cycloid of the spin arrangement [47].

1.2.3. Other Bi-Based Single Perovskites

Other Bi-based single perovskites have been proposed and proved to be multiferroic, but all of them are antiferromagnetic with only weak ferrimagnetism at most. First principles calculations predicted $BiCrO_3$ to be antiferroelectric and antiferromagnetic [48]. Experimental results show weak ferrimagneticordering (≈ 0.05 μ_B/f.u. saturated magnetization) below ≈ 120 K in bulk [49] and thin film samples [50], but contradictory conclusions were made about whether $BiCrO_3$ is antiferroelectric [51] or ferroelectric [50]. Another potential multiferroic is $BiCoO_3$, which was predicted to crystallize in tetragonal structure with large tetragonality c/a giving rise to huge polarization (≈ 150 $\mu C/cm^2$) [52]. The structure was confirmed by neutron diffraction [53], though the ferroelectric character has not been experimentally confirmed yet. Furthermore, $BiNiO_3$ has been considered, which is antiferromagnetic with a weak spin canted ferrimagnetic moment below 300 K as a consequence of the small Ni – O – Ni bond angles [54]. Due to the preferred Ni^{2+} valence state, $BiNiO_3$ displays an unusual charge disproportionation at the A-site of Bi^{3+} – Bi^{5+}. This is quite rare for Bi cations and affects the magnetic symmetry in $BiNiO_3$ [54]. Still, there is no experimental evidence of its ferroelectric character and indeed structural analysis reveals $BiNiO_3$ to be centrosymmetric [55].

1.3. Bi-Based Double-perovskites, $Bi_2BB'O_6$

Bi-based double-perovskites $Bi_2BB'O_6$ have been much less studied than their counterpart single perovskites, but recently more intense research has resulted in promising findings such as a small net magnetization at room temperature in Bi_2FeMnO_6 [56]. As mentioned before, the Goodenough-Kanamori's rules predict these oxides to allow engineering of ferromagnetic paths. In an ideal ferromagnetic double-perovskite the B – B' cation species may possess either d^3 - d^5 or d^3 - d^8 configuration, and an ideal 180º B – O – B' bonding angle [12]. However, due to the mixed valence states that most transition metal oxides can adapt, the desired electronic configuration of B - B'cations has been found difficult to be obtained. Moreover, long-range ferromagnetism requires alternating B - B' cation ordering along the fundamental perovskite directions [100], [010] and [001] as illustrated in Figure 8. Among all double-perovskites, Bi_2NiMnO_6 has been the most intensively studied due to the fact that B-site Ni^{2+} and Mn^{4+}cations have been proven to be long-range ordered [57, 58], thus allowing long-range ferromagnetism to occur. In the following, Bi_2NiMnO_6 will be discussed in detail and several alternative Bi-based double perovskites will be mentioned briefly.

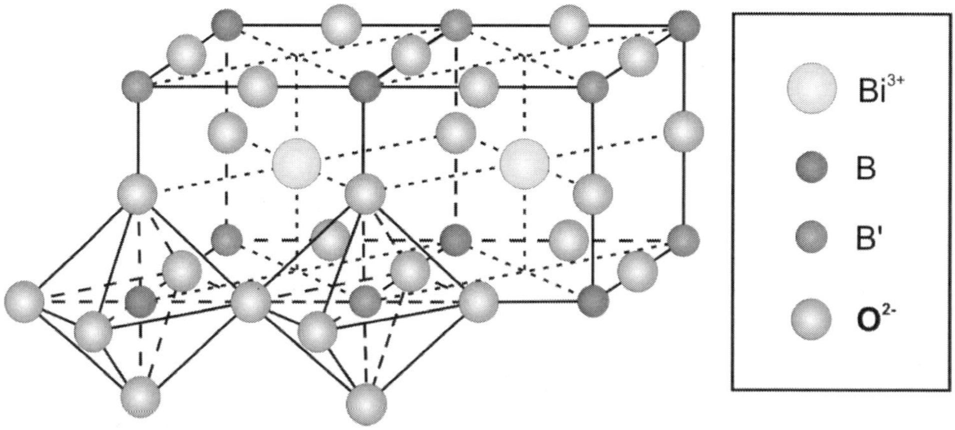

Figure 8. Scheme of the double-perovskite structure with long range B-site cation ordering.

1.3.1. Bi_2NiMnO_6

Equivalent to $BiMnO_3$, Bi_2NiMnO_6 is commonly synthesized in bulk form at ≈ 800ºC under high pressure of ≈ 6 GPa [59]. Yet epitaxial strain was also proven to be equally efficient, replacing the high synthesis pressures to produce single phase samples by pulsed laser deposition [60-62]. However, much alike $BiMnO_3$, the high Bi volatility and the multiphase tendency of these materials lead to narrow windows in the thin film deposition conditions for single phase stabilization [62]. On the other hand, partial replacement of Bi^{3+}cations by La^{3+}cations at the A-site has been proven to facilitate single-phase stabilisation in bismuth manganite compounds [30, 62-64] because of the slightly smaller ionic radius of La^{3+} with regard to Bi^{3+} (0.130 nm and 0.131 nm, respectively) [33]. It gives rise to a slightly reduced unit cell volume and exerts chemical pressure to prevent Bi^{3+}cation desorption during film growth.

Bi_2NiMnO_6 crystallizes in monoclinic *C2* structure, which is non-centrosymmetric allowing spontaneous polarization. The ferroelectric phase transition occurs at ≈ 485 K, above which the compound crystallizes in a centrosymmetric monoclinic $P2_1/n$ structure [59] and a dielectric anomaly can be observed across the transition (Figure 9a) [59].

The solid solution $(Bi_{1-x}La_x)_2NiMnO_6$ remains non-centrosymmetric at sufficiently low temperature for $x < 0.2$ [65] allowing ferroelectric order to be established. Conclusive ferroelectric domain switching current measurements on 10% La-doped $(Bi_{0.9}La_{0.1})_2NiMnO_6$ thin films have confirmed the ferroelectric character of this compound [66], although the ferroelectric Curie temperature (≈ 450 K) was inferred to be slightly inferior than that in bulk (≈ 485 K).

As stated above, ferromagnetism in Bi_2NiMnO_6 arises from the long-range order of Ni^{2+} – O – Mn^{4+} cations. The preferred Ni^{2+} valence state forces the Mncation to adapt 4+ valency, which not only favours the appropriate electronic configuration of the B cations for ferromagnetic interactions, but also the long range B-site cation order. Bi_2NiMnO_6 shows a large saturated magnetization (≈ 4 μ_B/f.u.) and orders ferromagnetically below 140 K in bulk (Figure 9b) [59], whereas a lower Curie temperature (≈ 100 K) was reported for thin films [58, 60].

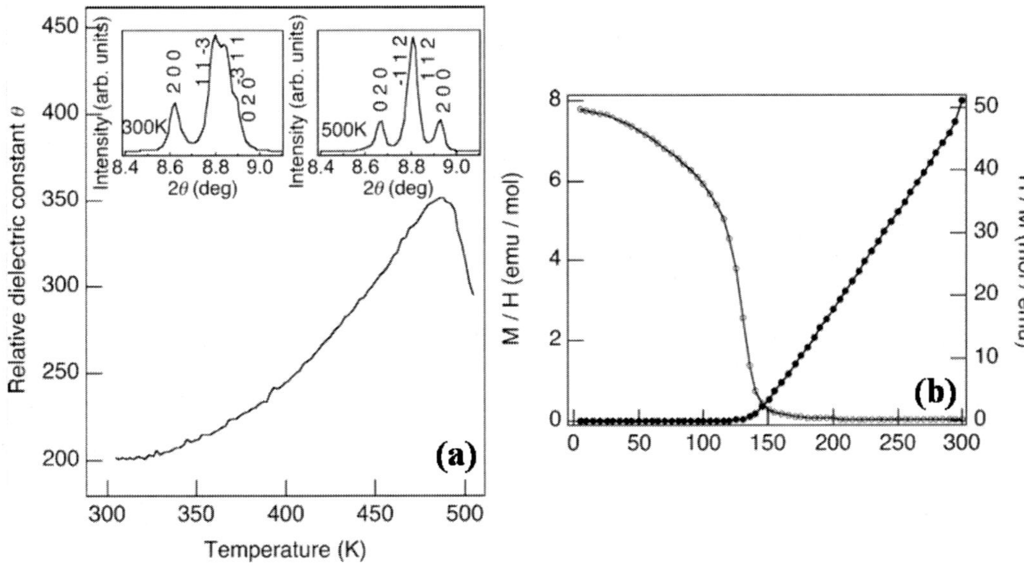

Figure 9. Temperature dependence of (a) the dielectric permittivity and (b) the magnetization of bulk Bi_2NiMnO_6. Reproduced from Reference [59], with permission from the American Chemical Society (ACS).

1.3.2. Other Bi-Based Double-perovskites

Other Bi-based double perovskites have been studied and proved to be multiferroic, but none of them show clear long-range B-site cation order like Bi_2NiMnO_6, which inhibits long-range ferromagnetism or ferrimagnetism.

As a first example, Bi_2FeMnO_6 is predicted to crystallize in monoclinic *C2* structure [67], but bulk samples are usually indexed as orthorhombic [68]. Both Fe and Mn are found to be mainly in a 3+ valence state, which hinders the B-site cation ordering [56, 67, 68]. A hypothetical B-site ordering would entail either ferromagnetic or more likely antiferromagnetic exchange with a net ferrimagnetic magnetization of $\approx$ 1 μ_B/f.u. (the magnetic moments of Fe^{3+} and Mn^{4+} are 5 μ_B and 4 μ_B, respectively). Yet the recorded magnetization values are much lower, thus confirming the lack of B-site order in this compound [56, 67, 68]. On the other hand, Bi_2FeMnO_6 was proven to be ferroelectric with a saturated polarization of $\approx$ 30 $\mu C/cm^2$ at 150 K [69].

Another example is Bi_2FeCrO_6. Ab-initiocalculation predicted Bi_2FeCrO_6 to show a large saturated polarization ($\approx$ 80 $\mu C/cm^2$) and a saturated magnetization of 2 μ_B/f.u. for a Cr^{3+} and Fe^{3+} B-site cation order [70]. Although Cr^{3+} and Fe^{3+} exhibit d^3 and d^5 orbital filling and ferromagnetic superexchange interaction would be expected, the magnetic $Fe^{3+} - O - Cr^{3+}$ interaction was in fact predicted to couple antiferromagnetically [70]. Still, a net magnetization of 2 μ_B/f.u. would be expected in case of long-range cation order due to the uncompensated magnetic moments (5μ_B and 3μ_B for Fe^{3+} and Cr^{3+}, respectively). However, long-range B-site order in Bi_2FeCrO_6 is not likely to occur due to the same charge of Fe^{3+} and Cr^{3+} and the similar ionic radii (0.645 Å and 0.615 Å, respectively) [33]. Indeed Bi_2FeCrO_6 is found to exhibit low magnetization (< 0.2 μ_B/f.u.) in both bulk samples [71] and thin films [72]. Nonetheless, some recent experimental results show a certain thickness dependence of the B-site order, pointing out that partial ordering of Fe^{3+} and Cr^{3+} can be improved by decreasing the thickness of Bi_2FeCrO_6 thin films [73]. The predicted ferroelectric properties were confirmed though experimentally with a large polarization of $\approx$ 60 $\mu C/cm^2$ along the [001] direction at $\approx$ 77 K, which would entail $\approx$ 100 $\mu C/cm^2$ along the polar axis [111], [72].

To date all known Bi containing multiferroic double perovskite oxides consist of 3*d* transition metals occupying the B-site, like Ni, Mn and Fe. Recently it was predicted though that combining 3*d* and 5*d* elements in Bi-based double perovskites allows robust room temperature ferrimagnetism to be attained [74], which is still a big challenge in multiferroics for the time being. In particular, Bi_2NiReO_6 and Bi_2MnReO_6 are expected to order ferrimagnetically below 360 K and 330 K, respectively. Although they would not be ferroelectric in the ground state, the ferroelectric phase is likely to be stabilized in thin films by small percentages of strain [74]. Still, these materials have not been synthesized so far, hence their predicted multiferroic properties remain to be confirmed experimentally.

2. EXPERIMENTAL

Functional properties in epitaxial thin films strongly depend on the film quality and the lattice strain. The latter arises due to the lattice mismatch between the substrate and the bulk crystal of the epitaxial phase under investigation. Therefore, the deposition technique used and the choice of an appropriate substrate are of paramount importancefor the production of high quality Bi containing multiferroic perovskites thin films, as will be demonstrated on the examples of $BiMnO_3$, $(Bi_{0.9}La_{0.1})_2NiMnO_6$ (BLNMO) and $BiFeO_3$.

2.1. Pulsed Laser Deposition of Bi-Based Perovskite Multiferroics

Among other physical vapor deposition techniques, pulsed laser deposition (PLD) is long known as the tool of choice for the growth of complex-oxide materials [75]. The technique is conceptually simple, even though involving rather complex physical processes: a pulsed laser beam leads to the ejection of particles from a solid target, where the particles range from atoms to electrons and clusters of particles. Such particles form of a spatially restricted plasma known as plume, which propagates towards the substrate and condensates on it. Figure 10 illustrates the PLD system used for growing $BiMnO_3$, BLNMO and $BiFeO_3$ films.

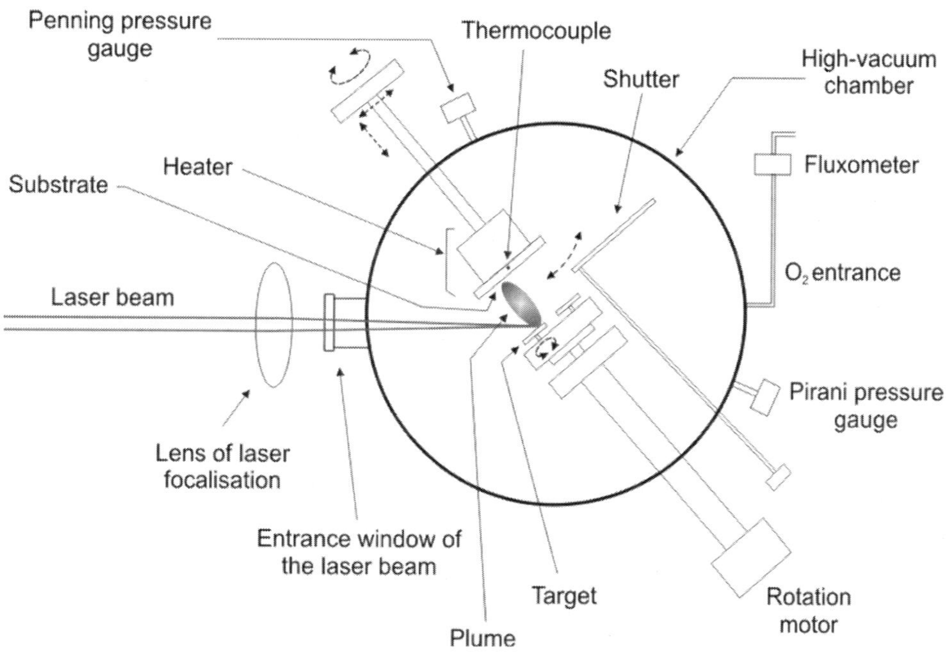

Figure 10. Schematic drawing of the PLD system used for this work.

The condensed material forms an epitaxial film on the substrate if grown under the correct deposition conditions, where the film often maintains the target stoichiometry virtually unaffected. This feature may be regarded the key factor for the large success of the PLD film growth technique in recent years. Yet, enriched targets are commonly used when volatile species such as Bi are involved, where stoichiometry is often not maintained.

The condensed or ablated products have higher kinetic energies (up to a few hundred eV) than in other conventional techniques such as radiofrequency magnetron sputtering or thermal evaporation. This high energy of impinging particles leads to a high ad-atom surface diffusion which enables epitaxial film growth [76]. In case of oxide films, O_2 gas is introduced into the vacuum chamber during the growth process to an appropriate amount to prevent oxygen vacancies forming. This is particularly important for the growth of insulating oxide films, where incomplete oxygenation often leads to undesired mixed cationic valence states and considerable conductivity. Furthermore, O_2 pressure during deposition plays a major role on the plume expansion velocity and therefore on film growth dynamics [77], where the latter is also critically effected by the substrate temperature.

The substrate selection is another issue of major importance, especially when strain engineering (i.e. replacement of mechanical pressure by epitaxial stress) is required to stabilize an otherwise unstable phase as it is the case for $BiMnO_3$ and $(Bi_{0.9}La_{0.1})_2NiMnO_6$. In order to facilitate epitaxial growth, Nb-doped (001)-oriented $SrTiO_3$ substrates were chosen with a cubic perovskite structure and a lattice parameter of 3.905 Å as shown in Table 1.

Table 1. Lattice parameters and mismatch ($f = a_{bulk}/a_{substrate} - 1$) with SrTiO3. a_{PC} stands for the pseudocubic lattice parameters in the bulk

Material	Crystal Structure	Lattice parameter	Mismatch with (001) $SrTiO_3$
$BiFeO_3$	Rhombohedral, R3c	$a_{PC} = 3.96$Å	$f = 1.41$ %
$BiMnO_3$	Monoclinic, C2	$a \approx c = 3.935$ Å $b = 3.99$ Å	$f_{a,c} = 0.77$ % $f_b = 2.15$ %
Bi_2NiMnO_6	Monoclinic, C2	$a_{PC} = 3.877$ Å	$f = -0.72$ %

In case that epitaxial thin films are coherently grown, the in-plane lattice parameter of film and substrate are identical when misfit dislocations are absent. For $BiMnO_3$, BLNMO and $BiFeO_3$ films the following mismatch would then be expected:

- The in-plane strain in $BiFeO_3$ films would be compressive and is relatively high (1.4%), which may lead to a tendency of the $BiFeO_3$ lattice to relax with the increase of film thickness by the creation of defects.
- In $BiMnO_3$ the ***a*** and ***c*** lattice parameters show a lower lattice mismatch to $SrTiO_3$ (0.77%) than that of ***b*** (2.15%), which suggests that it is the latter which stands out-of-plane in an epitaxial film. The in-plane strain would be compressive.
- The in-plane strain in BLNMO films would be tensile with a relatively low value of -0.71%, which may favour 2D growth.

Table 2. Optimum deposition conditions. The laser fluence was 1.8 J/cm^2 with a repetition rate of 5 Hz in each case

Material	% Bi target enrichment	O_2 pressure (mbar)	Substrate temperature (°C)
$BiFeO_3$	10	0.1	650
$BiMnO_3$	10	0.1	630
$(Bi_{0.9}La_{0.1})_2NiMnO_6$	0	0.5	620

To obtain single phase $BiMnO_3$, BLNMO and $BiFeO_3$ films all critical deposition parameters such as O_2 pressure, substrate temperature and film thickness need to be adjusted. For the work presented here a narrow window for the values of such parameters was encountered for all 3 species due to considerable Bi volatility as shown in Table 2 [31, 62].

3. RESULTS

3.1. Structural Characterization

3.1.1. X-Ray Diffraction

3.1.1.1. Phase Identification and Out-of-plane Orientation

X-ray diffraction (XRD) on $BiMnO_3$, BLMNO and $BiFeO_3$ films was carried out using a PANalytical Pro MRD 4-circle diffractometerusing Cu K_α radiation. Figure 11 shows representative θ/2θ diffractograms of $BiFeO_3$, $BiMnO_3$ and BLNMO.

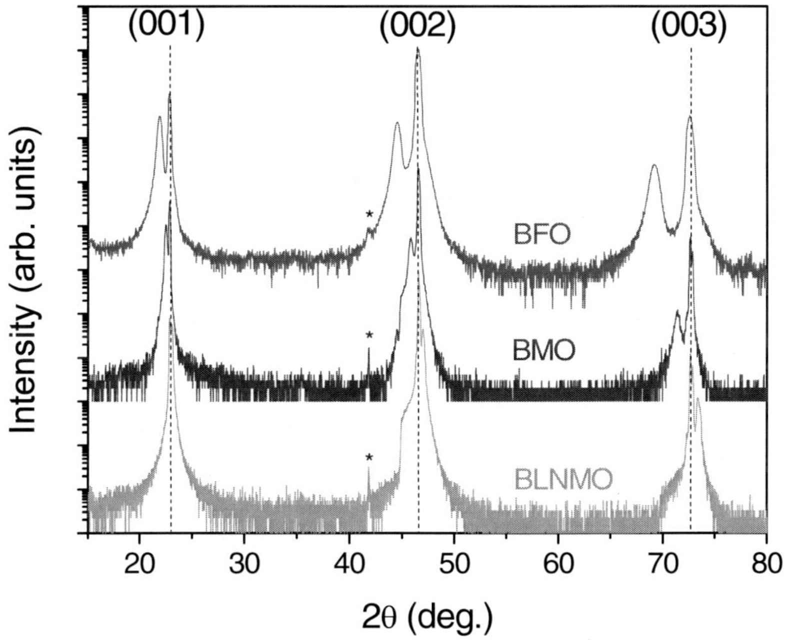

Figure 11. XRD symmetric θ-2θ diffractograms of the $BiFeO_3$ (top), $BiMnO_3$ (center) and BLNMO films (bottom). Substrate peaks are indicated by dashed lines. The asterisks (*) indicate the $SrTiO_3$ (002) peak caused by Cu K_β radiation.

In all samples the $SrTiO_3$ (00ℓ) substrate peaks are indicated by dashed lines, accompanied by the film peaks which correspond to the (00ℓ) reflections for BLNMO and $BiFeO_3$ and to the (0k0) for $BiMnO_3$ in pseudocubic notation. The film peaks are relatively narrow, which indicates good film crystallinity with the expected out-of-plane orientation, and no spurious phases are obvious.

Table 3. Out-of-plane lattice parameters obtained from θ/2θdiffractograms

Material	$BiFeO_3$	$BiMnO_3$	$(Bi_{0.9}La_{0.1})_2NiMnO_6$
Out-of-plane lattice parameter (Å)	4.04	3.99	3.87

As justified in section 2, $BiMnO_3$ has (010) out-of-plane orientation, whereas BLNMO and $BiFeO_3$ are (001)-oriented. Out-of-plane lattice parameters are listed in Table 3. In comparison to bulk values (Table 1) the out-of-plane lattice parameter is expanded for $BiFeO_3$, that of $BiMnO_3$ is unchanged and that of BLNMO is contracted.

3.1.1.2. ϕ-Scans and In-plane Orientation

The in-plane texture and in-plane orientation of the unit cells was characterized by pole figures and ϕ-scans of the (204) asymmetric reflection representing a lattice plane that is not parallel to the interface. Such reflection proved to be appropriate to discern between the film and substrate reflections, which is critical due to the similarity of the lattice parameters. A higher 2θ reflection is useful in such case, because the difference in 2θ for substrate and film is larger the higher the 2θ angle is.

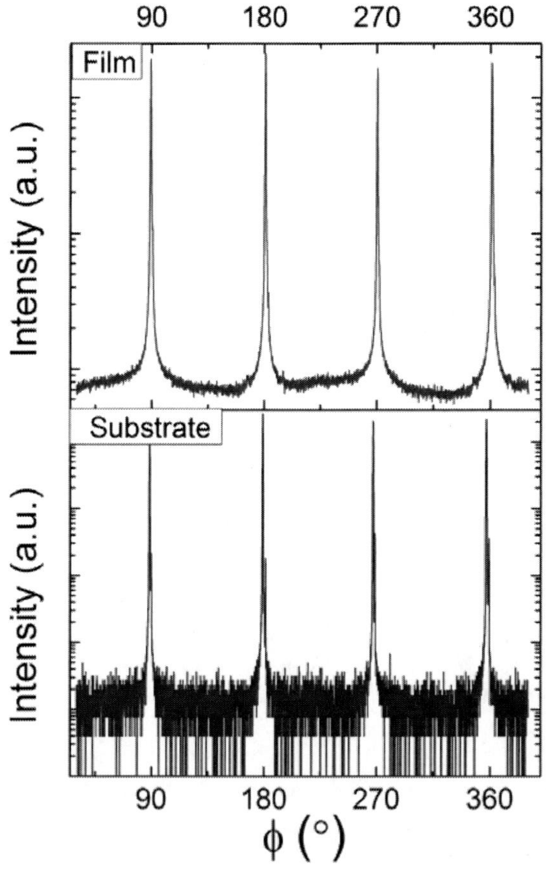

Figure 12. XRD ϕ-scans around the (204) reflection of both, 125-nm BLNMO film and $SrTiO_3$ substrate.

Figure 12 depicts the XRD ϕ-scans of the assymetric (204) reflection of both substrate and film of a 125-nm BLNMO film. Four 90º- ϕ-spaced peaks of the film are observed, which indicates that contributions from misoriented planes are absent. This, in turn, implies thatBLNMO films are coherently in-plane oriented. In order to confirm the in-plane texture, pole figures were obtained where ψ was scanned from 0 to 30º, as illustrated in Figure 13 for a 60-nm BLNMO film. In agreement with the ϕ-scans, Figure 13 reproduces the four 90°-ϕ-spaced peaks of the film, confirming the in-plane four-fold orthogonal symmetry.

Moreover, the fact that the values of the ϕ-peaks coincide with the ϕ-angles of the substrate confirms that BLNMO films have grown in a cube-on-cube fashion. In-plane film alignment with substrate and hence cube-on-cube epitaxy was also confirmed for $BiFeO_3$ and $BiMnO_3$ in the same way.

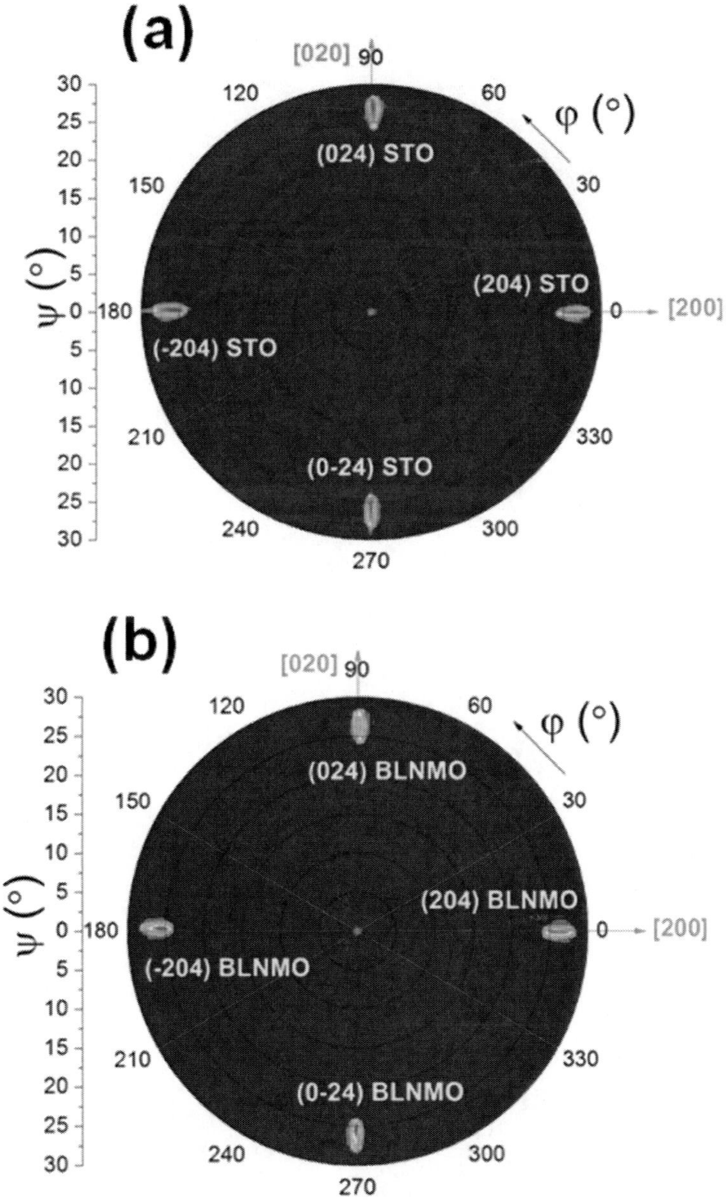

Figure 13. XRD pole figures around the (204) reflection of (a) $SrTiO_3$ substrate, and (b) 60-nm BLNMO film.

3.1.1.3. Reciprocal Space Maps

In reciprocal space maps (RSMs) the diffraction vector $\vec{q}$ is not directly measured, but several coupled $2\theta/\omega$ XRD scans are performed for different ω values, giving rise to a 2D 2θ–ω map of intensities, and q_{perp} and $q_{parallel}$ are obtained via trigonometric relationships. Figure 14 shows the RSMs around the (204) reflection for (a) $BiFeO_3$, (b) $BiMnO_3$, and (c) BLNMO. Both, the $SrTiO_3$ and the film reflections are displayed in each case, where $q_{parallel}$ of (204) is found to coincident with the substrate as indicated by the dashed lines.

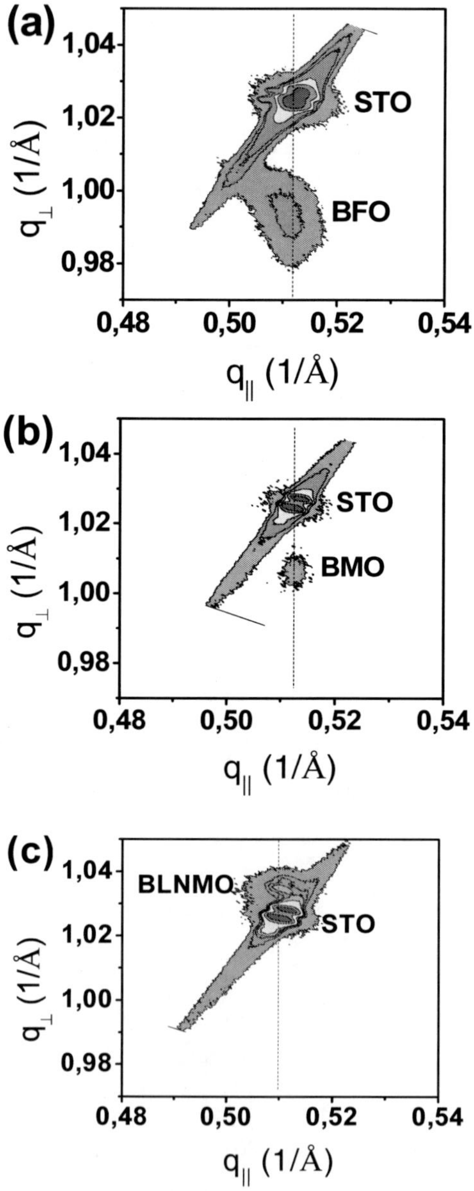

Figure 14. XRD reciprocal space maps of (a) $BiFeO_3$, (b) $BiMnO_3$ and (c) BLNMO around the (204) reflection of $SrTiO_3$. The dashed lines are a guide to the eye, indicating the in-plane q-value of the substrate reflection.

All samples adopt the in-plane lattice parameter of $SrTiO_3$ and, therefore, the films are fully strained. Unlike $BiFeO_3$, the $BiMnO_3$ and BLNMO peaks are narrow indicating an average low dispersion of lattice parameters in the film.

On the other hand, the high $BiFeO_3$ lattice mismatch (see Table 1) leads to a progressive relaxation of elastic energy, which is illustrated in form of a peak tail in Figure 14a extending towards lower q_{par} (higher in-plane lattice parameter) and higher q_{perp} (lower out-of-plane lattice parameter).The out-of-plane plane lattice parameters obtained from RSMs were coincident with those obtained from θ/2θ measurements (Table 2) as expected.

3.1.1.4. B-Site Order in $(Bi_{0.9}La_{0.1})_2NiMnO_6$ Detected by Synchrotron XRD

For a long-range ferromagnetic interaction, Ni and Mn cations are required to be located alternatingly along the [100], [010] and [001] crystallographic directions, which results in the alternation of Ni-planes and Mn-planes along the [111] direction (along the eight equivalent directions: [1-11], [11-1], [-111], [1-1-1], [-11-1], [-1-11], [-1-1-1], [111]). As a consequence, occurrence of the (1/2 1/2 1/2) superstructure XRD reflection should be expected. However, as the atomic scattering factor of Ni^{2+} and Mn^{4+} are very similar, the superstructure XRD reflection in Bi_2NiMnO_6 is very weak. Therefore, synchrotron radiation XRD was used, which provides an intense source of X-rays for detecting weak XRD reflections. Figure 15a and b depicts θ/2θ synchrotron XRD scans around (111) and (1/2 1/2 1/2) BLNMO reflections, respectively, of a 100-nm BLNMO film. Due to the similar lattice parameters of BLNMO and $SrTiO_3$, the perovskite (111) reflection of both film and substrate ($2\theta_{film}$ ~ 32.62º and $2\theta_{susbtrate}$ ~ 32.59º, respectively) can be seen in Figure 15a. Contrarily, as superstructure order is absent in $SrTiO_3$, the peak at $2\theta \approx 16.14°$ observed in Figure 15b solely corresponds to the BLNMO superstructure (1/2 1/2 1/2) reflection. Therefore, long-range B-site cation order in BLNMO thin films was confirmed unequivocally. The equivalent superstructure peaks were also observed on selected area electron diffraction patterns using high resolution electron microscopy [58].

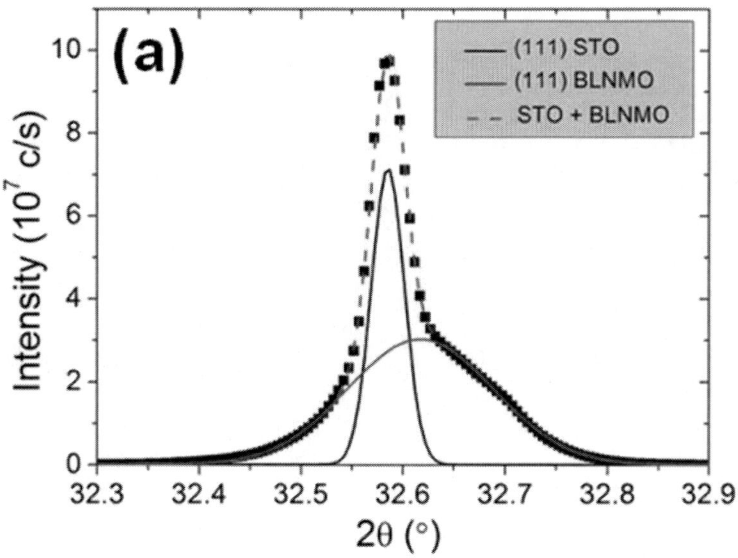

Figure 15. (Continued).

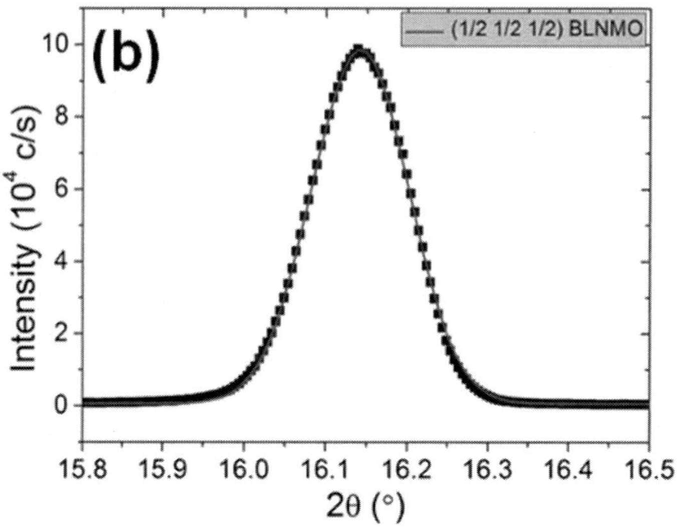

Figure 15. (a) θ/2θ synchrotron XRD scans around the perovskite (111) reflection for BLNMO and $SrTiO_3$. (b) θ/2θ synchrotron XRD scan around the (1/2 1/2 1/2) superstructure reflection for BLNMO. The diffraction intensity peaks were fitted using Gaussian functions.

3.1.2. HRTEM Microstructure Determination of $(Bi_{0.9}La_{0.1})_2NiMnO_6$

High-resolution TEM (HRTEM) images from BLNMO films are presented in Figure 16a and b. The high crystalline film quality indicated by X-rays is confirmed and no misfit dislocations are observed, as expected for a coherently strained film. The BLNMO/$SrTiO_3$ interface appears strikingly sharp with no cationic inter-diffusion across the interface.

Fast Fourier Transmforms (FFTs) of selected areas of the HRTEM images were computed in order to assess the periodicity of the crystal lattice. According to the FFTs depicted in Figure 17b and c the crystallographic directions of both substrate and film are well aligned in agreement with XRD measurements.

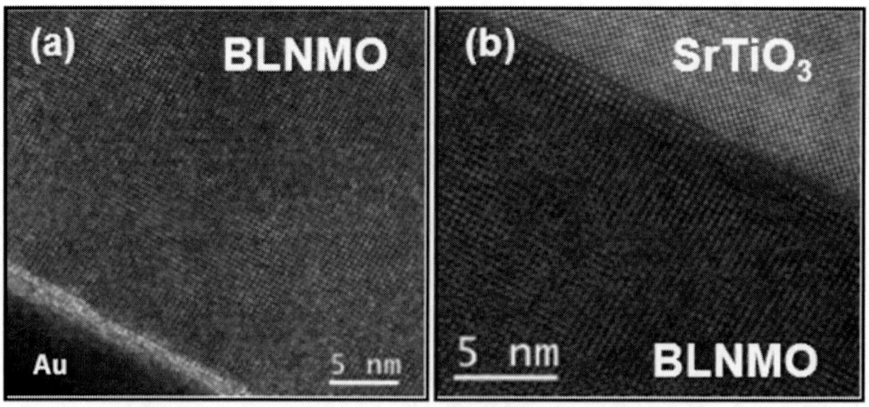

Figure 16. (a) HRTEM image of BLNMO film prepared in cross-section geometry along the [100] direction. (b) Higher magnification HRTEM image of the BLNMO/$SrTiO_3$ interface.

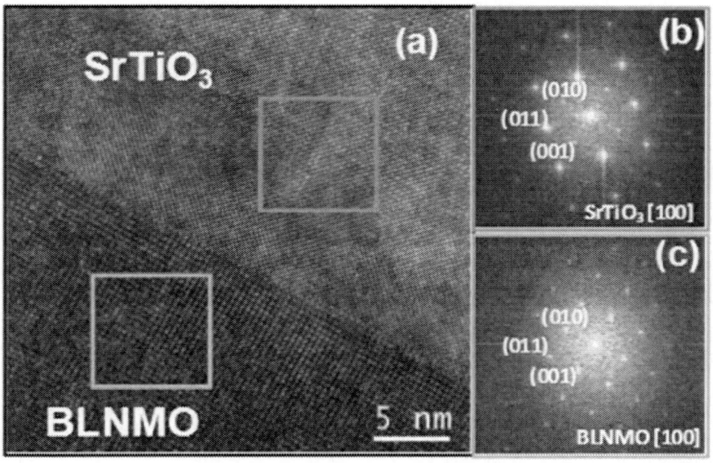

Figure 17. (a) HRTEM image of the BLNMO film and the $SrTiO_3$ substrate in cross-section geometry along the [100] direction. The green and blue squares mark the selected areas of the substrate and film where FFTs were taken. FFTs of (b) the $SrTiO_3$ substrate, and (c) of theBLNMO film.

3.1.3. Surface Topography

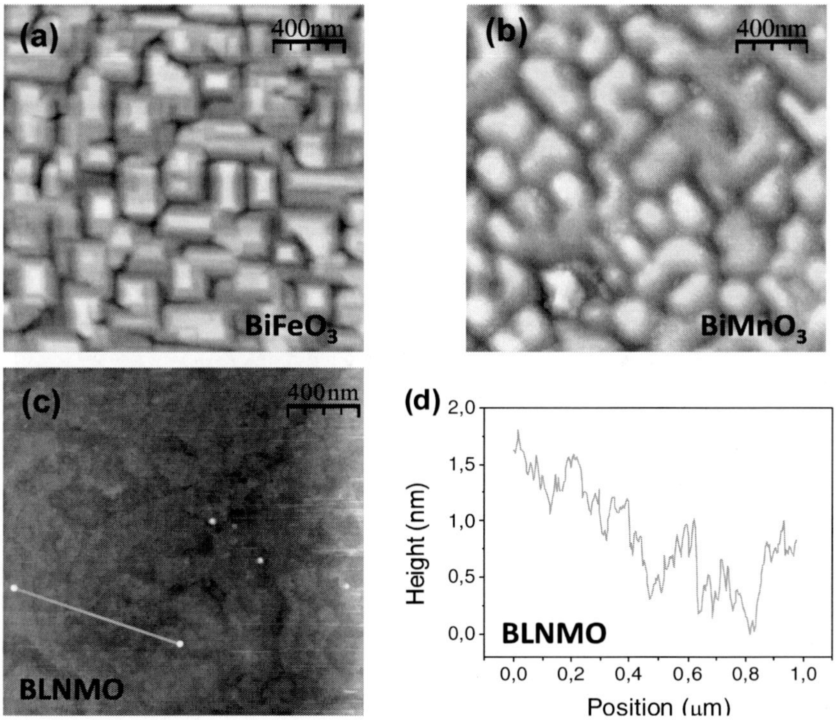

Figure 18. 2x2 μm^2AFM topography images of (a) $BiFeO_3$, (b) $BiMnO_3$ and (c) BLNMO. (d) corresponds to the BLNMO film height profile indicated by the solid (green) line in (c).

The surface topography of the films was studied by atomic force microscopy (AFM) using an Agilent 5420 instrument, aspresented in Figure 18.

Both $BiMnO_3$ and $BiFeO_3$ films developed faceted crystallites, evidencing 3D growth mode. Such facets can be deduced to represent coherently aligned crystal planes when studying the distribution of angles in the AFM data. Perfect alignment of these objects corroborates single-domain epitaxial growth and, as deduced from XRD analysis, their orientation coincides with the STO crystallographic axis. The root-mean-square roughness for $BiMnO_3$ and $BiFeO_3$ films was in the order of few nanometers not exceeding 5 % of the nominal film thickness.

On the contrary, the BLNMO films exhibits a quite different morphology where substrate steps and terraces are maintained. This is indicative of a 2D growth mode, in agreement with the higher epitaxial quality. Meandering can be observed as a consequence of anisotropic diffusion due to Ehrlich–Schwoebel terrace potentials [78].

3.2. Magnetic Properties

The magnetic properties of $BiMnO_3$, BLNMO and $BiFeO_3$ films were determined using SQUID magnetometry (Quantum Design MPMS XL) with magnetic field (H) applied in the film in-plane direction and the resulting film magnetization (M) being recorded.

3.2.1. $BiMnO_3$

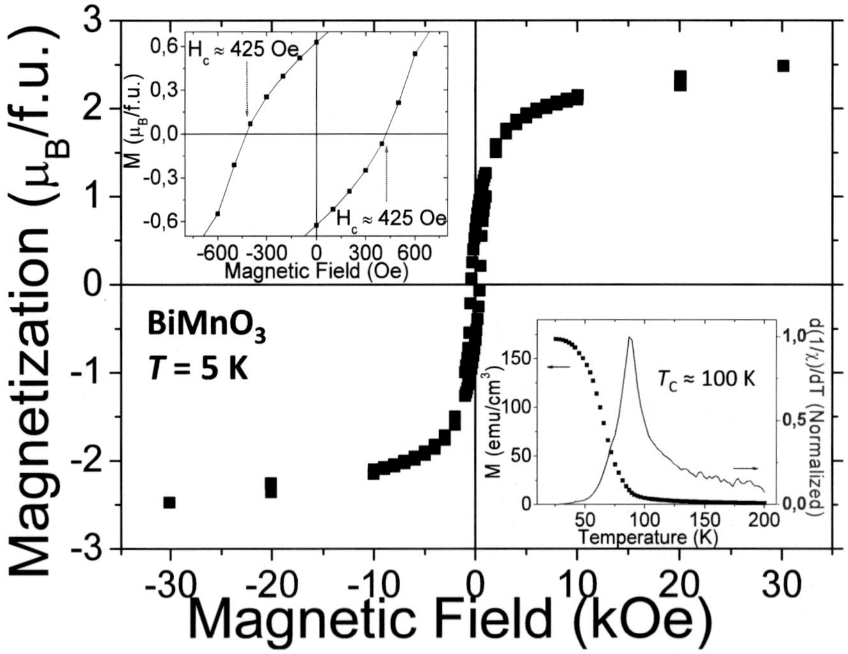

Figure 19. Magnetic field (H) dependence of magnetization (M) for a $BiMnO_3$ thin film at 5 K. Upper inset: Zoom of the low-field region. Lower inset: T dependence of M under field-cooled (FC) conditions using an external in-plane field of H = 1 kOe (left axis) and derivative of the inverse susceptibility (right axis).

The ferromagnetic moment in $BiMnO_3$ thin films is clearly observed in the hysteretic magnetization (M) vs applied magnetic field (H) curve recorded at 5 K (Figure 19). A saturated $BiMnO_3$ moment of $M_{sat.} \approx 2.5\ \mu_B$/f.u. was found, which is smaller than the bulk value (3.6 μ_B/f.u.) but close to the thin film value reported previously (2.2 μ_B/f.u.) [27]. A $BiMnO_3$ coercive field $H_{coerc.}$ of ≈ 425 Oe (upper inset of Figure 19) was found. The temperature dependence of M under 1 kOe applied H and the differentiated inverse of the magnetic susceptibility $1/\chi$ (lower inset of Figure 19) allowed determining the ferromagnetic Curie temperature $T_C \approx 100$ K, in close agreement to the bulk value [79-81].

3.2.2. $(Bi_{0.9}La_{0.1})_2NiMnO_6$

BLNMO films are B-site ordered, fully stoichiometric and the electronic configuration of Ni^{2+} and Mn^{4+}cations are those required for ferromagnetic coupling [58]. An ideal ferromagnetic ordering of the spins of the B-cations, $Ni^{2+}(t_{2g}{}^6e_g{}^2$, S=1) and $Mn^{4+}(t_{2g}{}^3$, S=3/2), implies a saturated magnetization of M_S=5 μ_B/f.u. Magnetization data of Figure 20 shows that the film does not saturate even at a field of 7 T, although the diamagnetic substrate contribution has been subtracted from the data.

The films have a magnetization M_S (7 T, 10 K) ≈ 3.5 μ_B/f.u., in close agreement with data reported for bulk $(Bi_{0.9}La_{0.1})_2NiMnO_6$samples ($M_S$ = 3.6 μ_B/f.u.), which do not saturate either at 7 T field and 5 K [65]. The presence of antisite defects and antiphaseboundaries, common defects in double perovskite structures [82], may account for both the hardness to saturation and the accompanying reduced magnetization.

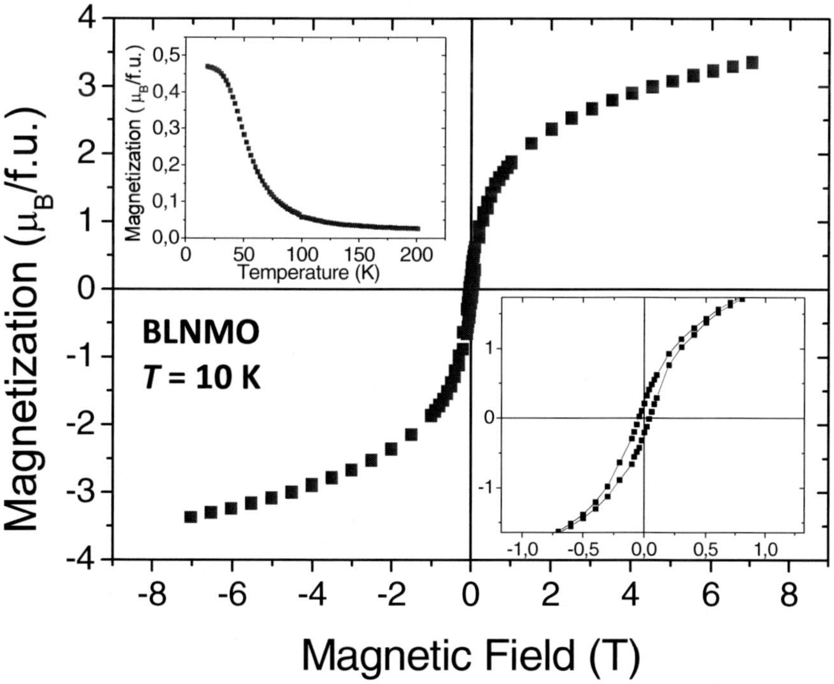

Figure 20. *H* dependence of *M* for a BLNMO thin film at 10 K. Lower inset: Zoom of the low-field region; units are identical to the main figure. Upper inset: Temperature dependence of the magnetization under an external in-plane magnetic field of 1kOe.

The magnetic remanence, as shown in Figure 20 (bottom panel), is rather low probably due to the small spin-orbit coupling of the magnetic cations.

On the other hand, inspection of the temperature dependent magnetization data (upper panel in Figure 20) indicates that the Curie transition occurs at $T_C \approx 100$ K. This value is lower than in the bulk (≈ 140 K) [59], but this is in agreement with the reduced T_C found previously for BLNMO thin films [60, 61]. The reduction in the magnetic transition temperature has been ascribed to the epitaxial stress exerted by the substrate [60, 62]. The rather broad transition temperature may result from a distribution of superexchange interactions within the sample, likely driven by a dispersion of the distinct length and angles of Ni–O–Mn bonds near the A-sites, which are occupied by either La or Bi cations.

3.2.3. $BiFeO_3$

The $BiFeO_3$ film showed signs of antiferromagnetism with a weak ferromagnetic moment up to room temperature (Figure 21). The $M_{sat.} \approx 0.05$ μ_B/Fe registered at 5 K agrees with previous reports [45, 83-85]. The appearance of a small ferrimagnetic component in epitaxial $BiFeO_3$ thin films may be ascribed to spin canting due to lattice or FE domain wall strain or oxygen vacancies [85].

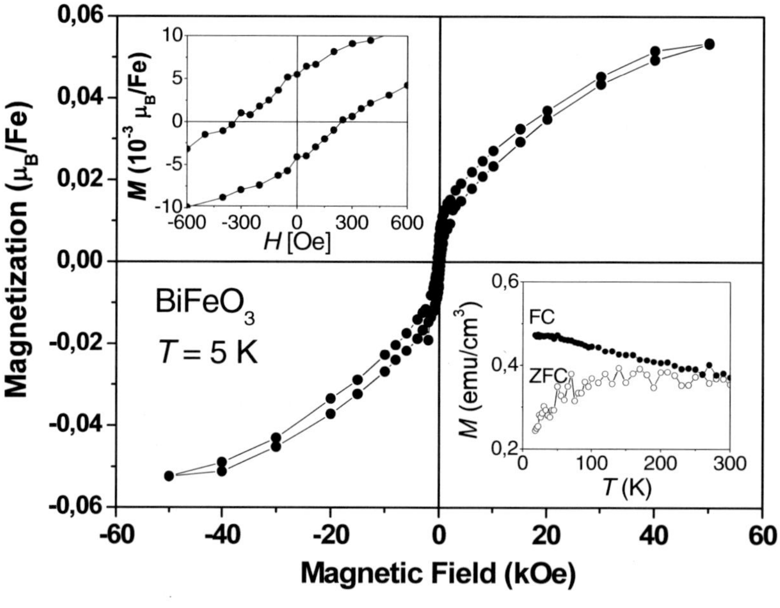

Figure 21. H dependence of M for $BiFeO_3$ thin films at 5 K. Upper inset: Zoom of the low-field region. Lower inset: Zero-field cooled (ZFC) and field cooled (FC) M vs T curves using an external in-plane magnetic field of 100 Oe.

3.3. Ferroelectric Properties

As mentioned in section 1, ferroelectricity in Bi containing perovskites commonly arises from the off-centering of the lone-pair electrons of Bi cations. If the displacement vectors of the lobe-distributed Bi^{3+} lone-pair electrons align spontaneously and coherently on a macroscopic scale, the conditions for ferroelectricity are fulfilled and ferroelectric domains form, which may be poled or switched by the application of sufficiently high electric fields.

3.3.1. Ferroelectric Hysteresis Loop Measurements

The most commonly used technique to detect ferroelectricity may be ferroelectric polarization measurements by using a time-dependent sinusoidal electric field to detect hysteretic behavior in form of a hysteresis loop. Since the poling or switching of ferroelectric domains by electric fields implies that Bi^{3+} lone-pair electrons change their lattice position, such electron movements can be picked up at the electrodes as a small current pulse at the moment when the switching occurs. Such ferroelectric testing can be ambiguous though [86], because applied electric fields may also cause leakage currents which can mask the switching current. Such leakage currents can be determined and subtracted from the ferroelectric switching, which then may lead to hysteretic curves consistent with ferroelectric polarization.

3.3.2. Piezo-force Microscopy

A more reliable technique to test for the switching of ferroelectric domains may be piezo-force microscopy, where ferroelectric domains are switched locally by applying a voltage between a scanning-probe microscopy (SPM) tip and the sample. The local switching can be monitored by the piezo-force response of the ferroelectric sample. This is demonstrated on 100 nm $BiFeO_3$ thin films deposited on conducting Nb-STO substrates as shown in Figure 22.

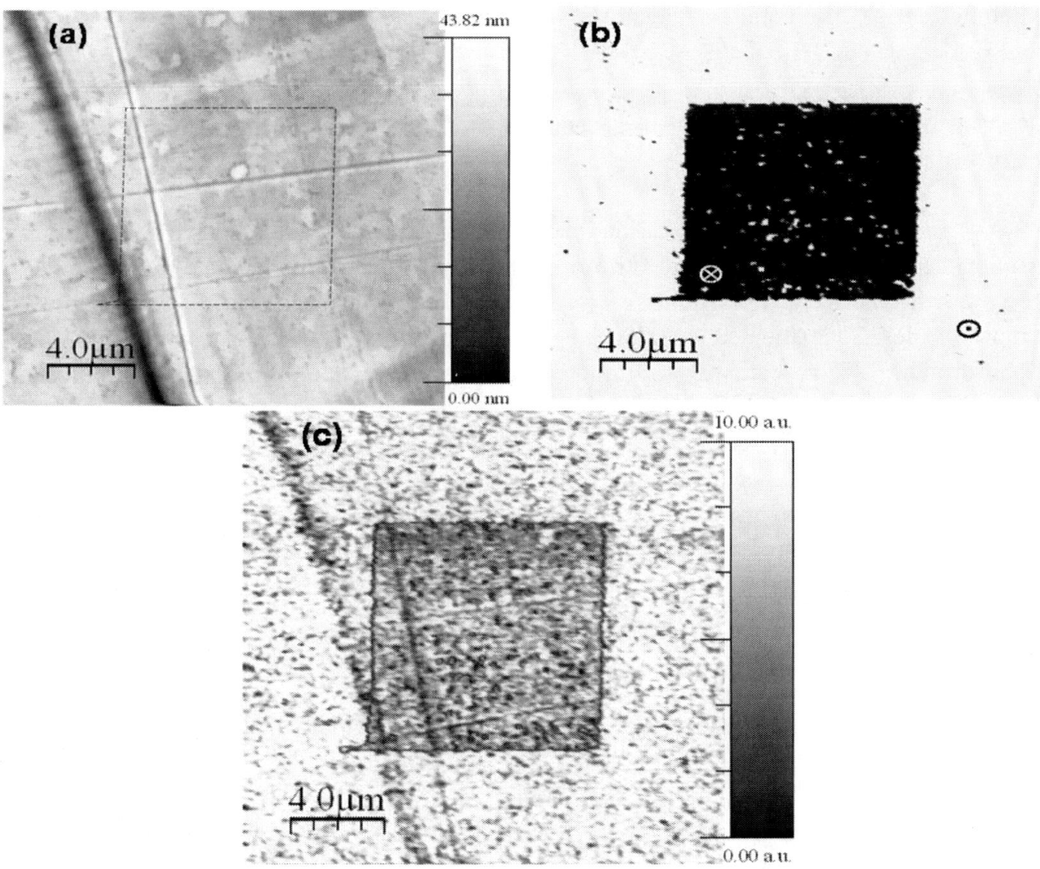

Figure 22. (a) Surface topography, (b) out-of-plane PFM phase, and (c) out-of-planePFMamplitude images (20x20 μm^2) for a $BiFeO_3$ thin film at room temperature. The marked area in (a) was previously scanned by applying a DC voltage (-5V) between tip and substrate.

Images were acquired using Si tips with PtIr coating at an AC voltage of 0.5 Vrms and a frequency of 50 kHz. In the set-up used here voltages were applied to the sample while the tip was grounded. Well-defined poled areas were patterned into the film by applying a DC voltage between tip and substrate, and subsequently checked by piezo-response force microscopy (PFM). Simultaneous topographic, out-of-plane PFM phase and amplitude images (Figures 22a, b, c) were recorded.

The topographic image in Figure 22a shows that switching the polarization within the marked area does not damage the $BiFeO_3$ film surface. The phase image in Figure 22b presents a 180° contrast between the poled area and the surrounding film, pointing to the switching of the perpendicular component of the polarization, with a corresponding minimum in the amplitude image at the domain wall (Figure 22c). Polarization can be switched back and forth by reversing the polarity of the applied voltage, confirming that the $BiFeO_3$ film was ferroelectric at room temperature (RT). PFM on $BiMnO_3$ and BLNMO films at RT is not feasible due to low film resistance.

A further possibility to detect for ferroelectricity is the use of dielectric spectroscopy to measure the dielectric permittivity ε' at various temperatures. In ferroelectric materials ε' vs T curves exhibit a maximum at the transition temperature T_C. However, these measurements can be troublesome in the presence of multiple dielectric relaxations and the direct observation of switching ferroelectric domains may always be preferable.

3.4. Magneto-electric Coupling Detected by Magnetic Field Dependent Impedance Spectroscopy

For potential industrial application of Bi containing multiferroic perovskites a strong coupling between magnetic and electric order parameters is desired. Such magneto-electric coupling (MEC) effects are difficult to detect, which is a major problem in the emerging field of multiferroics. Sophisticated techniques such as transmittance optical spectroscopy [87], X-ray absorption [88] or scattering [89] and magnetic field dependent piezo-force microscopy (PFM) [90] are expected to yield unambiguous results. Much discussion has arisen though over the more frequently used technique of dielectric spectroscopy, which, in multiferroics research, is often combined with applied magnetic fields. The use of such frequency (f), temperature (T) and magnetic field (H) dependent impedance spectroscopy (IS) allows detection of and discrimination between magneto-resistance (MR) and magneto-capacitance (MC) effects. It has been pointed out previously that such discrimination is a non-trivial task, because pure MR of a secondary and extrinsic dielectric relaxation in the sample can cause changes in capacitance [91, 92]. Therefore, it has been argued that all existing dielectric relaxations within one sample need to be identified and deconvoluted first, before a magnetic field can be applied. In polycrystals multiple dielectric relaxations may originate from grain interior (bulk) regions, grain boundaries (GBs) and electrode-sample interface (IF) layers [93], whereas in epitaxial thin films only bulk (or intrinsic film) and IF contributions are expected [94]. The MC and MR of each contribution in the sample must be determined separately by IS for a range of applied magnetic fields. In this way the true MC and MR of all separated relaxations can be obtained. Such analysis may then be repeated at various T to obtain a comprehensive picture of the intrinsic MC and MR behavior. This procedure is demonstrated here in this work, where first all dielectric relaxations occurring in the samples are deconvoluted.

Impedance spectroscopy measurements on all films were carried out in the out-of-plane direction, which leads to measurable thin film resistance in nominally insulating materials. For such purpose all films were deposited on highly conducting Nb-doped $SrTiO_3$ substrates, where the film surfaces would be covered with sets of Pt electrodes. In such way the impedance can be measured twice across the film, which was shown previously to be appropriate to obtain valid IS data [94].

3.4.1. Deconvolution of Dielectric Relaxations

IS data are usually represented on plots of the imaginary vs the real part of the impedance (-Z'' vs Z'): for each series dielectric relaxation ideally one semicircle appears in agreement with the RC element model [93, 95, 96]. In multiferroics research however, dielectric data are often represented in terms of dielectric permittivity ε' vs f or ε' vs T to directly display changes of ε' as a result of applied H, or ε' changes with T across a magnetic transition.

Impedance spectroscopy data from 50 nm $BiMnO_3$, 100 nm BLNMO and 100 nm $BiFeO_3$ thin films are represented in the ε' vs f notation in Figure 23a, b and c respectively. Clear signs of two dielectric relaxation processes were obvious in $BiMnO_3$ and BLNMO films as manifested by the existence of two ε' plateaus, ε'_{high} and ε'_{low}. Such plateaus corresponds to two series dielectric relaxations as represented by a series of two RC elements, where the low permittivity plateau ε'_{low} represents the intrinsic $BiMnO_3$/BLNMO film permittivity and ε'_{high} an extrinsic Maxwell-Wagner type electrode interface relaxation [92, 97-99]. The $BiMnO_3$ ε'_{low} plateau was found to be $\approx 35 - 40$, in a similar range as $BiMnO_3$ permittivity values of $\approx$ 30 reported before [22, 94, 97]. For BLNMO, ε'_{low} was found to be ≈ 100, where the increase as compared to $BiMnO_3$ may be a result of the B-site cation ordering. In the $BiFeO_3$ film no ε' plateaus are visible and at the T-range investigated only the intrinsic film contribution ε'_{low} is visible, displaying values in the range of ≈ 90 - 120. The strong f dependence of the intrinsic $BiFeO_3$ capacitance is a sign of a certain degree of non-ideality of the dielectric relaxation [100], which is also visible for the $BiMnO_3$ and BLNMO relaxations, where the ε'_{high} and ε'_{low} plateaus both show a slight f dependence too.

The $BiMnO_3$ and BLNMO data were fitted to an equivalent circuit of a series of two RC elements (see inset in Figures 22a and b), whereas the single relaxation in $BiFeO_3$ was fitted by one RC element.

In all spectra for $BiMnO_3$, BLNMO and $BiFeO_3$ the extrinsic parasitic contributions were accounted for by a series inductor L0 describing the inductance of the measurement wires and a series resistor R0 describing the resistance of the conducting substrate, the electrodes and measurement wires [94]. The non-ideality of the intrinsic $BiMnO_3$ and BLNMO film relaxations was accounted for by a parallel CPE (CPE1) and for the extrinsic $BiMnO_3$/BLNMO relaxations by the replacement of the ideal capacitor by CPE2. Such replacement of the ideal capacitor also gave good fits to the intrinsic $BiFeO_3$ film relaxation in terms of the R1-CPE1 element depicted in Figure 23c. The fact that in $BiMnO_3$ and BLNMO the intrinsic relaxations could only be fitted by a R1-CPE1-C1 element and not by a more simple R1-CPE1 element is debited to the fact that the ε' drop from ε'_{high} to ε'_{low} in ε' vs f is not as sharp as expected for an ideal intrinsic relaxation. Additionally, the ε'_{high} and ε'_{low} plateaus both exhibit a slight f-dependence. Such multiple signs of non-ideality can only be fitted with an R1-CPE1-C1 element for the intrinsic and an R2-CPE2 element for the extrinsic contribution.

In order to confirm the validity of each equivalent circuit presented in Figure 23a, b and c, and to get a clearer picture of the physics behind each deconvoluted dielectric relaxation it is advantageous to consider the temperature trends of all equivalent circuit components.

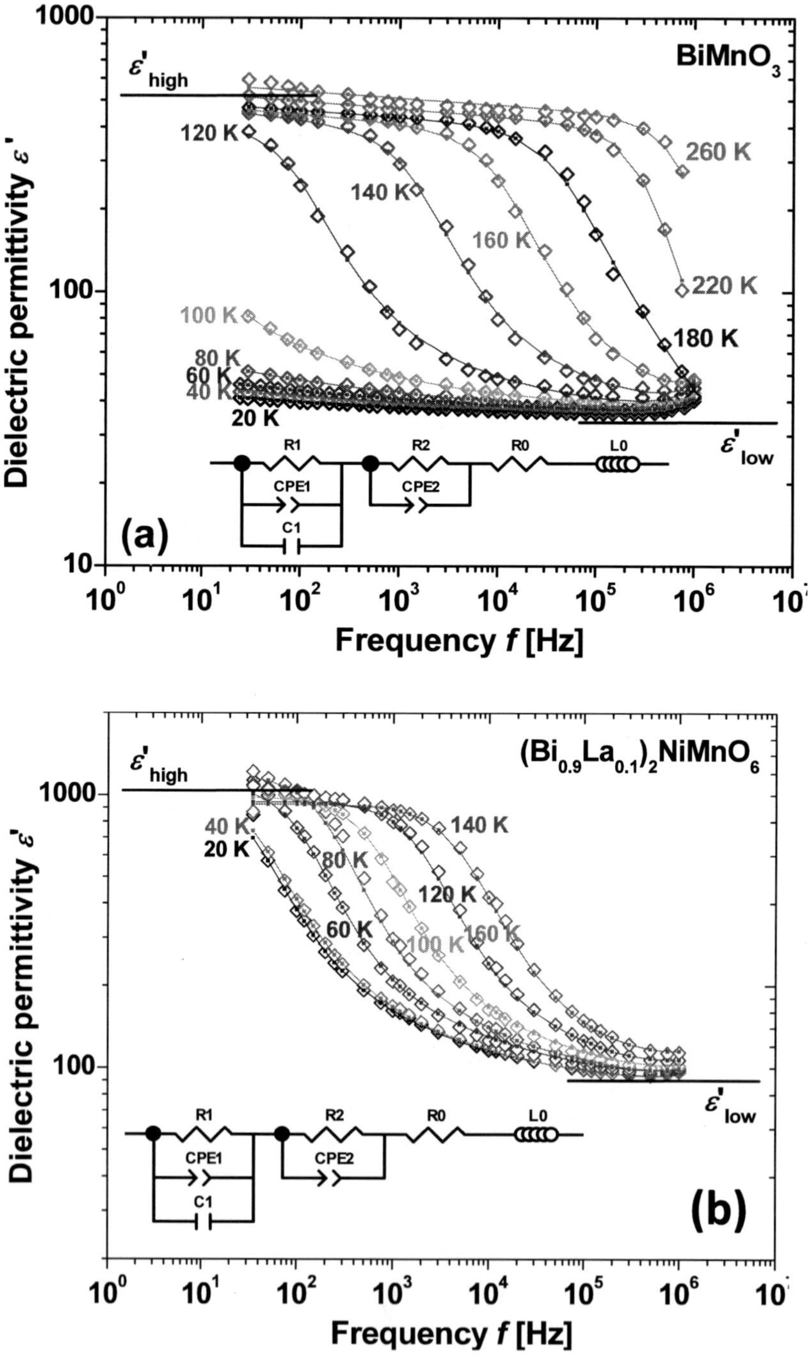

Figure 23. (Continued).

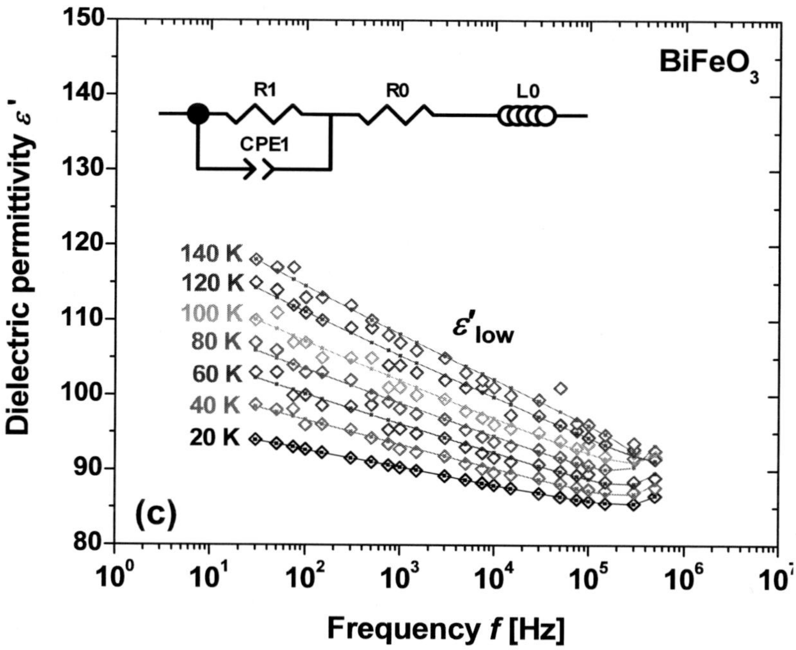

Figure 23. Dielectric permittivity ε' vs f for (a) 50 nm $BiMnO_3$ (b) 100 nm BLNMO and (c) 100 nm $BiFeO_3$ films at selected temperatures as indicated. Open symbols (◊) represent experimental data, full squares (■) and solid lines represent fits to the data using the equivalent circuit depicted below the curves. The occurrence of two ε' plateaus (ε'_{high} and ε'_{low}) for $BiMnO_3$ and BLNMO is consistent with two series dielectric relaxations, as represented by two series RC elements. The presence of one ε' plateau (ε'_{low}) for $BiFeO_3$ is consistent with one dielectric relaxation, as represented by one RC element. R0 and L0 represent parasitic contributions.

3.4.2. Equivalent Circuit Fitted Parameters for $BiMnO_3$, BLNMO and $BiFeO_3$ Films

Using the equivalent circuits shown in the insets of Figure 23a, b and c the $BiMnO_3$, BLNMO and $BiFeO_3$ intrinsic and extrinsic relative dielectric permittivity ε_1 (C1) and ε_2 (C2) respectively, the intrinsic and extrinsic resistivity ρ_1 (R1) and ρ_2 (R2) respectively, and the CPE exponents n were determined at various T. The resulting T dependences of such deconvoluted parameters are presented in Figures 24 and 25.

The $BiFeO_3$ intrinsic film permittivity ε_1 increases uniformly from 100 to 130 with T (Figure 24a), which is a clear sign for ferroelectricity. The intrinsic $BiMnO_3$ permittivity $\varepsilon_1 \approx 35 - 45$ shows a modest increase up to 75 K, again consistent with ferroelectricity. At higher T around the magnetic transition $T_C \approx 100$ K, the $BiMnO_3$ intrinsic ε_1 vs T curve displays a peak structure as demonstrated in the enlarged inset of Figure 24b. This may well represent an intrinsic MEC response of ε_1 to the $BiMnO_3$ magnetic transition at $T_C \approx 100$ K, since all extrinsic contributions had been deconvoluted and were accounted for by means of the equivalent circuit. This peak in ε_1 vs T is consistent with previous work [97], where the peak was stronger pronounced though. Since the peak structure observed here is rather weak, it can hardly be resolved from the experimental data in Figure 23a and only emerges from the fits. Surprisingly, the intrinsic BLNMO film permittivity ε_1 did not show any signs of a peak structure in the ε_1 vs T curve near the magnetic transition at $T_C \approx 100$ K and MEC coupling was not indicated. Obviously, the interruption of the Mn sublattice by Ni cations leads to an attenuation or disappearance of the MEC effect. The extrinsic $BiMnO_3$ and BLNMO interface relative permittivity ε_2 displayed in Figure 24a shows values of 550 – 600 and 800 -1100 respectively, which is slightly below the standard for such type of relaxation [93].

Figure 24c shows the CPE exponent n for the various $BiMnO_3$, BLNMO and$BiFeO_3$ relaxations. The intrinsic $BiMnO_3$ film relaxation shows a strongly increasing non-ideality above 75 K, which is manifested by a drop of n with increasing T.

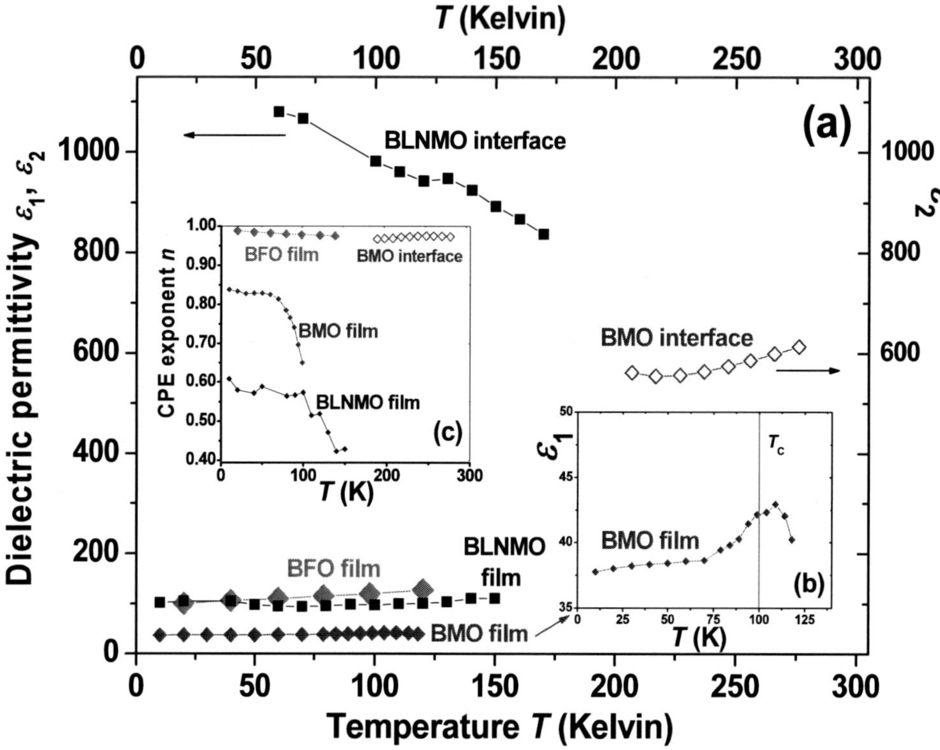

Figure 24. (a) $BiMnO_3$, BLNMO and $BiFeO_3$ intrinsic and extrinsic dielectric permittivity ε_1 and ε_2 vs T, obtained from the respective equivalent circuit fits. (b) Magnification of the intrinsic $BiMnO_3$ permittivity ε_1 displaying signs of MEC near T_C. (c) CPE exponents n, indicative of more ideal ($n \approx 1$) or less ideal ($n < 1$) character of the respective dielectric relaxation.

This may well be consistent with a broadening of the distribution of dielectric relaxation times τ, possibly as a reflection of the magnetic transition T_C. This, in turn, points towards increasing disorder in the $BiMnO_3$ film with increasing T due to the nearby onset of paramagnetism above 75 K. The intrinsic BLNMO relaxation shows a similar decrease in the CPE exponent n near T_C as the $BiMnO_3$ film and the equivalent interpretation may apply.

Figure 25 demonstrates the $BiMnO_3$ and BLNMO thin film and interface resistivity ρ_1 and ρ_2 respectively on plots of ρ vs reciprocal T. Both the intrinsic film $BiMnO_3$ and BLNMO charge transport curves (ρ_1 vs $1/T$) show correlations with T_C with clearly modified T-dependence above and below T_C:

- Above T_C electron hopping is indicated with a T independent activation energy E_A of 0.19 eV ($BiMnO_3$) and 70 meV (BLNMO). In case of $BiMnO_3$ this is in excellent agreement with previous work (0.2 eV) [97].
- Below T_C, the thermal activation of charge transport is weaker as reflected by a reduced slope of the ρ_1 vs $1/T$ curves implying a lower activation energy E_A.

The charge transport in $BiMnO_3$ and BLNMO seems to be based on similar mechanisms, which is most likely by electron hopping between B-site cations. The intrinsic resistivity ρ_1 of the $BiFeO_3$ film was high at all T investigated and could not be resolved. Therefore, the film quality may be regarded satisfactory with minimized leakage. In the equivalent circuit fits, the resistors R2 for the $BiMnO_3$/BLNMO interface and R1 for the $BiMnO_3$, BLNMO and $BiFeO_3$ film contributions were always set to infinity in case that the impedance data did not allow resolving high resistance.

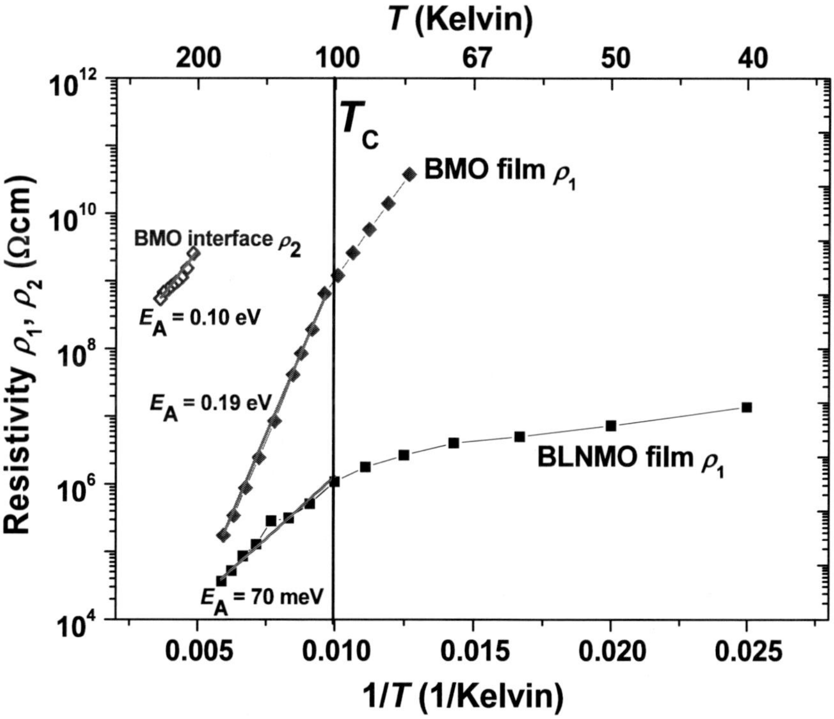

Figure 25. $BiMnO_3$ and BLNMO intrinsic and extrinsic resistivity ρ_1 and ρ_2 vs T, obtained from the respective equivalent circuit fits. Both the $BiMnO_3$ and BLNMO trends of ρ_1 vs T show correlation with T_C. The activation energies E_A are given in [eV/meV]. Solid (red) lines are linear Arrhenius fits.

3.4.3. Magnetic Field Dependent Dielectric Characterizations

After rationalizing the presence of extrinsic and intrinsic dielectric relaxations in $BiMnO_3$ and BLNMO, and the single intrinsic relaxation in $BiFeO_3$ films, the MR and MC in both types of film were determined. The f dependent $BiMnO_3$, BLNMO and $BiFeO_3$ film impedance was measured at selected fixed T repeatedly under various fixed applied H.

The complex impedance notation, $Z^* = Z' + i Z''$, is presented in Figure 26 in terms of $-Z''$ vs Z' curves for the $BiMnO_3$ film at 95 K close to T_C, where in such plots a semicircle is expected for each relaxation and the semicircle diameter corresponds to the resistivity of the respective relaxation. The partial semicircles in Figure 26 are a manifestation of the intrinsic relaxation, whereas the extrinsic interface resistance is large and not accessible at this *T* and *f* range. Although the intrinsic film semicircle is not fully displayed, the diameter may be extrapolated readily. On increasing *H* the semicircle diameter decreases, which entails that ρ_1 decreases as a result of MR. The *H* dependent R1 values allowed calculating the intrinsic MR defined by MR = [R(*H*=0) – R(*H*)]/R(0). The inset of Figure 26 shows rather high MR of up to 65% at 95 K and 100 K.

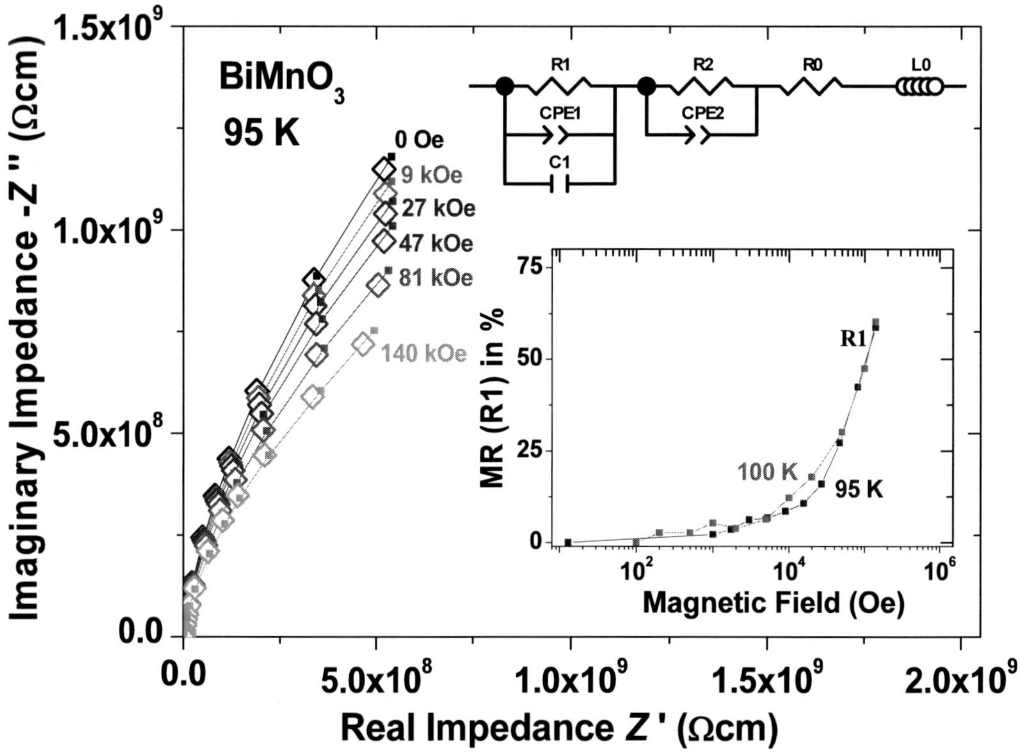

Figure 26. $BiMnO_3$ imaginary part of the impedance –*Z*'' vs real part *Z*' at 95 K for selected applied magnetic fields *H* as indicated. Open symbols (◊) represent experimental data, full squares (■) and solid lines represent fits to the data using the equivalent circuit depicted at the top of the curves. The decreasing size of the semicircle indicates MR of the intrinsic $BiMnO_3$ film relaxation. The inset displays MR vs *H* at 95 and 100 K, where MR = [R(*H*=0) – R(*H*)]/R(0).

No perceptible MR nor MC were detected in $BiMnO_3$ above the magnetic transition T_C. The BLNMO film showed similar MR like the $BiMnO_3$ film with a maximum in MR of ≈ 40% near the magnetic transition at ≈ 100 K [101].

In the $BiFeO_3$ films the resistance was too high to be resolved at all *T* investigated as mentioned before, and the MR could not be determined.

Figure 27 shows ε' vs *f* curves for $BiMnO_3$ films collected under various applied *H* at 95 K. At low *f*, ε' shows an upturn, which represents the onset of the extrinsic Maxwell-Wagner relaxation R2-CPE2 (see ε'_{high} in Figure 23a), where perceptible variations with *H* occur. It has been pointed out above that such variation in ε' with *H* in the vicinity of an extrinsic contribution is not necessarily a reflection of MEC, but can be caused entirely by the MR of an extrinsic Maxwell-Wagner relaxation [91]. Since the intrinsic $BiMnO_3$ film relaxation shows perceptible MR, the extrinsic relaxation may be expected to exhibit similar behavior and the variation in ε' at low f observed in Figure 27 may well be artificial and simply reflect the MR of the extrinsic Maxwell-Wagner relaxation. At higher f where the intrinsic relaxation is more dominant, only small MC can be seen. Intrinsic $BiMnO_3$ ε_1 values were obtained from the equivalent circuit fits by extracting C1 for various H and the intrinsic MC, defined as MC = [C(H=0) – C(H)]/C(0), is shown in the inset of Figure 27.

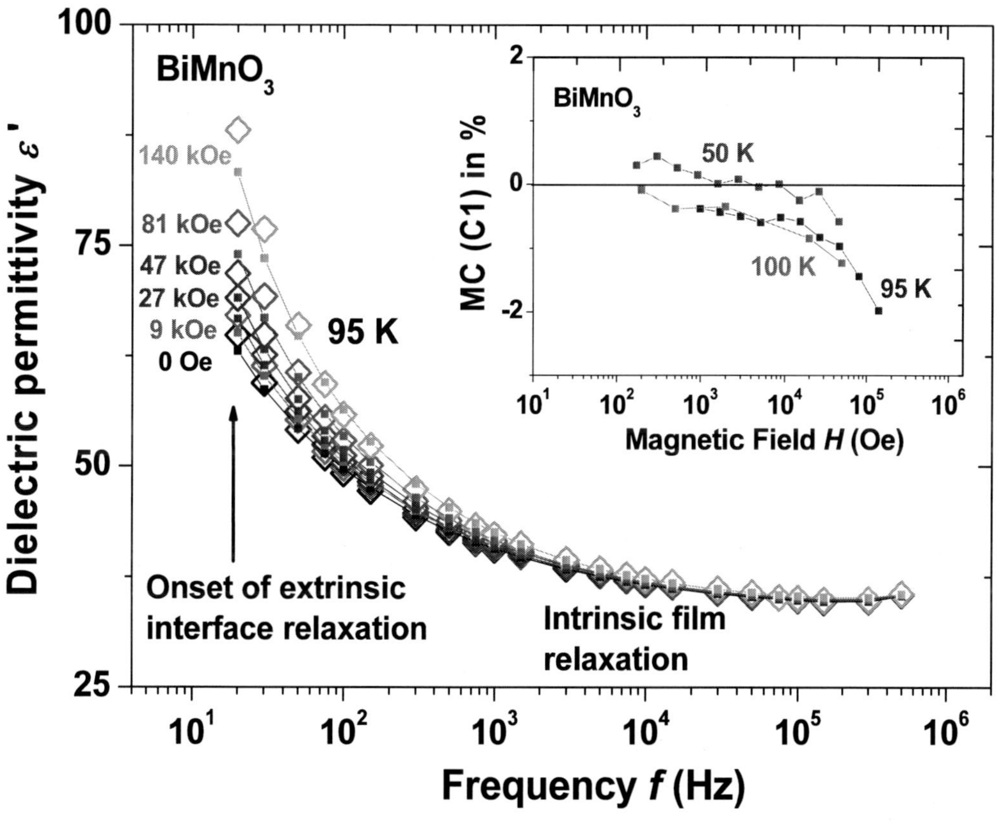

Figure 27. ε' vs f for a 50 nm $BiMnO_3$ thin film at 95 K for selected magnetic fields H as indicated. Open symbols (◊) represent experimental data, full squares (■) and solid lines represent fits to the data using the equivalent circuit depicted in Figure 23a and 26. The inset displays MC vs H at 50, 95 and 100 K, where MC = [C(H=0) – C(H)]/C(0).

Intrinsic MC and MEC appear to be rather small in the range of ≈ -1.5% at 90 kOe, which confirms the rather weak peak feature in ε_1 vs T (Figure 24b) and previous reports on polycrystals (MC ≈ -0.7% at 90 kOe) [22]. In the BLNMO film only signs of an extrinsic and artificial MEC due to MR effects were detected, whereas $BiFeO_3$ films exhibited no MEC at all. In $BiFeO_3$ the absence of any MEC may be debited to the fact that the antiferromagnetic transition temperature occurs at much higher T than the range investigated here, and only near such transition MC and MEC effects may be detectable.

CONCLUSION

It was shown that thin film deposition of multiferroic epitaxial thin films based on the Bi containing perovskite oxides $BiMnO_3$, $(Bi_{0.9}La_{0.1})_2NiMnO_6$, and $BiFeO_3$ can be performed adequately by pulsed laser deposition. Thorough structural film characterizations demonstrated the desired single film phases, single out-of-plane orientation and coherent in-plane strain. $BiFeO_3$ films were shown to be ferroelectric at room temperature and strong indications for ferroelectricity were observed in $BiMnO_3$. All films showed the desired ferromagnetic ($BiMnO_3$) or ferrimagnetic ($BiFeO_3$ and $(Bi_{0.9}La_{0.1})_2NiMnO_6$)) magnetism. Furthermore, it was shown that magnetic field dependent impedance spectroscopy is a powerful tool to deconvolute intrinsic and extrinsic contributions to the film dielectric properties, which concomitantly allows reliable determination of intrinsic magneto-electric coupling effects. Such effects were shown to be absent in $BiFeO_3$ and double-perovskite $(Bi_{0.9}La_{0.1})_2NiMnO_6$ at the temperature range investigated, whereas $BiMnO_3$ films showed weak MEC effects in the range of up to 2%.

ACKNOWLEDGMENTS

E. L., J. V. and M. V. acknowledge financial support from the Spanish Ministerio de Ciencia e Innovación (MICINN) under project numbers MAT2008-06761-C03, MAT2011-29269-C03 MAT2011-27470-C02-01, MAT2011-27470-C02-02, IMAGINE CSD2009-00013 and NANO-SELECT CSD2007-00041. RS acknowledges the Ramón y Cajal program from the MICINN. Many thanks go to SSSS Ltd. for support with the data analysis. Furthermore, the contributions from Carmen Munuera, Neven Biskup, Norbert Nemes, Alberto Rivera, Mar Garcia-Hernandez, Carlos Leon, Josep Fontcuberta and Jacobo Santamaria are acknowledged explicitly.

REFERENCES

[1] Eerenstein W., Mathur N. D. and Scott J. F., *Nature* 442, 759 (2006).
[2] Van Aken B. B., Rivera J.-P., Schmid H. and Fiebig M., *Nature* 449, 702 (2007).
[3] Spaldin N. A. and Fiebig M., *Science* 309, 391 (2005).
[4] Gajek M., Bibes M., Fusil S., Bouzehouane K., Fontcuberta J., Barthelemy A. and Fert A., *Nature Materials* 6, 296 (2007).
[5] Bea H., Gajek M., Bibes M. and Barthelemy A., *Journal of Physics-Condensed Matter* 20 (2008).
[6] Scott J. F., *Nature Materials* 6, 256(2007).
[7] Ju S., Cai T.-Y., Guo G.-Y. and Li Z.-Y., *Physical Review* B 75, 064419 (2007).
[8] Yang F., Tang M. H., Ye Z., Zhou Y. C., Zheng X. J., Tang J. X., Zhang J. J. and He J., *Journal of Applied Physics* 102, 044504 (2007).
[9] Hill N. A., *J. Phys. Chem.* B 104, 6694 (2000).
[10] Martin L. W., Chu Y. H. and Ramesh R., *Materials Science & Engineering R-Reports* 68, 89 (2010).

[11] Wang K. F., Liu J. M. and Ren Z. F., *Advances in Physics* 58, 321 (2009).

[12] Goodenough J. B., *Magnetism and the Chemical Bond.* (R. E. Krieger Pub. Co., Huntington, NY, 1976).

[13] Khomskii D., *Physics* 2, 20 (2009).

[14] Kimura T., Goto T., Shintani H., Ishizaka K., Arima T. and Tokura Y., *Nature* 426, 55 (2003).

[15] Hur N., Park S., Sharma P. A., Ahn J. S., Guha S. and Cheong S. W., *Nature* 429, 392 (2004).

[16] Van Aken B. B., Palstra T. T. M., Filippetti A. and Spaldin N. A., *Nature Materials* 3, 164-170 (2004).

[17] Lee J. H., Fang L., Vlahos E., Ke X., Jung Y. W., Kourkoutis L. F., Kim J.-W., Ryan P. J., Heeg T., Roeckerath M., Goian V., Bernhagen M., Uecker R., Hammel P. C., Rabe K. M., Kamba S., Schubert J., Freeland J. W., Muller D. A., Fennie C. J., Schiffer P., Gopalan V., Johnston-Halperin E. and Schlom D. G., *Nature* 466, 954 (2010).

[18] Seshadri R. and Hill N. A., *Chemistry of Materials* 13, 2892 (2001).

[19] Tokura Y. and Nagaosa N., *Science* 288, 462 (2000).

[20] Atou T., Chiba H., Ohoyama K., Yamaguchi Y. and Syono Y., *Journal of Solid State Chemistry* 145, 639 (1999).

[21] Moreira dos Santos A., Parashar S., Raju A. R., Zhao Y. S., Cheetham A. K. and Rao C. N. R., *Solid State Communications* 122, 49 (2002).

[22] Kimura T., Kawamoto S., Yamada I., Azuma M., Takano M. and Tokura Y., *Physical Review* B 67, 180401(R) (2003).

[23] Chi Z. H., Xiao C. J., Feng S. M., Li F. Y., Jin C. Q., Wang X. H., Chen R. Z. and Li L. T., *Journal of Applied Physics* 98, 103519 (2005).

[24] Montanari E., Righi L., Calestani G., Migliori A., Gilioli E. and Bolzoni F., *Chemistry of Materials* 17, 1765 (2005).

[25] Moreira dos Santos A. F., Cheetham A. K., Tian W., Pan X. Q., Jia Y. F., Murphy N. J., Lettieri J. and Schlom D. G., *Applied Physics Letters* 84, 91 (2004).

[26] Gajek M., Bibes M., Barthelemy A., Bouzehouane K., Fusil S., Varela M., Fontcuberta J. and Fert A., *Physical Review* B 72, 020406 (2005).

[27] Eerenstein W., Morrison F. D., Scott J. F. and Mathur N. D., *Applied Physics Letters* 87, 101906 (2005).

[28] Yang C.-H., Koo T. Y., Lee S.-H., Song C., Lee K.-B. and Jeong Y. H., *Europhysics Letters* 74, 348 (2006).

[29] Fujino S., Murakami M., Lim S. H., Salamanca-Riba L. G., Wuttig M. and Takeuchi I., *Journal of Applied Physics* 101, 013903 (2007).

[30] Gajek M., Bibes M., Wyczisk F., Varela M., Fontcuberta J. and Barthelemy A., *Physical Review* B75, 174417 (2007).

[31] Langenberg E., Varela M., Garcia-Cuenca M. V., Ferrater C., Sanchez F. and Fontcuberta J., Materials Science and Engineering B-Solid State *Materials for Advanced Technology* 144, 138 (2007).

[32] Havelia S., Wang S., Skowronski M. and Salvador P. A., *Journal of Applied Physics* 106, 123509 (2009).

[33] Shannon R. D., *Acta Crystallographica* A 32, 751 (1976).

[34] Belik A. A., Iikubo S., Kodama K., Igawa N., Shamoto S., Maie M., Nagai T., Matsui Y., Stefanovich S. Y., Lazoryak B. I. and E. Takayama-Muromachi, *Journal of the American Chemical Society* 128, 706 (2006).

[35] Yokosawa T., Belik A. A., Asaka T., Kimoto K., Takayama-Muromachi E. and Matsui Y., *Physical Review* B77, 024111 (2008).

[36] Baettig P., Seshadri R. and Spaldin N. A., *Journal of the American Chemical Society* 129, 9854 (2007).

[37] Sharan A., Lettieri J., Jia Y. F., Tian W., Pan X. Q., Schlom D. G. and Gopalan V., *Physical Review* B 69, 214109 (2004).

[38] Kubel F. and Schmid H., *Acta Crystallographica* B 46, 698 (1990).

[39] Prado-Gonjal J., Villafuerte-Castrejón M. E., Fuentes L. and Morán E., *Materials Research Bulletin* 44, 1734 (2009).

[40] Selbach S. M., Tybell T., Einarsrud M.-A. and Grande T., *Advanced Materials* 20, 3692 (2008).

[41] Wang J., Neaton J., Zheng H., Nagarajan V., Ogale S. B., Liu B., Viehland D., Vaithyanathan V., Schlom D. G., Waghmare U., Spaldin N. A., Rabe K. M., Wuttig M. and Ramesh R., *Science* 299, 1719 (2003).

[42] P. Fischer, M. Polomska, I. Sosnowska and M. Szymanski, *Journal of Physics C-Solid State Physics* 13, 1931 (1980).

[43] Sosnovska I., Peterlin-Neumaier T. and Steichele E., *Journal of Physics C-Solid State Physics* 15, 4835 (1982).

[44] Ederer C. and Spaldin N. A., *Physical Review* B71, 060401(R) (2005).

[45] Eerenstein W., Morrison F. D., Dho J., Blamire M. G., Scott J. F. and N. D. Mathur, *Science* 307, 1203a (2005).

[46] Zhao T., Scholl A., Zavaliche F., Lee K., Barry M., Doran A., Cruz M. P., Chu Y. H., Ederer C., Spaldin N. A., Das R. R., Kim D. M., Baek S. H., Eom C. B. and Ramesh R., *Nature Materials* 5, 823 (2006).

[47] Lebeugle D., Colson D., Forget A., Viret M., Bataille A. M. and Gukasov A., *Physical Review Letters* 100, 227602 (2008).

[48] Hill N. A., Battig P. and Daul C., *Journal of Physical Chemistry* B 106, 3383 (2002).

[49] Niitaka S., Azuma M., Takano M., Nishibori E., Takata M. and Sakata M., *Solid State Ionics* 172, 557 (2004).

[50] Murakami M., Fujino S., Lim S. H., Long C. J., Salamanca-Riba L. G., Wuttig M., Takeuchi I., Nagarajan V. and Varatharajan A., *Applied Physics Letters* 88, 152902 (2006).

[51] Kim D. H., Lee H. N., Varela M. and Christen H. M., *Applied Physics Letters* 89, 162904 (2006).

[52] Uratani Y., Shishidou T., Ishii F. and Oguchi T., *Japanese Journal of Applied Physics* 44, 7130 (2005).

[53] Belik A. A., Iikubo S., Kodama K., Igawa N., Shamoto S., Niitaka S., Azuma M., Shimakawa Y., Takano M., Izumi F. and Takayama-Muromachi E., *Chemistry of Materials* 18, 798 (2006).

[54] Ishiwata S., Azuma M., Takano M., Nishibori E., Takata M., Sakata and M. Kato K., *Journal of Materials Chemistry* 12, 3733 (2002).

[55] Carlsson S. J. E., Azuma M., Shimakawa Y., Takano M., Hewat A. and Attfield J. P., *Journal of Solid State Chemistry* 181, 611 (2008).

[56] Choi E. M., Patnaik S., Weal E., Sahonta S. L., Wang H., Bi Z., Xiong J., Blamire M. G., Jia Q. X. and MacManus-Driscoll J. L., *Applied Physics Letters* 98, 139903 (2011).

[57] Shimakawa Y., Kan D., Kawai M., Sakai M., Inoue S., Azuma M., Kimura S. and Sakata O., *Japanese Journal of Applied Physics* 46, L845 (2007).

[58] Langenberg E., Rebled J., Estrade S., Daumont C. J. M., Ventura J., Coy L. E., Polo M. C., Garcia-Cuenca M. V., Ferrater C., Noheda B., Peiro F., Varela M. and Fontcuberta J., *Journal of Applied Physics* 108, 123907 (2010).

[59] Azuma M., Takata K., Saito T., Ishiwata S., Shimakawa Y. and Takano M., *Journal of the American Chemical Society* 127, 8889 (2005).

[60] Sakai M., Masuno A., Kan D., Hashisaka M., Takata K., Azuma M., Takano M. and Shimakawa Y., *Applied Physics Letters* 90, 072903 (2007).

[61] Padhan P., le Clair P., A. Gupta and G. Srinivasan, *Journal of Physics-Condensed Matter* 20, 355003 (2008).

[62] Langenberg E., Varela M., Garcia-Cuenca M. V., Ferrater C., Polo M. C., Fina I., Fabrega L., Sanchez F. and Fontcuberta J., *Journal of Magnetism and Magnetic Materials* 321, 1748 (2009).

[63] Karimi S., Reaney I., Han Y., Pokorny J. and Sterianou I., *Journal of Materials Science* 44, 5102 (2009).

[64] Pradhan D. K., Choudhary R. N. P., C. Rinaldi and R. S. Katiyar, *Journal of Applied Physics* 106, 024102 (2009).

[65] Kobayashi Y., Shiozawa M., Sato K., Abe K. and Asai K., *Journal of the Physical Societyof Japan* 77, 084701 (2008).

[66] Langenberg E., Fina I., Gemeiner P., Dkhil B., Fabrega L., Varela M. and Fontcuberta J., *Applied Physics Letters* 100, 022902 (2012).

[67] Bi L., Taussig A. R., Kim H.-S., Wang L., Dionne G. F., Bono D., Persson K., Ceder G. and Ross C. A., *Physical Review* B78, 104106 (2008).

[68] Mandal P., Sundaresan A., Rao C. N. R., Iyo A., Shirage P. M., Tanaka Y., Simon C., Pralong V., Lebedev O. I., Caignaert V. and Raveau B., *Physical Review* B82, 100416(R) (2010).

[69] Miao J., Zhang X., Zhan Q., Jiang Y. and Chew K. H., *Applied Physics Letters* 99, 062905 (2011).

[70] Baettig P. and Spaldin N. A., *Applied Physics Letters* 86, 012505 (2005).

[71] Suchomel M. R., Thomas C. I., Allix M., Rosseinsky M. J., A. Fogg M. and Thomas M. F., *Applied Physics Letters* 90, 112909 (2007).

[72] Kim D. H., Lee H. N., M. D. Biegalski and H. M. Christen, *Applied Physics Letters* 91, 042906 (2007).

[73] Nechache R., Harnagea C. and Pignolet A., *Journal of Physics-Condensed Matter* 24, 096001 (2012).

[74] Lezaic M. and Spaldin N. A., *Physical Review* B 83, 024410 (2011).

[75] Christen H. M. and Eres G., *Journal of Physics-Condensed Matter* 20, 264005 (2008).

[76] Aziz M. J., *Applied Physics a-Materials Science & Processing* 93, 579 (2008).

[77] Trtik V., Perez A., Navarro J., Ferrater C., Sanchez F. and Varela M., *Applied Physics a-Materials Science & Processing* 69, S815 (1999).

[78] Hausser F. and Voigt A., *Journal of Crystal Growth* 303, 80 (2007).

[79] Chiba H., Atou T. and Syono Y., *Journal of Solid State Chemistry* 132, 139 (1997).

[80] Sugawara F., Iiida S., Syono Y. and Akimoto S.-I., *Journal of the Physical Societyof Japan* 25, 1553 (1968).

[81] Faqir H., Chiba H., Kikuchi M. and Syono Y., *Journal of Solid State Chemistry* 142, 113 (1999).

[82] Navarro J., Balcells L., Sandiumenge F., Bibes M., Roig A., Martinez B. and Fontcuberta J., *Journal of Physics-Condensed Matter* 13, 8481 (2001).

[83] Béa H., Bibes M., Barthélémy A., Bouzehouane K., Jacquet E., Khodan A., Contour J.-P., Fusil S., Wyczisk F., Forget A., Lebeugle D., Colson D. and Viret M., *Applied Physics Letters* 87, 072508 (2005).

[84] Bai M., Wang J. L., Wuttig M., Li J. F., Wang N. G., Pyatakov A. P., A. K. Zvezdin, L. E. Cross and D. Viehland, *Applied Physics Letters* 86, 032511 (2005).

[85] Prokhorov V. G., Kaminsky G. G., Kim J. M., Eom T. W., Park J. S., Lee Y. P., Svetchnikov V. L., Levtchenko G. G., Nikolaenko Y. M. and Khokhlov V. A., *Low Temperature Physics* 37, 129 (2011).

[86] Scott J. F., *Journal of Physics: Condensed Matter* 20, 021001 (2008).

[87] Choi W. S., Moon S. J., Seo S. S. A., Lee D., Lee J. H., Murugavel and P. Noh T. W., *Physical Review* B 78, 054440 (2008).

[88] Cho D.-Y., Kim J.-Y., Park B.-G., Rho K.-J., Park J.-H., Noh H.-J., Kim B.-J., Oh S.-J., Park H.-M., Ahn J.-S., Ishibashi H., Cheong S.-W., Lee J. H., Murugavel P., Noh T. W., Tanaka A. and Jo T., *Physical Review Letters* 98, 217601 (2007).

[89] Kajimoto R., Sagayama H., Sasai K., Fukuda T., Tsutsui S., T. Arima, K. Hirota, Y. Mitsui, H. Yoshizawa, A. Q. R. Baron, Y. Yamasaki and Y. Tokura, *Physical Review Letters* 102, 247602 (2009).

[90] Zavaliche F., Zheng H., Mohaddes-Ardabili L., Yang S. Y., Zhan Q., P. Shafer, E. Reilly, R. Chopdekar, Y. Jia, P. Wright, D. G. Schlom, Y. Suzuki and R. Ramesh, *Nano Letters* 5, 1793 (2005).

[91] G. Catalan, *Applied Physics Letters* 88, 102902 (2006).

[92] Schmidt R., Ventura J., Langenberg E., Nemes N. M., Munuera C., M. Varela, M. Garcia-Hernandez, C. Leon and J. Santamaria, *Physical Review* B86, 035113 (2012).

[93] Irvine J. T. S., Sinclair D. C. and West A. R., *Advanced Materials* 2, 132 (1990).

[94] Schmidt R., Eerenstein W., Winiecki T., Morrison F. D. and Midgley P. A., *Physical Review* B75, 245111 (2007).

[95] Barsukov E. and Macdonald J., *Impedance Spectroscopy: Theory, Experiment and Applications*. (John Wiley & Sons Inc., Hoboken, 2005).

[96] Schmidt R., in *Ceramic Materials Research Trends*, edited by P. B. Lin (Hauppauge, 2007), pp. 321.

[97] Schmidt R., Eerenstein W. and Midgley P. A., *Physical Review* B79, 214107 (2009).

[98] Liu G.-Z., Wang C., Wang C.-C., Qiu J., He M., Xing J., Jin K.-J., Lu H.-B.and Yang G.-Z., *Applied Physics Letters* 92, 122903 (2008).

[99] Srivastava A., Garg A. and Morrison F. D., *Journal of Applied Physics* 105, 054103 (2009).

[100] Ren P., Liu P., Xia B., Zou X., You L., Wang J. and Wang L., AIP *Advances* 2, 022133 (2012).

[101] Langenberg E., Fina I., Ventura J., Noheda B., Varela M. and Fontcuberta J., *Physical Review* B 86, 085108 (2012).

In: Perovskite
Editors: Jinghua Zhang and Huan Li
ISBN: 978-1-62417-800-9

Chapter 3

PEROVSKITES AS CATALYSTS FOR ENVIRONMENTAL REMEDIATION

Gina Pecchi* and Eduardo J. Delgado
University of Concepción, Concepción, Chile

ABSTRACT

The current awareness about environmental and health protection has increased the regulations of the chemical industrial processes. Consequently, the use of heterogeneous catalysts for pollutant abatement from vehicle emissions and industrial effluent is indispensable in order to attenuate the global warming and other environmental effects.

Currently, most of the heterogeneous catalysts are based on noble metals, due to the high activity; however, their main disadvantages are related to volatility, sulfur poisoning and cost. Therefore, much effort has been devoted to the substitution of noble metal catalysts with mixtures of selected metal oxides.

Transition metal perovskites, with the general formula ABO_3, are well known as very good oxidation catalysts due to thermal stability, poisoning resistance and low cost. Moreover, the oxidation on perovskite-type catalysts, the combustion is carried out much below the flame temperature, substantially reducing the formation of harmful species such as CO and NO_x. In addition, perovskite-like oxides can be tailored to create a wide family of catalysts by varying either the A-site or the B-site metal ion, or both. Indeed, the catalytic activity of a perovskite can be modified by inserting proper transition metal ions. Partial substitution at the A-site with another metal cation, doping, can strongly affect the catalytic activity, due to stabilization of unusual oxidation states of the B cation and simultaneous formation of structural defects created by such a substitution. Structural defects are responsible not only for part of the catalytic activity, but also for oxygen mobility within the crystal lattice of the solid due to non-stoichiometry created by the substitution. The presence of ionic vacancies affects the catalytic activity by favouring (or not) reactant adsorption from the gas phase. Thus, an oxygen vacancy in the surface layer is not only the place where oxygen can enter into the solid state structure, but also

* Department of Physical Chemistry, Chemical Science Faculty, University of Concepción, Concepción, Chile. E-mail: gpecchi@udec.cl.

the site with a missing oxygen ion surrounded by a maximum number of cations represents the most acidic site, much more acidic than the regular surface cations.

In this chapter, the synthesis, characterization and catalytic activity of pure and doped perovskites for the abatement of some of the most hazardous pollutants delivered to the atmosphere as volatile organic compounds (VOCs) and particulate soot, is addressed. In particular, the performance of pure and doped lanthanum ferrites and cobaltites in the catalytic oxidation of some VOCs as toluene, ethanol, acetyl acetate and n-hexane is discussed. Finally, the catalytic activity of alkali metal niobates in the soot combustion and the relation between defects and catalytic activity of calcium doped lanthanum ferrites in the combustion of methane is also discussed.

INTRODUCTION

The current awareness about environmental and health protection has increased the regulations of the chemical industrial processes. Important air pollutants emitted by vehicles, in industrial processes, and when employing solvents are grouped under the acronym of VOCs (volatile organic compounds). Since these compounds can be easily oxidized, their catalytic combustion is an effective way to control air pollution. The use of heterogeneous catalysts for pollutant abatement effluent is indispensable in order to attenuate the global warming and other environmental effects. Currently, most of the heterogeneous catalysts are based on noble metals, due to the high activity; however, their main disadvantages are related to volatility, sulfur poisoning and cost. Supported precious metals, such as Pt and Pd, are well established as efficient catalysts for VOC combustion. However, due to cost and environmental protection reasons, alternative materials have been considered. Much effort has been devoted to the substitution of noble metal catalysts with mixtures of selected metal oxides [1]. Transition metal perovskites-type oxides, general formula ABO_3, offer good catalyst activity, high thermal stability and low cost and can be considered to be use as catalysts in exothermic reactions such as the total combustion of VOCs [2]. They have been investigated extensively as catalysts in terms of [3]:

(i) Transition metal ion contribution in unusual oxidation state;
(ii) Amount of non-stoichiometric oxygen;
(iii) Effect of lattice defect structures [4].

In the ABO_3 structures, A is a large cation and B a small cation of the d-transition series. These oxides consist of close-packed, ordered AO_3 layers which are stacked with the B cations occupy all the interlayer oxygen octahedra. The ideal structure of the perovskite-type oxide is a cubic type, and when some distortions of the cubic phase are present, the cation size mismatch and cation or anion non-stoichiometry is induced. For these highly oxidative systems it is of interest to relate bulk and surface defect chemistry to catalytic activity. Perovskite-type oxides can be tailored to create a wide family of catalysts by varying either the A-site or the B-site metal ion, or both forming solid solutions that favour the alteration of the physical-chemical properties. The catalytic activity of a perovskite can be modified by partial substitution at the A-site with another metal cation, doping, due to stabilization of unusual oxidation states of the B cation and simultaneous formation of structural defects. Structural defects are responsible not only of part of the catalytic activity, but also for oxygen mobility within the crystal lattice of the solid due to the non-stoichiometry created by the substitution. The presence of ionic vacancies affects the catalytic activity by favouring reactant adsorption from the gas phase. An oxygen vacancy in the surface layer is not only the place where oxygen can enter into the solid state structure, but also the site with a missing oxygen ion surrounded by a maximum number of cations represents the most acidic site, much more acidic than the regular surface cations [5].

In this chapter, the synthesis, characterization and catalytic activity of pure and doped perovskites for the abatement of some of the most hazardous pollutants delivered to the atmosphere as volatile organic compounds (VOCs), methane and particulate soot are addressed. In particular, the performance of pure and doped lanthanum ferrites and cobaltites in the catalytic oxidation of some VOCs is discussed. Finally, the catalytic activity of alkali metal niobates in the soot combustion and the relation between defects and catalytic activity of calcium doped lanthanum ferrites in the combustion of methane is also discussed.

CATALYTIC ACTIVITY IN TOTAL COMBUSTION OF VOCS

Total Combustion of Acetyl Acetate

The catalytic activity evaluation was performed in a flow reactor at atmospheric pressure with the catalysts diluted with silica as an inert. Before the activity test, the catalyst was activated with dry air at 500°C. The temperature was linearly increased up to the required temperature and maintained constant for 30 min. Several isothermal steps were performed until complete conversion. The reactant gaseous mixture was $C_4H_8O_2$:O_2:He=1:10:89 (molar). Reactor effluents were analyzed using an on-line gas chromatograph Hewlett Packard model HP 4890D with thermal conductivity detector. The column was 30-m capillary Supelco 25462.

A-Site Substituted $La_{1-x}Ca_xFeO_3$ (x= 0.0, 0.1, 0.2, 0.3, 0.4)

The $La_{1-x}Ca_xFeO_3$ perovskites [6] prepared by the citrate method, show S_{BET} values with no dependence on Ca substitution with surface areas between 20–38 m^2g^{-1} [7]. The XRD patterns of the solids calcined at 700°C indicates that unsubstituted $LaFeO_3$ corresponds to an orthorhombic structure with absence of extra phases. When La is partially substituted for Ca the observed lattice distortions towards lower 2θ angles are explained considering that the substitution of La^{3+} with a lower valence action like Ca^{2+} entails either the oxidation of the cation at the B-site ($Fe^{3+}\rightarrow Fe^{4+}$) or the formation of oxygen vacancies. The presence of Fe^{4+} in the perovskite structure, is in agreement with previous reported for other iron-based perovskites [8].

Only CaO is detected as a segregated phase for $x\geq0.2$. Pure $LaFeO_3$ perovskite is a non reducible solid, and upon Ca substitution two reduction peaks appears as consequence of the cation exchange of Ca^{2+} by La^{3+} in the lattice, with the necessary oxidation of Fe^{3+} to Fe^{4+} to compensate the lower charge of Ca^{2+} [9, 10]. For pure $LaFeO_3$, he evolution of oxygen desorption experiments show the expected physisorbed species at low temperatures and presence of the so-called α-oxygen as a shoulder. The desorption of α- and β-oxygen show a noticeable increase in intensity upon Ca substitution. As this peak is associated with the lattice oxygen or with oxygen species occupying inner vacancies, the evolution of these oxygen species is a measurement of the replacement of La^{3+} for Ca^{2+} in the perovskite structure. The binding XPS energies of the Ca $2p_{3/2}$ component has two peaks at binding energy of 346.2 and 346.8 eV corresponding to surface Ca-O and carbonated surface Ca^{2+} ions, respectively. The Ca substitution x=0.2 shows the largest extent of Ca incorporated into the perovskite structure. The percentage of O1s components show that the intensity of the peaks at 529.2 and 531.0 eV change almost linearly in an opposite direction with Ca substitution, indicative of the incorporation of Ca in the structure. Since carbonation degree is related to the extent of segregated phases, the low (CO_3^{2-}/La+Ca) surface ratio for x=0.2 indicates the one with the lowest extent of segregated phases. For $La_{1-x}Ca_xFeO_3$ perovskite-type oxide the substitution of calcium ions into the perovskite lattice leads to a solid with structural and electronic defects [11, 12].

The catalytic results show the typical S-shaped curves, in which the reaction starts at about 180°C, then the conversion increases rapidly upon increasing temperature and complete conversion is reached at 370°C. The catalytic activity results in terms of the ignition temperature (T_{50}), reaction rate and intrinsic activity at 250°C are shown in Table 1. It can be observed an enhancement of the catalytic activity, a decrease in the ignition temperature and an increases in the reaction rate and intrinsic activity with Ca substitution. Considering that characterization results indicates that the extent of Fe^{4+} increases with Ca substitution, it is proposed that the higher activity and stability are attributed to a synergy between the pair Fe^{4+}/Fe^{3+} with the oxygen vacancies generated by the calcium substitution. The large extent of α-oxygen for the higher x values is in agreement with the proposed Mars-van Krevelen type oxidation-reduction cycle for these type of solids [13].

Table 1. Ignition temperature and reaction rate at conversion level below 10% of acetylcetate at 250°C for $La_{1-x}Ca_xFeO_3$ perovskites

	T_{50}, °C	Reaction rate μmolg^{-1}h^{-1}	Intrinsic activity μmolm^{-2} h^{-1}
$LaFeO_3$	340	0.4	0.02
$La_{0.9}Ca_{0.1}FeO_3$	325	1.3	0.07
$La_{0.8}Ca_{0.2}FeO_3$	320	2.6	0.06
$La_{0.7}Ca_{0.3}FeO_3$	316	3.0	0.08
$La_{0.6}Ca_{0.4}FeO_3$	314	3.1	0.10

Table 2. Ignition temperature and reaction rate at conversion level below 10% of acetylcetate at 250°C for $La_{1-x}Ca_xNiO_3$ perovskites

	T_{50}, °C	Reaction rate μmolg^{-1}h^{-1}	Intrinsic activity μmolm^{-2} h^{-1}
$LaNiO_3$	300	1.8	0.13
$La_{0.9}Ca_{0.1}NiO_3$	305	1.7	0.13
$La_{0.8}Ca_{0.2}NiO_3$	307	1.7	0.11
$La_{0.7}Ca_{0.3}NiO_3$	305	1.6	0.10
$La_{0.6}Ca_{0.4}NiO_3$	306	1.6	0.10

A-Site Substituted $La_{1-x}Ca_xNiO_3$ (x= 0.0, 0.1, 0.2, 0.3, 0.4)

The Ca substituted lanthanum $La_{1-x}Ca_xNiO_3$ perovskites [6] prepared by the citrate method show specific surface areas between 15 – 20 m^2g^{-1}. The XRD patterns indicates that unsubstituted $LaNiO_3$ corresponds to an rhombohedral structure with NiO present as segregated phase. The XRD of the Ca substituted solids indicates that as Ca increases the diffraction line at 2θ=32.752°, corresponding to $LaNiO_3$ structure, decreases in intensity and new diffractions lines corresponding to a brownmillerite-type structure, $La_2Ni_2O_5$, and NiO and CaO as segregated phases appears. This result indicates the difficulty to substitute La by Ca in $LaNiO_3$ even to a limited extent. TPR experiments of pure $LaNiO_3$ and the substituted $La_{1-x}Ca_xNiO_3$ show two well defined reduction peaks, with different intensities. Hydrogen consumption for the first peak is consistent with the reduction of Ni^{3+} to Ni^{2+}, yielding a $La_2Ni_2O_5$ phase [14] and the second peak is an evidence for the formation of metallic nickel deposited on lanthanum oxide [15]. The O_2-TPD profiles for the pure $LaNiO_3$ and the Ca substituted solids show a high number of desorption peaks, indicative of the presence of segregated metal oxide phases in a high dispersion degree. XPS results indicates presence of Ni^{2+} ions on the surface and no clear tendency of the Ca 2p3/2 and O 1s components is observed. The (Ni/La+Ca) surface ratio greater than the nominal indicates that the surface is Ni enriched and the almost constant (CO_3^{2-}/La+Ca) surface ratio the same carbonation degree. From this finding is conclusive that upon Ca substitution, the Ni^{3+} ions are not stabilized, therefore it is not possible to obtain the perovskite structure and a new $La_2Ni_2O_5$ structure with a large extent of segregated phase as mixed oxides as NiO and CaO in a high dispersion degree appears.

The catalytic results also show the typical S-shaped curves, the reaction starts at about 180°C, then the conversion increases rapidly upon increasing temperature up to 340°C. The catalytic activity results in terms of the ignition temperature (T_{50}), reaction rate and intrinsic activity at 250°C are shown in Table 2.

It can be seen that Ca substitution have not a beneficial influence in the catalytic activity of $LaNiO_3$, being the highest catalytic activity for pure $LaNiO_3$ perovskite and the lower for the Ca substituted perovskites. This result may be explained in terms of the defects formed in the solid upon the substitution of La^{3+} by Ca^{2+} in $LaNiO_3$. The ionic radious of Ni^{4+} makes not possible to substitute La^{3+} by a lower oxidation state cation as Ca^{2+} and a new compounds with definite composition and structure result from network rearrangement. The Brownmillerite-type structure $La_2Ni_2O_5$, considered as an oxygen-deficient perovskite [16], entails the change of the oxidation state of nickel from 3+ to 2+, suggesting the hypothesis that Ni^{3+} ions are the active sites for acetylacetate oxidation.

B-Site Substituted $LaFe_{1-x}Ni_xO_3$ (x= 0.0, 0.1, 0.2, 0.3, 1.0)

Pure $LaFeO_3$ perovskite prepared by the citrate method shows the largest surface area (21 m^2g^{-1}) while $LaNiO_3$ shows the lowest (8 m^2g^{-1}) BET area. Partial substitution of iron by nickel results in a progressive decrease in the value of BET area [17]. The XRD patterns of $LaFe_{1-x}Ni_xO_3$ oxides calcined at 700°C show that $LaFeO_3$ and $LaNiO_3$ samples belong to orthorhombic and rhombohedral phases and a shift towards larger 2θ angles upon substitution of Fe^{3+} by Ni^{3+} is detected. As Ni substitution increases, the unit cell volume decreases due to the smaller size of Ni^{3+} with the shift in the most important peaks of the perovskite structure. This finding suggests the complete insertion of nickel for substitution x=0.1, and segregated simple oxides as NiO_x for substitutions x≥0.2. The infrared spectra of pure $LaFeO_3$ and substituted $LaFe_{1-x}Ni_xO_3$ perovskites are consistent with orthorhombic structures, unlike from the rhombohedral $LaNiO_3$ structure, in which a large band attributed to bulk carbonates is detected. This result is conclusive that Ni^{3+} ions have been incorporated into the perovskite structure only in a lower degree. The TPR profiles of $LaFeO_3$ show no reduction peaks, meanwhile two large reduction peaks for $LaNiO_3$ are detected. The substituted perovskites (x≥0.2) exhibit a similar behavior of $LaNiO_3$ and the lowest substituted perovskite (x=0.1) only one small and broad peak in the TPR profiles. This is indicative of the higher stability of the $LaFeO_3$ and substituted samples in reductor atmosphere. The O_2-TPD profiles show a significant enhancement in the intensity of the denoted α-oxygen with Ni substitution, in close relation with the catalytic activity in an oxidation processes.

XPS study of C 1s shows high binding energy component due to surface carbonates. The percentage of the peaks at 529.2 and 531.0 eV of O1s components change almost linearly in a perfect opposite way, respect to the Ni substitution. The increase in the lattice oxygen component in the sample with larger Ni substitution can be related to higher catalytic activity in combustion reactions. The Fe/La and Ni/La surface ratios are higher than the nominal compositions, indicative of a Ni- and Fe- enriched surface of the substituted perovskites. The Fe/La surface ratio for $LaFeO_3$ is 0.96 and for Ni/La in $LaNiO_3$ is 1.13, the same of the greater substituted $LaFe_{1-x}Ni_xO_3$ samples. This finding is also conclusive that Ni^{3+} ions are more difficult to be stabilized than Fe^{3+} ions in the perovskite lattice, which results in surface segregation of NiO. The x=0.1 substituted sample shows only smaller differences between the bulk and surface ratios.

The characterization results of $LaFe_{1-x}Ni_xO_3$ perovskite-type oxides reveal that for substitution x=0.1, almost a total insertion of nickel into the perovskite structure with no segregation of single oxides is observed. At higher substitution degree (x≥0.2) segregated phases of NiO are detected [18, 17].

The catalytic results show the typical S-shaped curves, indicates that the reaction starts at 160°C and complete conversion is reached at 360°C. The catalytic activity results in terms of the ignition temperature (T_{50}), reaction rate and intrinsic activity at 200°C are shown in Table 3. Carbon dioxide and water were the only observed products at complete conversion, while trace amounts of acetaldehyde for lower conversion level were detected. An increase in the reaction rate and intrinsic activity, in parallel with a decrease in the ignition temperature of acetyl acetate combustion were observed upon increasing the nickel substitution from 0 to 0.3 and then it drops for the $LaNiO_3$. These results can be explained in terms of the cooperative effect of a $LaFe_{1-x}Ni_xO_3$ and NiO phases, whose relative concentration determines the oxygen activation capability and hence their reactivity for the oxidation of this organic compound. The active sites are related to the generation of sites associated to the presence of α–oxygen, in agreement with previous work on lanthanum substituted manganites [9] and with segregated NiO phase.

B-Site Substituted $LaMn_{1-x}Co_xO_3$ (x= 0.0, 0.1, 0.3, 0.5, 0.7, 0.9, 1.0)

The BET surface areas of the $LaMn_{1-x}Co_xO_3$ perovskite indicates that manganite exhibits the largest BET value of 40 m^2g^{-1} and upon substitution, remains almost constant till x<0.5. For higher x values, a progressive decrease occurs until reaching the cobaltite surface value of 16 m^2g^{-1} [19, 20].

Table 3. Ignition temperature and reaction rate at conversion level below 10% of acetylcetate at 250°C for $LaFe_{1-x}Ni_xO_3$ perovskites

	T_{50}, °C	Reaction rate $\mu molg^{-1}h^{-1}$	Intrinsic activity $\mu molm^{-2}\,h^{-1}$
$LaFeO_3$	340	0.2	0.01
La $Fe_{0.9}Ni_{0.1}O_3$	325	1.1	0.08
La $Fe_{0.8}Ni_{0.2}O_3$	293	1.0	0.12
La $Fe_{0.7}Ni_{0.3}O_3$	282	1.2	0.23
$LaNiO_3$	300	1.6	0.19

Table 4. Ignition temperature and reaction rate at conversion level below 10% of acetylcetate at 230°C for $LaMn_{1-x}Co_xO_3$ perovskites

	T_{50}, °C	Reaction rate $\mu molg^{-1}h^{-1}$	Intrinsic activity $\mu molm^{-2}\,h^{-1}$
$LaMnO_3$	243	2.3	0.06
$LaMn_{0.9}Co_{0.1}O_3$	246	2.5	0.06
$LaMn_{0.7}Co_{0.3}O_3$	252	1.6	0.05
$LaMn_{0.5}Co_{0.5}O_3$	252	2.5	0.08
$LaMn_{0.3}Co_{0.7}O_3$	258	1.1	0.05
$LaMn_{0.1}Co_{0.9}O_3$	252	1.2	0.15
$LaCoO_3$	260	1.1	0.07

The XRD shows that $LaMnO_3$ and $LaCoO_3$ perovskites belong to cubic and rhombohedral phases, respectively. For the substituted perovskites different tendencies are clearly observed for values of $x<0.5$, $x>0.5$ and $x=0.5$. For x=0.1 and 0.3 an orthorhombic structure is detected, for x=0.5, the diffractogram is similar to the one for x=0.0, and for x= 0.7, 0.9 and 1.0, consistent with a rombohedral structure.

The absence of diffraction lines belonging to single manganese or cobalt oxides is indicative that these oxide phases are highly dispersed or in an amorphous form. Presence of several ionic states (Mn^{3+},Mn^{4+},Co^{2+},Co^{3+}) makes difficult to decide which is the combination of oxidation states to comply with the electroneutrality requirement. Upon Co substitution in manganite, this ion enters the solid solution as Co^{2+}, thus the formation of an equivalent amount of Mn^{4+} is proposed. During the transformation of Mn^{3+} to Mn^{4+}, there is a progressive decrease of the amount of the Jahn-Teller cations, responsible of the lattice deformation. The IR spectra indicates that the most important band of the perovskites structure at 604 cm^{-1}, are more intense for the perovskites x=0.0 and 0.5, which correspond to more symmetrical structures. The x=0.3 and 0.5 present vibration modes associated to a low symmetry in the network and x=0.7 and 0.9 to a rhombohedral structure of $LaCoO_3$. In the perovskites with xCo=0.7 and 0.9 also appears bands associated to CO_3^{2-} indicative of segregated phases. TPR profiles indicates the cobaltite reduction associated only to surface oxygen species, and pure manganite higher stability in reductor atmosphere. The paramagnetic and the magnetically ordered regimes properties show two well identified regions separated by the equimolar composition x=0.5. For cobalt contents between x=0.0 and $x\leq0.3$, the Curie-Weiss temperature increases and the magnetic moment of the paramagnetic regime decrease, indicative of a progressive transformation of Mn^{3+} into Mn^{4+} ions, for the x>0.5 values the Curie-Weiss temperature sharply decreases, reaching negative values at x=1.0.

The magnetic results indicates that during the first half of the series (x≤0.5) strong Co^{2+}-Mn^{4+} ferromagnetic interactions are triggered when Co^{2+} is inserted into the lattice, while antiferromagnetism is favored when cobalt is introduced as Co^{3+} (x>0.5). These results allow to conclude that manganite tolerates Co insertion without large structural changes up to values of x≤0.3. The insertion of Mn into the cobaltite structure produces notable distortions in the cobaltite highly symmetric rhombohedral structure. In the substituted perovskite with x=0.5, an highly symmetric orthorhombic structure appears due to the presence of the same quantity of Mn^{4+} and Co^{2+} ions.

The catalytic results show that the reaction starts at 190°C and complete conversion is reached at 270°C. The catalytic activity results in terms of the ignition temperature (T_{50}), reaction rate and intrinsic activity at 230°C are shown in Table 4. The reaction products were carbon dioxide, water and traces of acetaldehyde [20] with no formation of CO. The substituted perovskites display ignition temperatures between these two values, indicating that changes in xCo values produce modification in the active sites required for organic compound oxidation. The reaction rate shows the same trend that the intrinsic activity. Manganite and cobaltite show almost the same values without significant differences with x values of 0.1, 0.3 and 0.7. The changes in the structural properties of perovskite-type oxides by isomorphic substitution of cation B with B' are related to four parameters, the first two geometric and the last two electronic:

i) Mass of cation B';
ii) Ionic radius of cation B', and consequently the length of the B'-O bond;
iii) Electronic structure of cation B';
iv) Cation B'-O bond strength.

The geometric factors are considered in the calculation of the tolerance factor t, which present very similar values for the total series. Consequently, the catalytic characterization and activity results should be related with electronic factors. The first stage of every heterogeneous catalytic reaction is an adsorption step, consequently, the higher reaction rate values for x=0.1 and x=0.5 must be attributed to the adequate crystalline structure of these two perovskites. The x=0.5 perovskite, the one that presents the two previously mentioned effects, adequate crystalline structure and electronic effect due to the optimized Mn^{4+}-Co^{2+} interactions show the large intrinsic activity than the rest of the series.

Total Combustion of Toluene

The catalytic tests were carried out in a fixed flow reactor system under atmospheric pressure with the catalysts (20-40 mesh) diluted with SiC, of the same particle size in a quartz reactor. Gas flow rates were adjusted by mass flow controllers (Brooks 5850E). The helium stream, bubbled through two thermostated and pressurized saturators containing toluene, was mixed with a mixture of oxygen/He to obtain a total composition of 4000 ppm of toluene and 20% of oxygen reheated before entry into the reactor. Prior to the activity test, the catalyst was activated with 100 $mLmin^{-1}$ of dry air for 1 h at 500°C. The temperature was linearly increased up to the required temperature, and maintained constant for 30 min. The activity tests were performed at different temperatures, ranging from 50°C until the complete conversion.

The exit gas mixture from the reactor was analyzed using an online gas chromatograph system equipped with TCD and FID detector using He as a carrier gas.

A-Site Substituted $La_{1-x}Ca_xFeO_3$ (x= 0.0, 0.1, 0.2, 0.3, 0.4)

The catalytic results show that the reaction starts at 220°C and complete conversion is reached at 360°C. The catalytic activity results in terms of the ignition temperature (T_{50}), reaction rate and intrinsic activity at 260°C are shown in Table 5. It can be observed a large decreases in T_{50} from pure $LaFeO_3$ till x=0.2.

Considering that characterization results indicates large extent of α-oxygens and segregated CaO phases for the x=0.3 and 0.4 perovskites, it is concluded that the decrease in the total oxidation of toluene for the two higher substituted perovskites is the presence of CaO.

A-Site Substituted $La_{1-x}Ca_xNiO_3$ (x= 0.0, 0.1, 0.2, 0.3, 0.4)

The catalytic results show that the reaction starts at 250°C and complete conversion is reached at 390°C. The catalytic activity results in terms of the ignition temperature (T_{50}), reaction rate and intrinsic activity at 200°C are shown in Table 6.

The substitution of calcium ions produces a structure modification leading to an increase in T_{50} and a decrease in the intrinsic activity. For these perovskites the electronic perturbation cannot be compensated by nickel ions and thus a change in the crystalline structure and segregation of crystallites occurs, resulting in a non-stoichiometric perovskite and NiO and CaO as segregated phases.

Total Combustion of Ethanol

The catalytic activity evaluation was performed in a conventional flow reactor at atmospheric pressure using the catalysts diluted with silica.

Table 5. Ignition temperature and reaction rate at conversion level below 10% of toluene at 260°C for $La_{1-x}Ca_xFeO_3$ perovskites

	T_{50}, °C	Reaction rate $\mu molg^{-1}h^{-1}$	Intrinsic activity $\mu molm^{-2}h^{-1}$
$LaFeO_3$	323	14	0.70
$La_{0.9}Ca_{0.1}FeO_3$	310	25	1.4
$La_{0.8}Ca_{0.2}FeO_3$	289	58	1.5
$La_{0.7}Ca_{0.3}FeO_3$	290	53	1.4
$La_{0.6}Ca_{0.4}FeO_3$	300	46	1.5

Table 6. Ignition temperature and reaction rate at conversion level below 10% of toluene at 260°C for $La_{1-x}Ca_xNiO_3$ perovskites

	T_{50}, °C	Reaction rate μmolg^{-1}h^{-1}	Intrinsic activity μmolm^{-2} h^{-1}
$LaNiO_3$	297	44	3.1
$La_{0.9}Ca_{0.1}NiO_3$	305	32	2.5
$La_{0.8}Ca_{0.2}NiO_3$	333	10	1.0
$La_{0.7}Ca_{0.3}NiO_3$	342	12	0.8
$La_{0.6}Ca_{0.4}NiO_3$	346	10	0.6

Table 7. Ignition temperature and reaction rate at conversion level below 10% of ethanol at 200°C for $La_{1-x}Ca_xFeO_3$ perovskites

	T_{50}, °C	Reaction rate μmolg^{-1}h^{-1}	Intrinsic activity μmolm^{-2} h^{-1}
$LaFeO_3$	247	0.4	0.02
$La_{0.9}Ca_{0.1}FeO_3$	243	0.9	0.05
$La_{0.8}Ca_{0.2}FeO_3$	239	1.4	0.04
$La_{0.7}Ca_{0.3}FeO_3$	241	1.3	0.03
$La_{0.6}Ca_{0.4}FeO_3$	242	1.2	0.04

Table 8. Ignition temperature and reaction rate at conversion level below 10% of ethanol at 200°C for $La_{1-x}Ca_xNiO_3$ perovskites

	T_{ign}^{50}, °C	Reaction rate μmolg^{-1}h^{-1}	Intrinsic activity μmolm^{-2} h^{-1}
$LaNiO_3$	250	2.7	0.19
$La_{0.9}Ca_{0.1}NiO_3$	249	2.7	0.21
$La_{0.8}Ca_{0.2}NiO_3$	247	2.7	0.18
$La_{0.7}Ca_{0.3}NiO_3$	248	2.8	0.18
$La_{0.6}Ca_{0.4}NiO_3$	248	3.4	0.21

Table 9. Ignition temperature and reaction rate at conversion level below 10% of ethanol at 200°C for $LaFe_{1-y}Ni_yO_3$ perovskites

	T_{50}, °C	Reaction rate μmolg^{-1}h^{-1}	Intrinsic activity μmolm^{-2} h^{-1}
$LaFeO_3$	247	0.3	0.01
La $Fe_{0.9}Ni_{0.1}O_3$	222	0.5	0.03
La $Fe_{0.8}Ni_{0.2}O_3$	218	1.0	0.12
La $Fe_{0.7}Ni_{0.3}O_3$	220	1.2	0.22
$LaNiO_3$	250	1.1	0.13

The reactant mixture was fed into the reactor, the temperature linearly increased up to the required temperature and maintained constant for 30 min. Several isothermal steps were performed up to reach complete conversion. The reaction was carried out under excess of oxygen, molar mixture of C_2H_5OH:O_2:He=1:21:79. A set of valves allowed bypass the reactor and feed directly the stream to the gas chromatography sampling loop, which provided a direct measurement of the VOC concentrations fed using an on-line gas chromatograph Hewlett Packard model HP 4890D with thermal conductivity detector. Helium as carrier gas and a Supelco 25462 and Betadex 225 columns (30m-0.5 mm) were used for permanent gas and VOC analysis, respectively.

A-Site Substituted $La_{1-x}Ca_xFeO_3$ (x= 0.0, 0.1, 0.2, 0.3, 0.4)

The catalytic results show that the reaction starts at 180°C and complete conversion is reached at 370°C. The catalytic activity results in terms of the ignition temperature (T_{50}), reaction rate and intrinsic activity at 200°C are shown in Table 7. The increasing substitution of calcium for lanthanum in the $La_{1-x}Ca_xFeO_3$ series is beneficial for total ethanol combustion due to the electronic perturbation compensated by oxidation of part of Fe^{3+} to Fe^{4+}. Eventhough, for the ethanol combustion as for the toluene combustion, the presence of CaO segregated phases in the larger Ca substituted perovskites has an opposite effect, the total effect is an increases in the catalytic activity respect to $LaFeO_3$.

A-Site Substituted $La_{1-x}Ca_xNiO_3$(x= 0.0, 0.1, 0.2, 0.3, 0.4)

The catalytic results indicates that the reaction starts at 170°C and complete conversion is reached at 300°C. The catalytic activity results in terms of the ignition temperature (T50), reaction rate and intrinsic activity at 200°C are shown in Table 8. It can be seen that the catalytic activity do not depend on calcium content, in agreement with the combustion behavior of polar molecules [21].

B-Site Substituted $LaFe_{1-x}Ni_xO_3$ (x= 0.0, 0.1, 0.2, 0.3, 1.0)

The catalytic results indicates that the reaction starts at 150°C and complete conversion is reached at 350°C. The catalytic activity results in terms of the ignition temperature (T_{50}), reaction rate and intrinsic activity at 200°C are shown in Table 9. Carbon dioxide and water are the main products at high conversion levels, while trace amounts of methanol were also detected [18]. A significant enhancement in the catalytic activity respect to $LaFeO_3$ with nickel substitution is observed, however, if the substitution of iron by nickel is complete the reaction rate drops. These results can be explained in terms of the insertion of Ni^{3+} ions into the $LaFeO_3$ structure having a part of the nickel content as segregated NiO phase is crucial to enhance the performance in ethanol combustion. The cooperative effect of $LaFe_{1-x}Ni_xO_3$ and NiO phases, is due to their relative concentration which determine the oxygen activation capability and hence their activity for the oxidation of ethanol.

Total Combustion of n-Hexane

The catalytic activity for the total combustion of n-hexane was performed in a conventional flow reactor under atmospheric pressure using the catalysts (20-40 mesh) diluted with 1.5 g of SiC, of the same particle size. Gas flow rates were adjusted by mass flow controllers (Brooks 5850E). A nitrogen stream was bubbled through two thermostated and pressurized saturators containing n-hexane at –5° C. This stream was further mixed with air to obtain a mixture gas of 2 vol% C_6H_{14} with air reheated before entry into the reactor. The temperature was linearly increased up to the required temperature, and maintained constant for 30 min.

Reactor effluents were analyzed using an on-line gas chromatograph Hewlett Packard model HP 4890D with thermal conductivity detector. Helium was used as a carrier gas and the column used was a 30-m capillary Supelco 25462.

A- and B-Site Substituted $La_{1-x}Ag_xMn_{0.9}Co_{0.1}O_{3.5}$(x= 0.0, 0.1, 0.2, 0.3)

The characterization results shows almost the same specific area about 28 m^2g^{-1}, and a rhombohedral structure is obtained for x≤ 0.2, while an orthorhombic structure is observed for x=0.3 [18]. Upon Ag substitution, surface oxygen species associated to oxygen vacancies in the crystal lattice are formed. XPS results indicates an almost complete substitution for x=0.1 and that further Ag loading (x≥0.2) increase silver segregated species.

The catalytic results show the typical S-shaped curves, the reaction starts at 150°C then the conversion increases rapidly and complete conversion is reached at 350°C. The catalytic activity results in terms of the ignition temperature (T_{50}), reaction rate and intrinsic activity at 180°C are shown in Table 10.

Carbon dioxide and water were the only detected products at the total conversion level at temperatures lower than 320°C. It can be seen a noticeable increase in the catalytic activity for x=0.1. For larger Ag-substitution the activity increases more slowly. By comparing catalyst performance in the region of low conversions (<10%), samples x=0.2 and 0.3 displayed similar activity behavior, eventhough the silver surface segregation is much lower in x=0.2.

This result is explained considering the differences in the crystalline structure. Only for the rhombohedral structure of x=0.1, almost no silver segregation occurs, and the relation between catalytic activity and La substituted by Ag appear close to the non-stoichiometric oxygen ratio, suggesting the predominance of an intrafacial mechanism. The enhancement of the catalytic activity for x=0.1 comes from the change of electronic sites of adsorbed O-species itself, a weak interaction between O^- and low coordinated metal site results in higher reactivity. The high reaction rate for x=0.3 is explained considering the larger amount of segregated silver phase which acts as additional active sites for n-hexane combustion.

Table 10. Ignition temperature and reaction rate at conversion level below 10% of n-hexane at 180°C for $La_{1-x}Ag_xMn_{0.9}Co_{0.1}O_3$ perovskites

	T_{50}, °C	Reaction rate μmol$g^{-1}h^{-1}$	Intrinsic activity μmolm^{-2} h^{-1}
$LaMn_{0.9}Co_{0.1}O_3$	275	0.04	0.001
$La_{0.9}Ag_{0.1}Mn_{0.9}Co_{0.1}O_3$	243	2.6	0.09
$La_{0.8}Ag_{0.2}Mn_{0.9}Co_{0.1}O_3$	233	1.9	0.07
$La_{0.7}Ag_{0.3}Mn_{0.9}Co_{0.1}O_3$	231	3.3	0.11

Total Combustion of Methane

A-Site Substituted $La_{1-x}CaFeO_3$ (x= 0.0, 0.1, 0.2, 0.3, 0.4)

Defects are defined as any deviation from the ideally perfect crystal and the main effect of defect results from their difference in charge state relative to the ion they replace. The presence of these charge compensating defects can have a deep effect on transport and catalytic properties. The relation between defects and catalytic activity of $La_{1-x}Ca_xFeO_3$ in the methane combustion can be related with the higher extent of reducibility and increasing α-oxygen upon Ca substitution [22]. The catalytic activity measured in terms of the ignition temperature T_{50} and apparent activation energy are summarized in Table 11.

Table 11. Ignition temperature and apparent activation energy for the catalytic combustion of methane for $La_{1-x}Ca_xFeO_3$ perovskites

	T_{50}, °C	E_{act}, kJmol^{-1}
$LaFeO_3$	512	106
$La_{0.9}Ca_{0.1}FeO_3$	508	98
$La_{0.8}Ca_{0.2}FeO_3$	502	96
$La_{0.7}Ca_{0.3}FeO_3$	494	95
$La_{0.6}Ca_{0.4}FeO_3$	487	95

It is observed that T_{50} decreases, as the level of doping goes up and the doped perovskites show lower activation energies in agreement with results reported by Ciambelli et al[7]. It is also observed that the activation energy is practically not affected by the level of Ca-doping. This finding seems to support the so-called compensatory effect, i.e. the higher reaction rate is due to an increase in the number of active sites rather than an increase in the activity of the separate sites, which is reflected in a larger pre-exponential factor. The higher Ca content, the higher number of active sites, whose effectiveness does not seem to be affected by the presence of a higher number of neigh bouring active sites [23]. According to the principle of electroneutrality, the replacement of metal A by a lower oxidation state cation either increases the oxidation state of metal B or decreases the number of oxygen anions (oxygen vacancies), or both as represented in Figure 1 and the following equations:

Oxygen vacancy formation by substitution of Ca for La in the lattice:

$$2\,CaO + 2\,La_{La} + O_O^x \rightarrow 2\,Ca_{La}^{/} + V_O^{\bullet\bullet} + La_2O_3 \quad (1)$$

Hole formation by equilibrium between gaseous oxygen and oxygen vacancy:

$$V_O^{\bullet\bullet} + \frac{1}{2}O_2 \longleftrightarrow O_O^x + 2h^{\bullet} \quad (2)$$

Fe^{4+} formation by oxidation of Fe^{3+}:

$$Fe_{Fe}^{x} + h^{\bullet} \longleftrightarrow Fe_{Fe}^{\bullet} \tag{3}$$

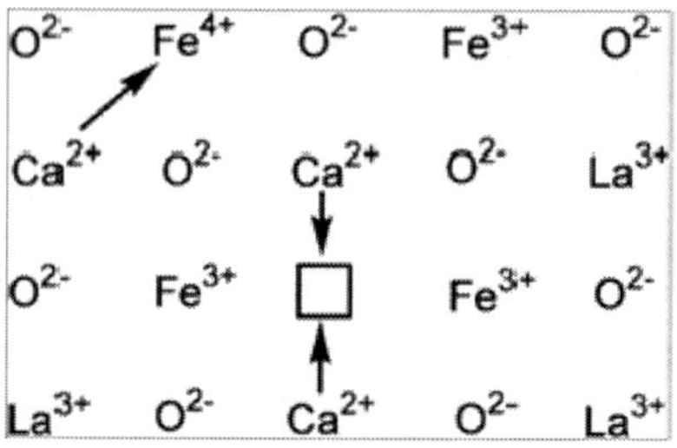

Figure 1. Schematic representation of defect patterns in $La_{1-x}Ca_xFeO_3$.

The empirical evidence, shows that the quantity of weakly adsorbed oxygen species increases with the substitution of La by Ca. Such phenomenon can be explained by the oxygen adsorption into the oxygen vacancies formed by the partial substitution of La^{3+} by Ca^{2+}. It seems to be clear that reactions (1) and (2) support the O_2-TPD results. On the other hand, XRD analyses show some lattice distortions upon Ca substitution. Thus, a shift to higher 2θ angles when the calcium doping increases is observed, Figure 2, indicating a lower interplanar distance.

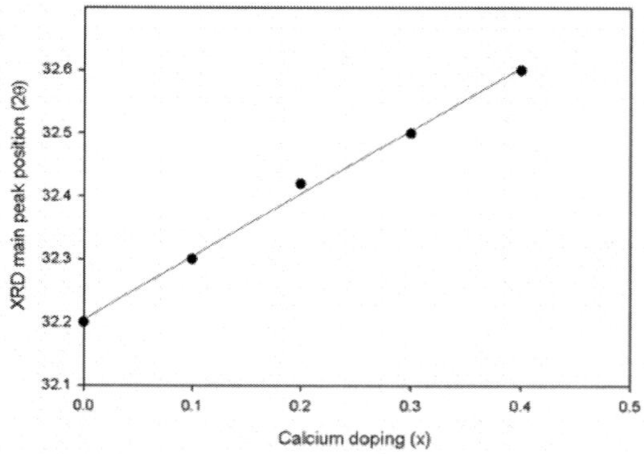

Figure 2. XRD main peak position (2θ) as function of the calcium doping.

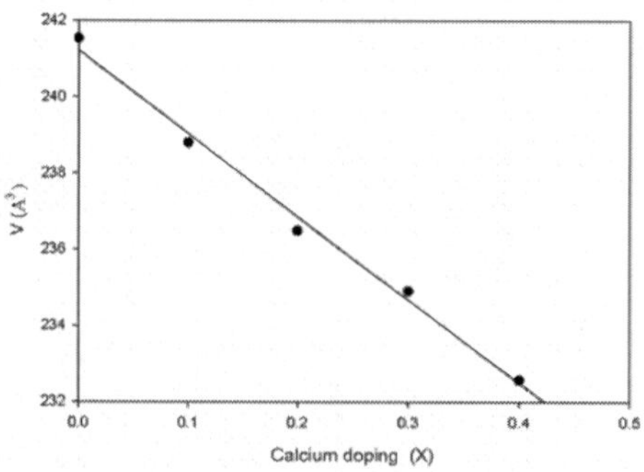

Figure 3. Unit cell volume as function of calcium doping.

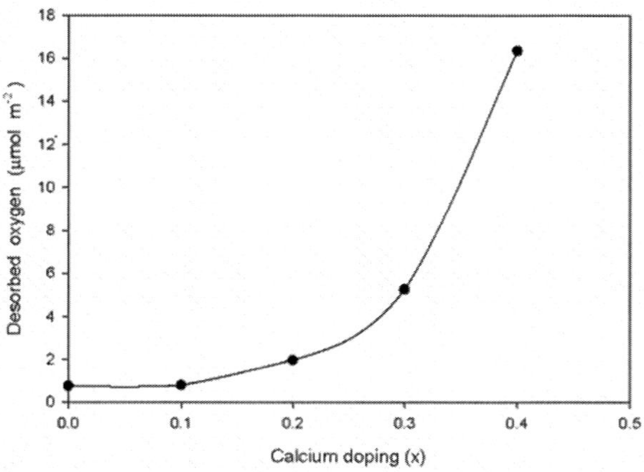

Figure 4. Desorbed oxygen as function of calcium doping.

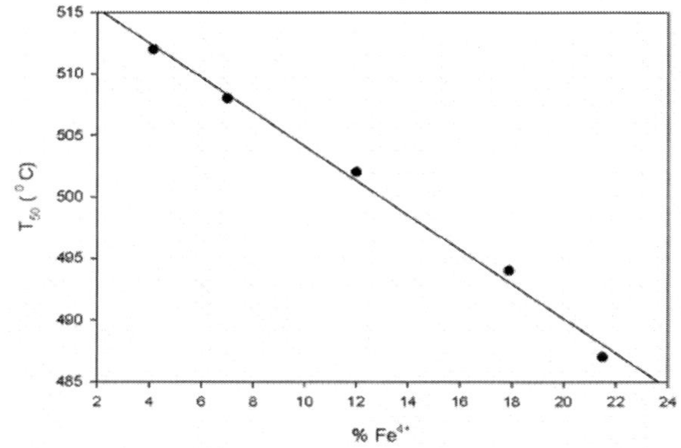

Figure 5. Ignition temperature T_{50} as function of amount of Fe^{4+} (%).

This modification cannot be attributed only to the partial substitution of La^{3+} with Ca^{2+} since the ionic radius are very similar, 0.136 and 0.134 nm, respectively. The cell volume should be very slightly affected by this substitution and an increase of a fraction of the iron oxidation state from $Fe^{3+} \rightarrow Fe^{4+}$ must occur and/or oxygen vacancies must be generated. The differences in the ionic radius of iron Fe^{4+} (0.0585 nm) and Fe^{3+} (0.0645 nm) explain the change in the cell volume shown in Figure 3. The changes in the cell volume is a direct effect of the existence of some iron in oxidation state (IV) and consequently confirms its presence. This finding supports reaction (3), in agreement with previous results reported for other iron-based perovskites [8, 9]. In the O_2-TPD experiments, Figure 4, show that at low substitution levels, $x \leq 0.1$, the amount of desorbed oxygen keeps nearly constant indicating that the charge compensation comes almost from the oxidation state increase of a part of the iron ions from Fe^{3+} to Fe^{4+}. At higher substitution levels ($x \geq 0.2$) the amount of desorbed oxygen increases, suggesting that charge compensation comes mainly from the formation of oxygen vacancies. Eventhough the amount of Fe^{4+} also increases, this increment is always lower than that necessary to charge balance. The obtained results indicates that the T_{50} values do not correlate well with the oxygen vacancies, meanwhile correlate pretty well, Figure 5, with the amount of

Fe^{4+}(%) with a squared correlation coefficient of 0.99 and standard error of 0.92, according the following equation:

$$T_{50} = (5181 \pm 0.9) - (1.4 \pm 0.1)(\%Fe^{4+}) \tag{4}$$

This finding suggests that methane combustion over these perovskites proceed through a redox mechanism where the pair Fe^{4+}/Fe^{3+} plays an important role and the active sites would be associated to the Fe^{4+} ions. Such condition promotes an important increase of the catalytic activity, likely due to the enabled possibility of the lattice oxygen close to Fe^{4+} that can be easily reacted away by methane with the simultaneous reduction of iron to Fe^{3+}. Fresh oxygen can then adsorb on the generated oxygen vacancies, restoring in this way the original condition.

Total Combustion of Soot

The catalytic activity on soot was studied using carbon black, CB, as model soot [24]. The catalyst-carbon black (CB) mixture 4:1 was prepared macerated for 3 min in an agata mortar to get tight contact. The combustion of the mixture was carried out in a Thermobalance (Netzch 409 PC) in air. The combustion rate was normalized to the initial weight of CB in each sample.

A-Site Substituted $La_{1-x}Ca_xFeO_3$

The catalytic activity was measured by means of the temperature of the maximum reaction rate (Tm). Table 12 show the obtained Tm values. For comparison reason the value of the non catalyzed reaction was also included. A significant decreases in more than 180°C respect to the rate of the uncatalized combustion of CB was detected with an increase of the catalytic activity for larger Ca^{2+} substitution degree. Characterization results confirmed that larger Ca substitution produces α-oxygen in a larger extent, modification that has a favorable effect over its capacity to enhance the combustion of carbonaceous substances. It activates the redox mechanism associated with the reduction of Fe^{4+} to Fe^{3+} forming at the same time oxygen vacancies that favor the production of highly mobile surface oxygen species with high oxidative capacity. This result is in agreement with the increase in the catalytic activity upon La^{3+} substituted by K^+ in La_2CuO_4-type oxides [25].

Table 12. Maximum reaction rate temperature of CB combustion on $La_{1-x}Ca_xFeO_3$ perovskites

	Tm, °C
Non-catalyzed	656
$LaFeO_3$	470
$La_{0.9}Ca_{0.1}FeO_3$	462
$La_{0.8}Ca_{0.2}FeO_3$	458
$La_{0.7}Ca_{0.3}FeO_3$	450
$La_{0.6}Ca_{0.4}FeO_3$	445

Pure Alkaline Niobates: $LiNbO_3$, $NaNbO_3$, $KNbO_3$, $RbNbO_3$

The specific areas of the alkaline niobates prepared by a sol-gel method an calcinated at 650°C is about 15 m^2g^{-1} for $NaNbO_3$ and about 10 m^2g^{-1} for the other alkaline niobates. XRD indicates that $LiNbO_3$ is consistent with a rhombohedrical structure with hexagonal axis; $NaNbO_3$ with an orthorhombic structure and $KNbO_3$ is obtained along with a mixture of complex oxides of $K_2Nb_{16}O_{41}$ and $K_3Nb_7O_{19}$. For Rb, it was not possible to obtain the stoichiometry $RbNbO_3$ structure and an mixture of an amorphous material of $Rb_4Nb_6O_{17}$ and $Rb_4Nb_6O_{17}{\cdot}3H_2O$ was obtained.

Table 13. Maximum reaction rate temperature of CB combustion in the first and fourth combustion cycles and apparent activation energy of the alkaline niobates

	Tm, °C		Eact, kJmol^{-1}
	cycle 1	cycle 4	
$LiNbO_3$	636	634	104
$NaNbO_3$	523	522	61
$KNbO_3$	460	460	86
$RbNbO_3$	612	---	---

A larger extent of α-oxygens, compared to Li and Rb counterparts, are detected for Na and K niobates. The temperatures of the maximum reaction of CB combustion, Tm are compiled in Table 13.

The Tm values indicate that all niobates show catalytic activity for soot combustion with dependence on the nature of the alkaline metal. The highest catalytic activity, is shown by $KNbO_3$ following an atomic number dependent order, namely: $KNbO_3 > NaNbO_3 > LiNbO_3$. $RbNbO_3$ do not follow this trend due presumably to the perovskite structure was not reached at the studied calcination temperature. Considering that only $KNbO_3$ and $NaNbO_3$ formed the perovskite structure, this crystalline structure is determining in the catalytic activity and in the extent of desorbed oxygen.

The differences in the catalytic activity among the alkaline niobates cannot be explained only in terms of differences in the surface area, but also in the chemical nature of the alkaline metal which also determines the structure of the solid. The catalytic activity is not the only prerequisite for the applicability of a catalyst in soot combustion, its thermal stability is of great importance. It is well known that K impregnated catalysts suffer deactivation because potassium loss by volatilization during the process.

In order to assess the stability of the niobates, several consecutive combustion cycles with the same sample were carried out. To avoid thermal changes of the powders, the successive combustion cycles were stopped at 650°C, the calcination temperature. Interestingly, no deactivation is observed, in contrast to other potassium impregnated catalysts [26], as a consequence of the stabilization of potassium inside the perovskite structure. In order to obtain experimental evidence to confirm this conclusion, AAS and DRX for the solid after the last cycle were carried out.

No differences are observed, consequently no loss of Na and K is concluded, due to the stabilization of the alkaline metal inside the perovskite structure avoiding volatilization of the metal.

This study opens interesting perspectives to be explored for future catalyst development. For one hand, it is worth to investigate the formation the α-type chemisorbed oxygen species in non-stoichiometric of $KNbO_3$ which should entail the improvement of the catalytic activity. For the other hand, it can be expected that $RbNbO_3$ could follow the trend of Li, Na and K, and consequently it could have higher catalytic activity than $KNbO_3$. Further research focused on the synthesis and catalytic assessment of $RbNbO_3$ should be also highly desirable.

CONCLUSION

The most important properties of perovskites to be used as catalysts for total combustion reactions are textural (specific area), structural (crystalline structure) and surface (surface oxygenated species, segregated phases and reduction ability). These properties are close related with the synthesis method, which in turn determines the composition, vacancies and redox properties of the perovskite.

- *Specific area.* Catalyst is a surface science, and the first stage of every heterogeneous catalytic reaction is the adsorption on the adsorbent´s surface, consequently it is highly desirable to use perovskites with larger S_{BET} values. Eventhough, it is not always in close relation with the catalytic activity.
- *Crystal structure.* Empirical evidences show that orthorhombic and rhombohedral structures are the most appropriate structures of perovskite-type oxides to be used in catalytic combustion reactions.
- *Nature of the surface oxygenated species.* The presence of α-oxygens improves the catalytic activity of VOCs combustion at lower temperatures, meanwhile β-oxygens are active for catalytic combustion at higher temperatures such as methane. These oxygen species determines the activation capability and the reactivity. The desorption of these oxygen produces the important oxygen vacancies always in close relation with the catalytic activity.
- *Nature of the segregated phases.* Larger extent of catalytic active segregated phases in a high dispersion degree contributes to an enhancement in the catalytic activity.
- *Reduction ability.* The reduction ability in consecutives reduction cycles, without loss of the perovskite structure, is a measured of the thermal stability in reductor atmosphere. Eventhough, this property is not always in close relation with the catalytic activity in oxidation reactions, may be of great importance in catalytic hydrogenation reactions.

The catalytic behavior of a perovskite is closely related to the nature of the substrate to be combusted, therefore to tailor the catalytic properties of perovskite-type oxides, it is absolutely necessary to chose correctly the A- and the B-site metal ion. Moreover, a correct election of the A'- or B'- substituted cations with a metal with lower oxidation state and similar ionic radii make it possible the substitution in a higher extent.

REFERENCES

[1] Pecchi, G., et al., Doping of lanthanum cobaltite by Mn: thermal, magnetic, and catalytic effect. *Journal of Materials Science*, 2008. 43(15): p. 5282-5290.

[2] O'Connell, M., et al., Catalytic oxidation over lanthanum-transition metal perovskite materials. *Catalysis Today*, 1999. 47(1-4): p. 123-132.

[3] Irusta, S., et al., Development and application of perovskite-based catalytic membrane reactors. *Catalysis Letters*, 1998. 54(1-2): p. 69-78.

[4] Zhang, H. M., et al., Oxygen sorption and catalytic properties of la1-xsrxco1-yfeyo3 perovskite-type oxides. *Journal of Catalysis*, 1990. 121(2): p. 432-440.

[5] Langmuir, I., The constitution and fundamental properties of solids and liquids. Part i. Solids. *Journal of the American Chemical Society*, 1916. 38(11): p. 2221-2295.

[6] Pecchi, G., et al., Effect of B-site cation on the catalytic activity of La1-xCaxBO3 (B = Fe, Ni) perovskite-type oxides for toluene combustion. *Journal of Chemical Technology and Biotechnology*, 2011. 86(8): p. 1067-1073.

[7] Ciambelli, P., et al., La, Ca and Fe oxide perovskites: preparation, characterization and catalytic properties for methane combustion. *Applied Catalysis B-Environmental*, 2001. 33(3): p. 193-203.

[8] Wu, Y., et al., A comparative-study on perovskite-type mixed-oxide catalysts Ax'A1-xBo3-lambda (A' = Ca, Sr, A = La, B = Mn, Fe, Coo) for NH3 oxidation. *Journal of Catalysis*, 1989. 120(1): p. 88-107.

[9] Barbero, B. P., J. A. Gamboa, and L. E. Cadus, Synthesis and characterisation of La1-xCaxFeO3 perovskite-type oxide catalysts for total oxidation of volatile organic compounds. *Applied Catalysis B-Environmental*, 2006. 65(1-2): p. 21-30.

[10] Ciambelli, P., et al., AMnO(3) (A=La, Nd, Sm) and Sm1-xSrxMnO3 perovskites as combustion catalysts: structural, redox and catalytic properties. *Applied Catalysis B-Environmental*, 2000. 24(3-4): p. 243-253.

[11] Utsumi, S., et al., Preparation and characterization of inexpensive heterogeneous catalysts for air pollution control: Two case studies. *Catalysis Today*, 2007. 123(1-4): p. 208-217.

[12] Pecchi, G., et al., Effect of the preparation method on the catalytic activity of La1-xCaxFeO3 perovskite-type oxides. *Catalysis Today*, 2008. 133: p. 420-427.

[13] Stojanovic, M., et al., Synthesis and characterization of LaCr1-xNixO3 perovskite oxide catalysts. *Journal of Catalysis*, 1997. 166(2): p. 315-323.

[14] Provendier, H., et al., Stabilisation of active nickel catalysts in partial oxidation of methane to synthesis gas by iron addition. *Applied Catalysis a-General*, 1999. 180(1-2): p. 163-173.

[15] Kawasaki, S., M. Takano, and Y. Takeda, Physical properties of Co-substituted iron perovskite systems. *Solid State Ionics*, 1998. 108(1-4): p. 221-226.

[16] Rivas, M. E., et al., H(2) production from CH(4) decomposition: Regeneration capability and performance of nickel and rhodium oxide catalysts. *Journal of Power Sources*, 2008. 184(1): p. 265-275.

[17] Pecchi, G., et al., Surface properties and performance for VOCs combustion of LaFe1-yNiyO3 perovskite oxides. *Journal of Solid State Chemistry*, 2008. 181(4): p. 905-912.

[18] Pecchi, G., et al., Effect of additive Ag on the physicochemical and catalytic properties of LaMn0.9Co0.1O3.5 perovskite. *Applied Catalysis a-General*, 2009. 371(1-2): p. 78-84.

[19] Pecchi, G., C. Campos, and O. Pena, Thermal stability against reduction of LaMn1-yCoyO3 perovskites. *Materials Research Bulletin*, 2009. 44(4): p. 846-853.

[20] Pecchi, G. and C. Campos, Structure and activity of LaMn1-yCoyO3 perovskites. *Reaction Kinetics and Catalysis Letters*, 2007. 91(2): p. 353-359.

[21] Trikalitis, P. N. and P. J. Pomonis, Catalytic activity and selectivity of perovskites la1-xsrxv1-x3+vx4+o3 for the transformation of isopropanol. *Applied Catalysis a-General*, 1995. 131(2): p. 309-322.

[22] Pecchi, G., et al., Relation between defects and catalytic activity of calcium doped LaFeO3 perovskite. *Solid State Ionics*, 2011. 187(1): p. 27-32.

[23] Saracco, G., et al., Methane combustion on Mg-doped LaCrO3 perovskite catalysts. *Applied Catalysis B-Environmental*, 1996. 8(2): p. 229-244.

[24] Jimenez, R., et al., Effect of Ca-substitution in La1-xCaxFeO3 perovskites on the catalytic activity for soot combustion. *Fuel Processing Technology*, 2010. 91(5): p. 546-549.

[25] Fino, D., et al., Studies on kinetics and reactions mechanism of La2-xKxCu1-yVyO4 layered perovskites for the combined removal of diesel particulate and NOx. *Applied Catalysis B-Environmental*, 2003. 43(3): p. 243-259.

[26] Lopez-Suarez, F. E., et al., Potassium Stability in Soot Combustion Perovskite Catalysts. *Topics in Catalysis*, 2009. 52(13-20): p. 2097-2100.

In: Perovskite
Editors: Jinghua Zhang and Huan Li
ISBN: 978-1-62417-800-9

Chapter 4

Microwave-Assisted Synthesis and Characterization of Perovskite Oxides

***Jesús Prado-Gonjal*[1], *Rainer Schmidt*[2] *and Emilio Morán*[1]**
[1]Dpto. Química Inorgánica I, Facultad C. Químicas,
Universidad Complutense, Madrid, Spain
[2]Dpto. Física Aplicada, GFMC, Facultad C. Físicas,
Universidad Complutense, Madrid, Spain

Abstract

The use of microwave irradiation is a promising alternative heat source for the synthesis of inorganic materials such as perovskite oxides. The method offers massive energy and time savings as compared to the traditional ceramic method. In this work we review the basic principles of the microwave heating mechanism based on interactions between dipoles in the material and the electromagnetic microwave.

Furthermore, we comment on and classify all different sub-types of microwave synthesis such as solid-state microwave and microwave assisted hydrothermal synthesis. In the experimental part of this work we present a large range of materials that were synthesized in our laboratories by one of the microwave techniques, where such materials include superconducting, ferromagnetic, ferroelectric, dielectric and multiferroic perovskite systems.

1. Introduction

Perovskite oxides are a fascinating class of inorganic materials [1], which are usually prepared by means of the conventional ceramic method. Such method is especially well suited for oxides since reactions are carried out in air at ambient pressure. It involves the homogeneous grinding and mixing of stoichiometric amounts of solid precursor oxide reactants followed by heating at elevated temperatures for long periods of time (typically 1000 ºC and several hours or days), where the grinding and heating cycle is repeated until the desired pure phase is obtained. Although there can be no doubt about the usefulness and easiness of this universal method of synthesis which works in thermodynamic equilibrium conditions, there are some intrinsic drawbacks such as the need for high temperatures to accelerate diffusion -which prevents metastable phases to appear-, the volatility of some reactants, the need of regrinding several times or the high energy and time consumptions. On these grounds, there is a permanent stimulus for searching alternative routes such as decomposition of alternative precursors, sol-gel methods and many more, where the reaction times can often be notably reduced. Synthesis is achieved in out-of-equilibrium conditions and such methods may be called "Fast Chemistry"; among them combustion, sonochemistry, "spark plasma" and microwave-assisted methods of synthesis, where the latter is the focus of this work. Many perovskite oxides can be prepared by these means as will be demonstrated in the following.

Microwave research is indebted to British scientists J. Randall and H. A. Boot, who designed a magnetron valve to generate microwaves for radar systems back in 1940. Six years later, Percy Spencer, an engineer working on magnetrons for the Raytheon Company, accidentally discovered the rapid effect of microwaves on heating a chocolate bar and after some proofs being made with other kinds of food it was realized that microwave cooking was feasible and the path for a new kind of domestic furnaces, nowadays of general use, was open [2].

Nevertheless, it was not until 1975 that the heating effect of microwaves on ceramic materials was observed, described and studied by Sutton [3]. At a first glance, these heating effects are quite surprising since microwave radiation ranges in the low-energy zone of the electromagnetic spectrum (microwave frequencies range between 0.3 and 300 GHz), (Figure 1) [4].

Most microwave heating effects can be understood in terms of electric dipoles in a material which follow the alternating electromagnetic field associated to the microwave, i. e. dipolar molecules such as water do rotate, and the resistance to that movement generates a considerable amount of heat [5]. For domestic purposes the conventionally used frequency is 2.45 GHz equivalent to 0.0016 eV, which is far below the energy of even a weak bonding energy such as the hydrogen bonding (0.21 eV) [4].

Microwave assisted synthesis flourished firstly in the pharmaceutics and organic industry but in solid state chemistry and materials science it is still in its infancy; there are still many questions to be answered such as fundamental details on the reaction mechanisms and kinetic aspects of the reactions. The lack of such knowledge turns the method into a trial and error one at the current stage [6]. Therefore, the aims of this review article are:

i) To comment on fundamental aspects of the interactions between microwaves and solid matter.
ii) To compare conventional heating versus microwave heating.
iii) To demonstrate that microwave assisted synthesis is a fast and efficient method to produce perovskite oxide materials.
iv) To compare several ways to perform microwave assisted synthesis: solid state, single-mode and solvothermal routes.
v) To give relevant examples from our own work or taken from the literature of important perovskite materials in different areas such as energy materials (components for solid oxide fuel cells SOFC's), magnetic, ferroelectric, multiferroic materials and superconductors.

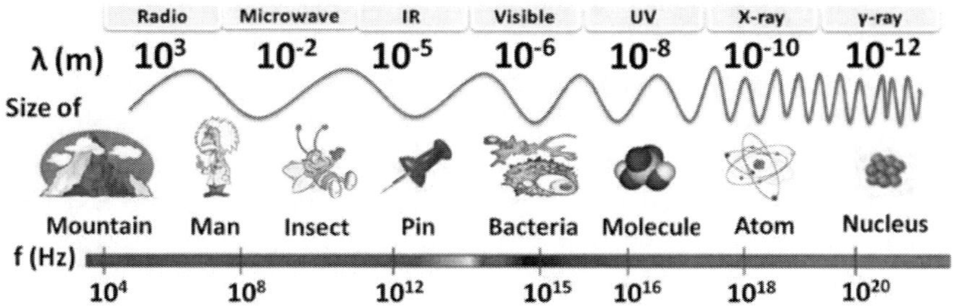

Figure 1. Electromagnetic spectra. Microwave frequencies range between 0.3 and 300 GHz.

2. On the Interaction between Microwaves and Matter

Any electromagnetic radiation has two components: an electric and a magnetic one. An electromagnetic microwave may be reflected, transmitted or absorbed by condensed matter depending on the nature of such matter. Therefore, in this context of condensed matter interacting with microwaves three classes of materials can be defined [7]:

i) *Reflecting materials,* which do not allow penetration by the electromagnetic wave; materials with free electrons such as certain metals behave in this way.
ii) *Transparent materials,* which allow the wave to propagate all the way through with low attenuation, such as Teflon and silica glass. Generally speaking, insulators or materials with low dielectric permittivity and low losses are concerned here.
iii) *Absorbing materials,* which transform the electromagnetic energy into heat; this is the case for dipolar liquids such as water and dielectric or polar materials such as ferroelectrics with high dielectric permittivity. Dielectric losses favor the absorption of the microwave energy by the material. Therefore, an ideal microwave absorber has high dielectric permittivity and an appropriate level of losses.

Since electric dipoles in condensed matter can strongly couple to microwave heating sources, it is usually the electric component of the microwave and its interaction with the sample which governs the microwave heating process of polar materials. The microwave heating process of such polar materials critically depends on both the dielectric permittivity and the dielectric losses as mentioned above. The dielectric permittivity describes the "polarizability" of the material where the overall polarization is termed (α_1). This term includes several accumulative components: electronic polarization (α_e), atomic polarization (α_a), dipolar polarization (α_p), ionic polarization(α_{ion}) and interfacial polarization (α_{if}):

$$(\alpha_1) = (\alpha_e) + (\alpha_a) + (\alpha_p) + (\alpha_{ion}) + (\alpha_{if})$$

- Electronic polarization α_e: when the atom is subjected to an external electric field, redistribution of the charges occurs and the atomic electron cloud shifts away from the equilibrium position with respect to the positive nuclei resulting in an induced dipole moment. This effect occurs in all solid matter containing localized electrons.

- Atomic polarization α_a: This is equivalent to the electronic polarization, except that it accounts for the shift of the atomic cores as a response to the applied electric field. It is usually small as compared to α_e.
- Dipolar polarization α_p: In thermal equilibrium, electric dipoles in the material will be randomly oriented and thus carry no net polarization. An external field can align these dipoles to some extent and thus induces a polarization to the material. This contribution only exists in polar materials such as dielectrics with a strong ionic character of the chemical bonds.
- Ionic polarization (or ionic displacement) α_a: This occurs due to relative displacements of positive and negative ions from their equilibrium lattice sites. The positively and negatively charged ions in the crystals are displaced spontaneously from their equilibrium lattice positions and the centers of negative and positive charge within one unit cell do not coincide. This constitutes a net dipole moment, which can be switched by high applied electric fields. This effect only occurs in piezo- and ferroelectrics.
- Interfacial polarization α_i: Surfaces, grain boundaries, interphase boundaries (including the surface of precipitates) may be charged, i.e. they contain dipoles which may become oriented to some degree in an external field and thus contribute to the total polarization of the material. Such effects can occur in all kind of condensed matter, but the quantitative contribution to the total polarization is usually small due to the generally small volume fraction of interfaces in bulk material.

Polarization highly depends on the frequency of the applied electric field as illustrated in Figure 2. At low frequencies all dipoles can follow the applied field, but with increasing frequency the different contributions to the overall polarization consecutively "relax out", which involves that the respective dipoles cannot follow the alternating applied electric field anymore and do not polarize. This is reflected in a consecutive reduction of the dielectric permittivity ε'. Atomic and electronic contributions show a double-peak structure in the dielectric permittivity, which occurs when the frequency of the applied field matches the resonance frequency of the switching dipoles.

In materials with high dielectric losses the microwave gets attenuated. Such attenuation can be improved by the addition of small amounts of metallic powder, which reflects the microwave and hinders loss-free propagation of the wave. The power density of absorbed microwaves may be expressed as:

$$P\ (W/m^3) = \text{electrical losses} + \text{magnetic losses} = \omega\, \varepsilon_o\, \varepsilon''\, E^2_{rms} + \omega\, \mu_o\, \mu''\, H^2_{rms}$$

where $\omega = 2\pi f$ = angular frequency, ε_o = dielectric permittivity of vacuum, ε" = relative dielectric losses (polarization + conduction), E_{rms} = internal electrical field of the microwave, μ_o.= magnetic permittivity in vacuum, μ" = relative magnetic losses and H_{rms} = internal magnetic field of the microwave. The magnetic losses μ" are usually less relevant and the susceptibility of a polar material to microwave heating is mainly determined by the dielectric losses. Materials with losses $\varepsilon'' < 10^{-2}$ are quite difficult to be heated up, while for those with $\varepsilon''_{eff} > 5$ the surface is heated rather than the bulk.

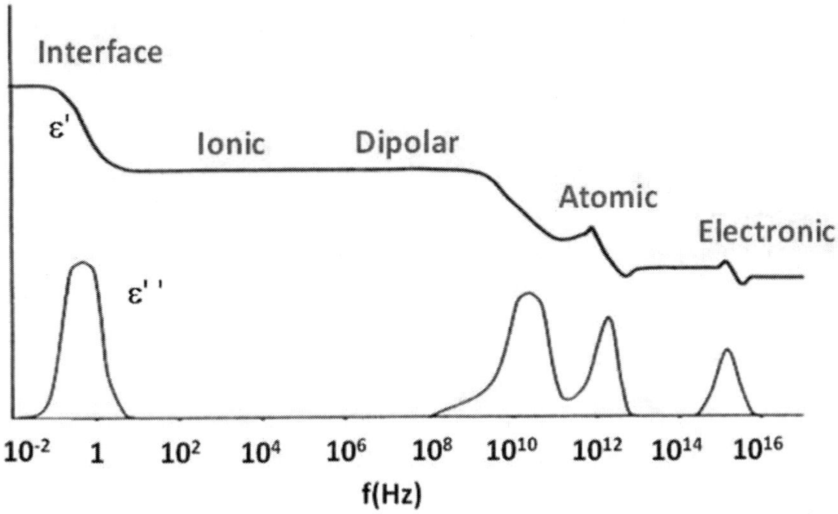

Figure 2. Frequency dependence of the polarization mechanisms.

Polar materials with intermediate values, $10^{-2} < \varepsilon''_{eff} < 5$, are best candidates for microwave heating [8-9]. The ability of a particular material to increase its temperature when exposed to microwaves can be described by the "loss tangent" $\tan\delta = (\varepsilon''/\varepsilon')$. In most aqueous and organic media $\tan\delta$ diminishes upon heating and microwave heating is less effective [10].

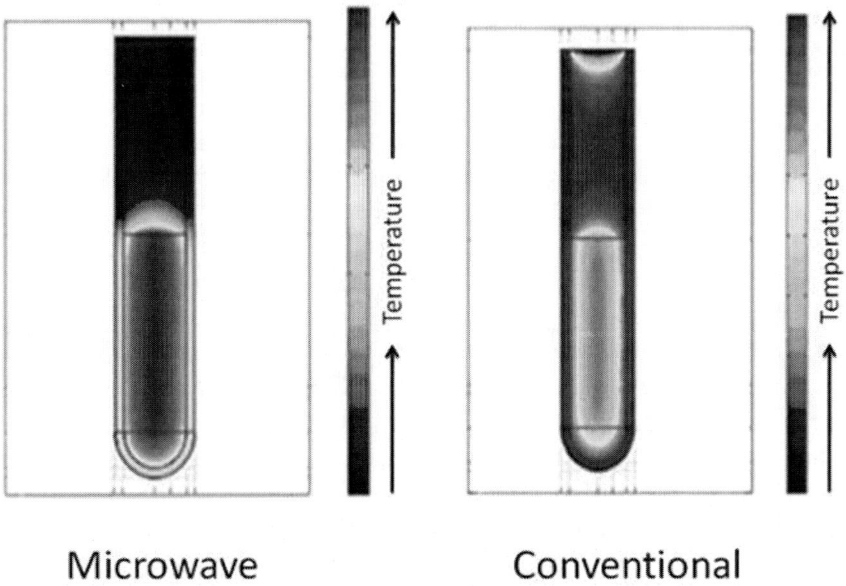

Figure 3. Differences in the temperature distribution in microwave and conventional processes (image downloaded from the web: http://www.biotage.com/DynPage.aspx?id=22052).

3. Microwave Heating

The main features which distinguish microwave heating from conventional methods are:

a) *Energy transfer*: In conventional heat treatments energy is transferred to the material through heat conduction and convection, creating thermal gradients. In the case of microwave heating, energy is directly transferred to the material through an interaction at the molecular level with the electromagnetic field (Figure 3). It has been argued that the way of energy transfer is in fact crucial in the synthesis of materials [11]. The depth that the radiation reaches varies depending on the material and other factors such as the dielectric and magnetic properties, the frequency and power of the microwave, temperature, conductivity, the size and the density of the material.
b) *Rapid heating*: Microwaves massively reduce the processing time as compared to conventional synthesis but significant qualitative changes in the crystal structure and the resulting material properties are usually not encountered. In some cases even quantitative improvements can be achieved. Some authors claim that chemical reactions, when microwave-assisted, are accelerated between 10 - 1000 times compared with conventional method [12]. A problem with such fast heating is the formation of "hot spots" and the temperature profile may not be homogeneous, which could possibly affect the reaction container [13-14].
c) *Selective heating of materials*: Microwaves can be used to carry out selective heating, which is not possible by using conventional heat treatments. Depending on the characteristics of the specific material (dielectric properties, size and structure) and its ability to couple to the electric and/or magnetic field of the microwaves, heating is located in particular areas. Since some materials do not couple well with the most frequently used 2.45 GHz microwaves at room temperature, a radiation susceptor (such as SiC, graphite or activated carbon) can be added to the precursor mixture. The susceptor strongly couples with the radiation causing a considerable increase in temperature to facilitate a more efficient microwave heating process [7, 15]. However, most polar materials do couple to the electromagnetic wave sufficiently strong to reach high temperatures, no additives are necessary and the reaction time is just a few minutes. In Table 1 the highest temperatures reached in selected materials are summarized.

Non-thermal effects: Many research groups are currently studying a number of anomalies involved with microwave heating that have been broadly termed "non-thermal effects", which include any differences of the microwave method as compared to conventional methods which cannot be explained by differences in the temperature and the temperature profile [16]. Today's debate focusses on anomalous effects between the electric field and the interfaces of the particles which possibly involves the formation of a plasma and the increase of the solid diffusion due to second-order effects [17-18].

Table 1. Maximum temperatures reached within different materials after few minutes of exposure to microwaves [8]

Material	t (min)	T (°C)	Material	t (min)	T (° C)
C(amorphous)	1	1283	V_2O_5	11	714
C(graphite)	2	1100	Cr_2O_3	7	130
Ti	1	1150	MnO_2	6	1287
V	1	557	V_2O_5	11	714
Fe	7	768	MnO_2	6	1287
Co	3	697	Mn_2O_3	6	1180
Ni	1	384	Fe_3O_4	3	1258
Zn	3	581	Co_2O_3	3	1290
Zr	6	462	NiO	6	1305
Nb	6	358	CuO	6	1012
Mo	4	660	ZnO	5	326
W	6	690	WO_3	6	1270

4. Microwave Assisted Synthesis

Microwave assisted synthesis of perovskite oxides may be performed by several alternative methodologies:

i) Direct irradiation of a mixture of the solid reactants;
ii) Irradiation of a solution in an autoclave;
iii) Combining microwaves with other synthetic procedures such as sol-gel or combustion [4].

4.1. Solid State Microwave Synthesis

There are many materials with various applications that have been synthesized by microwave irradiation of solid precursors in recent years. The most common synthetic procedure consists in mixing the precursors and packing them in a pellet, which is then deposited in an adequate crucible (usually porcelain, alumina or SiC) and placed in a microwave oven (a domestic one may be used). As mentioned above, in some cases the presence of a radiation susceptor is needed in order to generate a sufficient initial temperature for the reaction to start [19]. It has been shown that solid state microwave assisted synthesis is a very fast and effective method for the production of oxides.

4.1.1. "Simple" ABO_3 Perovskites

Stoichiometric perovskite oxides have the ideal chemical formula ABO_3, where A and B are cations of different sizes. A is a large cation with a similar size than the O^{2-} anion. Therefore, the A-cations and O^{2-} anions are cubic close-packed, whereas the smaller B-site cations reside on octahedral interstices. A-site cations are usually alkali, alkaline earth or

Lanthanide rare-earth, and B is a medium-sized cation, normally a transition metal with preference for the 6-fold octahedral coordination. In our laboratory different perovskites families have been prepared by solid state microwave synthesis, namely $LaMO_3$ (M = Al, Cr, Mn, Fe, Co) [20] and $RECrO_3$ (RE = Rare Earth). The first family of materials, $LaMO_3$ can be used in very different technological applications such as components of solid oxide fuel cells (SOFCs), separation membranes, magneto-optical or magneto-resistant materials and substrates for thin films [21-23]. On the other hand, RE chromites have become recently the subject of much interest as possible multiferroic materials due to potential magneto-electric coupling between the RE^{3+} and Cr^{3+} cations [24-25].

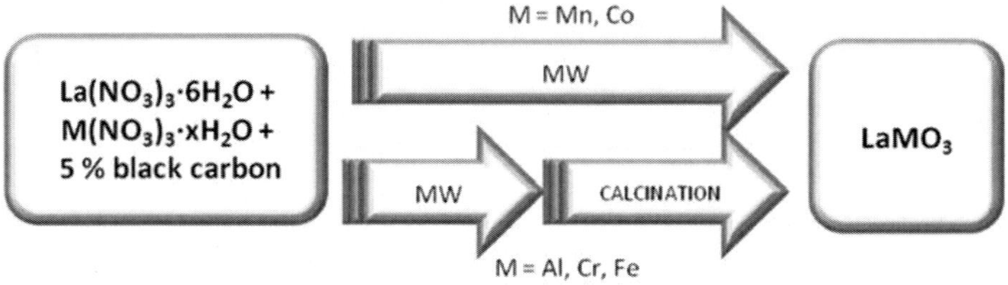

Figure 4. Microwave synthesis of $LaMO_3$ (M = Al, Cr, Mn, Fe, Co). Image reproduced from reference [20] with permission from Elsevier.

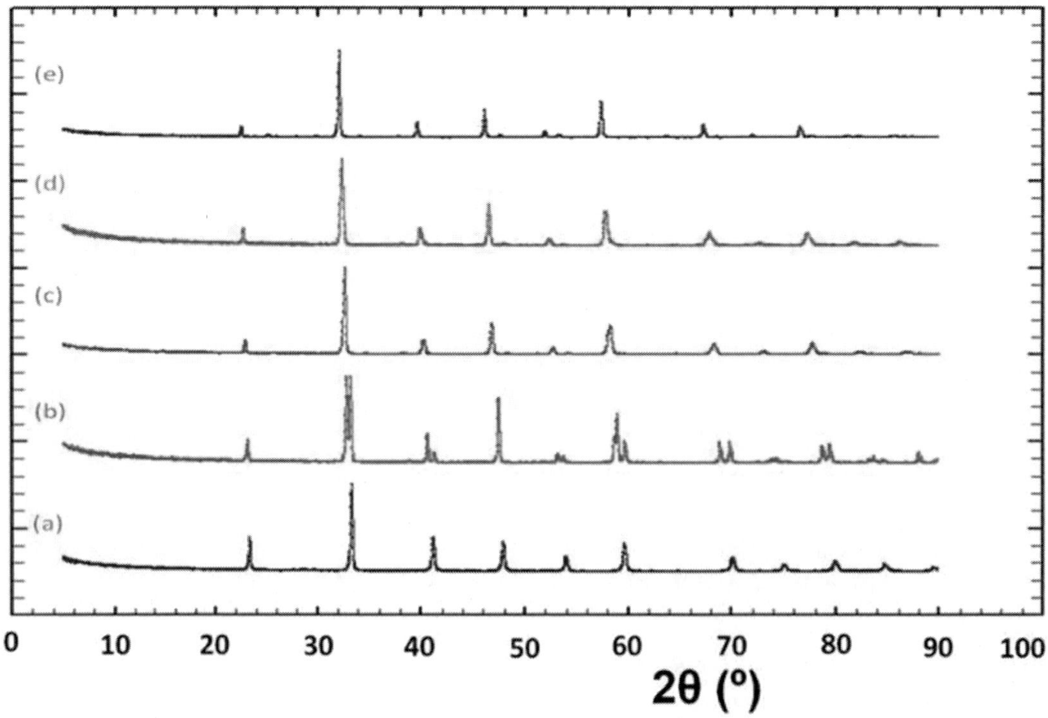

Figure 5. X-ray diffraction patterns of $LaMO_3$ M= (a) Al, (b) Co, (c) Cr, (d) Mn, (e) Fe.

Figure 6. Scanning electron microscopy (SEM) image of $LaCoO_3$.

The synthetic scheme for the $LaMO_3$ family is depicted in Figure 4. In the case of M = Al and Fe amorphous materials are produced in just 10 minutes. After heat treatment at 500 ºC in a conventional furnace the corresponding $LaAlO_3$ and $LaFeO_3$ perovskite phases can be obtained. It should be noted that the heat treatment is done at a much lower temperature than the usual ceramic synthesis temperature ($\geq$ 1000 ° C).

For both $LaCoO_3$ and $LaMnO_3$, 30 minutes of microwave treatments are sufficient to obtain pure and crystalline products, without any further heating. This is due to the metallic conduction of these materials, which turns the first particles being produced into radiation susceptors and the reaction becomes self-propagated [20]. The X-ray diffraction patterns are shown in Figure 5 where the phase purity of the microwave synthesized $LaMO_3$ family is demonstrated.

Figure 6 shows a scanning electron micrograph (SEM) for the $LaCoO_3$ powder, where agglomeration of particles of different sizes can be clearly seen. The excellent crystallinity of these materials enables to perform an accurate Rietveld refinement adjusting both the cell parameters and the atomic positions for each one of the synthesized materials. In Figure 7 the magnetic properties of microwave assisted and conventionally synthesized $LaCoO_3$ are compared. It can be seen that the curves of magnetization vs temperature are qualitatively very similar, only small quantitative differences exist mainly at low temperature as demonstrated in the Figure 7 insets. Such differences may be ascribed to magnetic defects, which are well known to dominate the low temperature magnetism of $LaCoO_3$.

The qualitative agreement of the magnetism is a clear indication that microwave assisted synthesis can successfully reproduce the $LaCoO_3$ crystal structure and the corresponding magnetic interactions between Co^{3+} cations. The broad peak at $\approx$ 100 Kelvin displayed in the magnetization curves is commonly understood in terms of a spin-state transition, where the Co^{3+} spins undergo a transformation from a low-temperature low-spin to an intermediate- or high-spin state at higher temperatures.

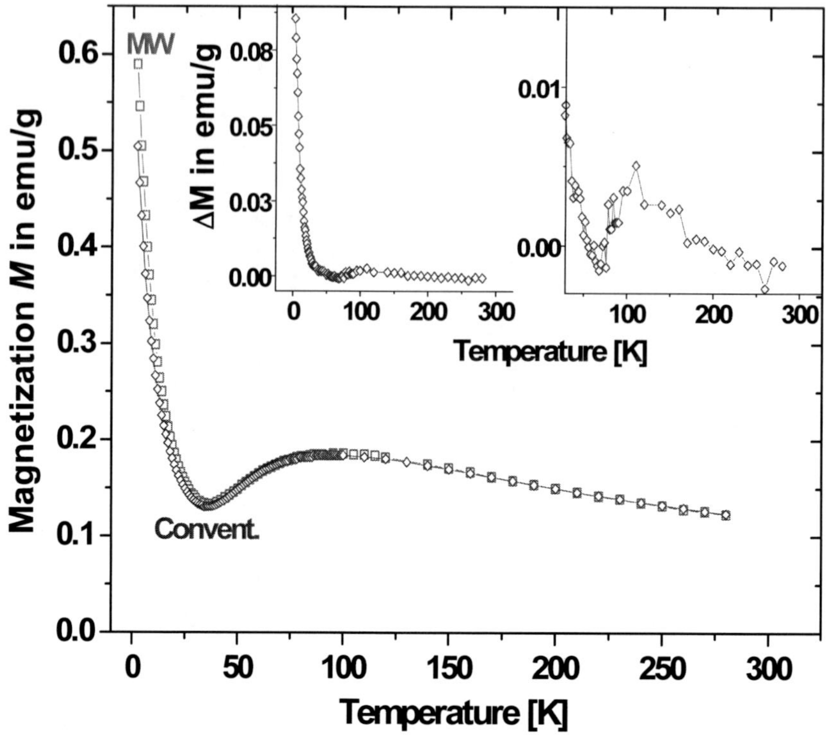

Figure 7. Magnetization vs temperature curves for microwave (MW) and conventionally (Convent.) synthesized $LaCoO_3$. In the Figure insets the differences in magnetization (ΔM) between the two curves are displayed at various temperature ranges.

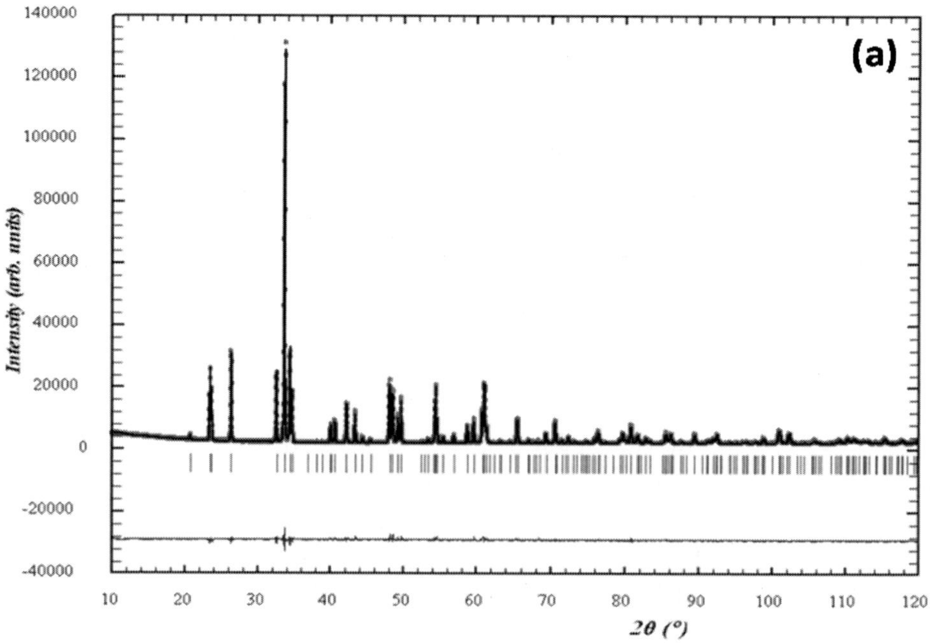

Figure 8. (Continued).

Figure 8. (a) Rietveld refinement of $ErCrO_3$; Space group Pbnm. Cell parameters: $\boldsymbol{a}$ = 5.21389(4) Å, $\boldsymbol{b}$ = 5.50302(4) Å and $\boldsymbol{c}$ = 7.50391(6) Å; R_{exp} = 1.74 (b) Scanning electron microscopy image of $ErCrO_3$.

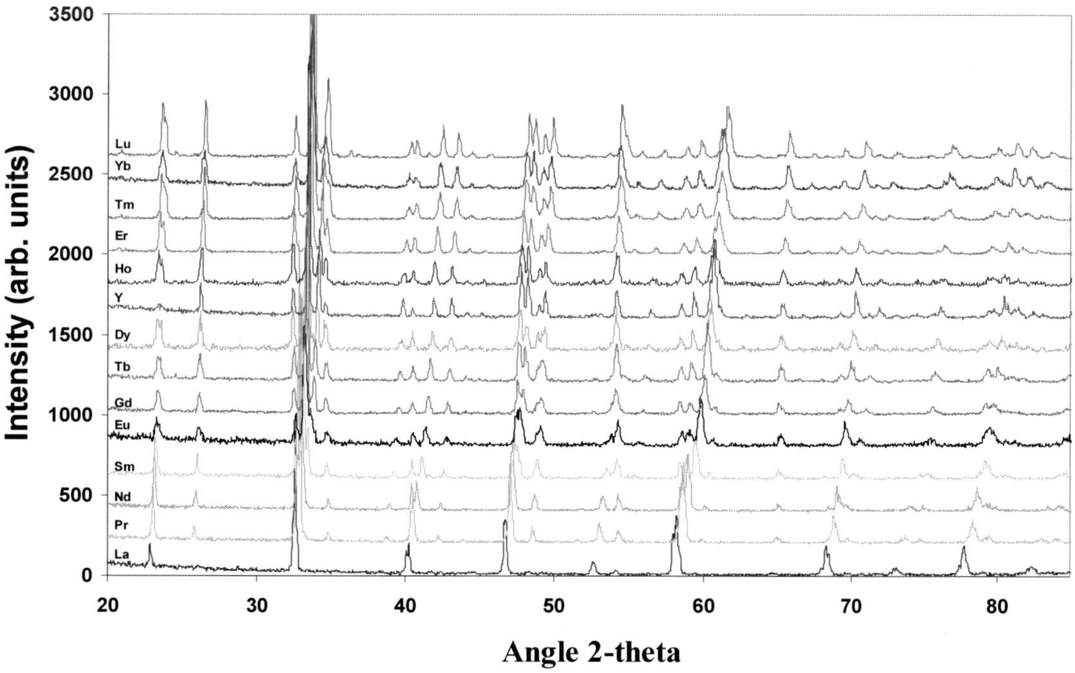

Figure 9. XRD patterns of microwave synthesized $(RE)CrO_3$ powders.

Solid state microwave synthesis was also applied to the Rare-Earth chromite series. As a representative example Figure 8a shows the Rietveld refinement performed for $ErCrO_3$ powder (space group Pbnm). The SEM micrograph (Figure 8b) of $ErCrO_3$ powder shows the agglomeration of large particles with spherical form which are made up of smaller particles near the nano-range [26]. For the full RE chromite series amorphous materials are formed during MW irradiation, which crystallize in the zircon structure $RCrO_4$ at 500 ºC. Upon increasing the temperature to 800 ºC the materials transform into the corresponding perovskites $RCrO_3$ within 2 hours [26]. Figure 9 shows a summary of the X-ray patterns of the microwave synthesized RE chromite series, where phase purity is indicated in all compositions. The X-ray peak positions move to higher angles with decreasing RE ionic radius and increasing atomic number, which is a reflection of the decreasing unit cell size. The decrease in the ionic radius also leads to increasing octahedral tilting in the perovskite cell and decreasing Neel temperature for Cr^{3+}-Cr^{3+} antiferromagnetic ordering, which occurs in all species.

Table 2. Synthesis conditions for some niobates and titanates by microwave or conventional routes

Material	Reactants	Microwave	Conventional
$LiNbO_3$	Li_2CO_3, Nb_2O_5	15 minutes 800 W	12 h 500ºC
$NaNbO_3$	Na_2CO_3, Nb_2O_5	17 minutes 800 W	t>>>, grindings 1250ºC
$KNbO_3$	K_2CO_3, Nb_2O_5	12 minutes 800 W	30 h 1000ºC
$BaTiO_3$	$BaCO_3$, TiO_2	25 minutes 1000 W	t>>>, grindings 1400º C
$PbTiO_3$	$PbNO_3$, TiO_2	9 minutes 600 W	8 h 360ºC

A wealth of further perovskite structures can be formed by solid state microwave synthesis, where a selected few ones are summarized in Table 2. Vaidhyanathan et al. [27] showed a decade ago that the synthesis of some niobates and titanates, among them ferroelectrics, piezoelectrics and insulators, can be performed by microwaves in a controlled fashion.

4.1.2. Complex Perovskites

The term "complex perovskites" is used here in reference to compositions that are more complicated than the ABO_3 archetype such as quaternary oxides and doped perovskites, or more complex structures such as the Ruddlesden-Popper series $A_{n+1}B_nO_{3n+1}$ or Sillen-Aurivillius phases. Perovskite superstructures are found when different species become ordered (either in the A or B sites and/or oxygen vacancies), double perovskites or some cuprate superconductors providing good examples for that behavior.

It can be stated generally that the more complex the perovskites are the more difficult it is to prepare them by microwave assisted synthesis. In more complex perovskites better diffusion processes are required to uniformly disperse 3 or more cations across the sample during the synthesis. At the current stage this still poses a major challenge for the experimental solid state chemist; nevertheless there are several cases worth mentioning of successful microwave synthesis of complex perovskites.

4.1.2.1. Cuprate Superconductors

In 1986 A. Müller and G. Bednorz discovered superconductivity above 30 K in $Ba_{0.15}La_2CuO_4$ [28], which has triggered a large surge of research into these high temperature superconductors (HTSC) with many of them being cuprates with perovskite-related structures. $YBa_2Cu_3O_7$ (YBCO) may be one of the most relevant examples, because it is the first HTSC material with a superconducting transition temperature (90 K) above the boiling point of liquid nitrogen. Shortly after its discovery, the microwave-assisted synthesis of YBCO was reported [29] and the microstructure and transport properties of YBCO zone melted samples processed in a microwave cavity have been studied [30].

Nevertheless, the literature on microwave-assisted synthesis of cuprate superconductors is scarce. Here, the microwave synthesis of superconducting La_2CuO_4 in a "single-step" process is presented using solid-state microwave assisted synthesis (as described previously). It is found that a microwave power threshold exists to successfully synthesize the phase pure compound at 800 W (Figure 10a). The resulting powder exhibited particles sized in the range of 100 - 300 nm (Figure 10b). The superconducting transition temperature in this compound was found to be at ≈ 30 K upon further oxidation [31].

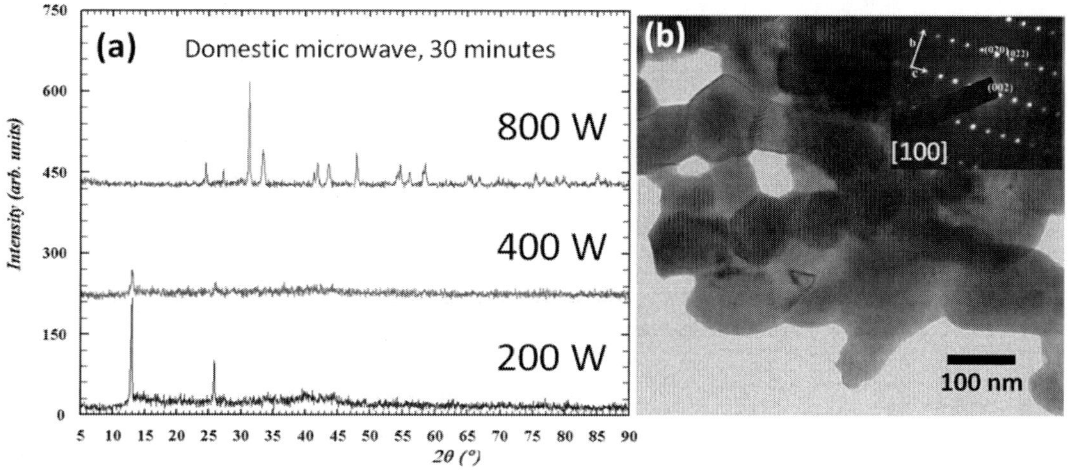

Figure 10. (a) XRD patterns of microwave synthesized La_2CuO_4 using different microwave power. The pure phase is obtained when using 800 W. (b) TEM and SAED picture (inset) of La_2CuO_4 sample after 800 W power and 30 minutes treatment.

4.1.2.2. $CaCu_3Ti_4O_{12}$

$CaCu_3Ti_4O_{12}$ (CCTO) is a further complex perovskite material that can be produced by solid state microwaves synthesis. This 1:3 A-site ordered perovskite related structure is strongly tilted and is commonly indexed with space group Im-3. It has recently attracted much attention due to its giant dielectric permittivity, which may be interesting for capacitors in microelectronic devices. It is quite well established though that the giant permittivity is not intrinsic but debited to a core-shell structure of conducting bulk and insulating grain boundaries, thus forming an Internal barrier layer (IBLC) structure [32-34]. It has been reported that solid microwave synthesis reduces the processing time and, more importantly, the IBLC structure is preserved and even higher values of the giant dielectric permittivity can be obtained [35-36]. Furthermore, it is possible to achieve higher densities in pellets and lower dielectric losses by microwave sintering.

Figure 11a shows the X-ray pattern of microwave synthesized CCTO oxide powders at different temperature for 30 min. The formation of single phase CCTO is indicated for heating at 800 °C and diffraction peaks are quite sharp indicating that the powder is well crystallized. The dielectric properties of ceramics obtained from two kinds of such microwave synthesized CCTO powders were investigated. After ceramic sintering at 1100 °C for 3 h, the relative dielectric constant from microwave (MS) powder (≈ 21400, at 1 kHz) is higher than that from conventional (CS) powder (≈ 10240, at 1 kHz) at room temperature as shown in Figure 11b. This can be attributed to a larger grain size of the ceramics obtained from MS powder as compared to the CS powder [35].

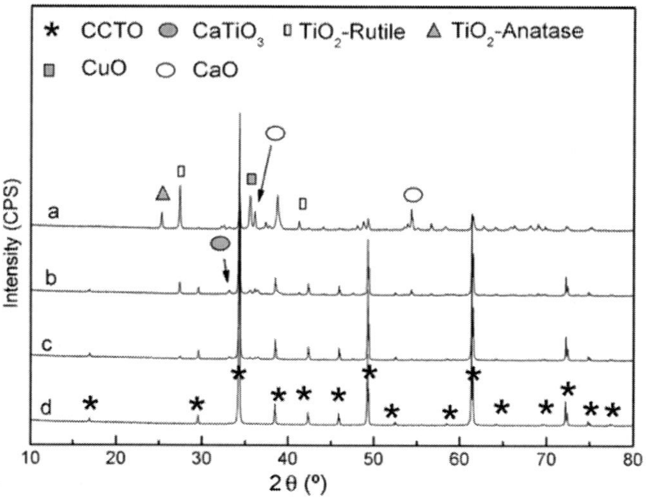

a

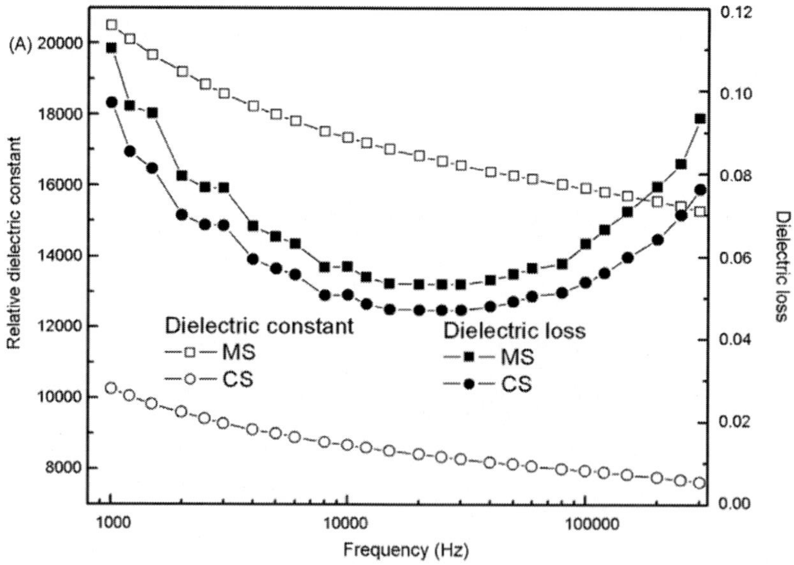

b

Figure 11. (A) XRD patterns of CCTO powder synthesized at different temperature for 30 min by microwave heating: (a) 750 C; (b) first, 800 °C; (c) second, 800 °C; (d) third, 800 °C. (B) Dielectric properties of sintered pellets from microwave and conventional powder. Images reproduced from reference [35] with permission from Elsevier.

4.2. Single Mode Solid State Microwave-Assisted Synthesis

Conventional microwave ovens constitute a multi-mode cavity configuration, where the electric (E) and magnetic (H) components of the microwave cannot be separated. Contrarily, in a single-mode cavity the E and H distributions are utilized separately by carefully choosing the position of the sample during synthesis with respect to the maxima and minima of the E and H components of the microwave (Figure 12). The single-mode microwave equipment used in this study is based on a microwave generator working at the standard 2.45 GHz frequency with a variable power up to 2 kW and a TE10p microwave cavity. The experimental set-up includs a microwave generator, circulator (water cooled magnetron), 3-stub tuner (impedance agreement accord) and the cavity. Tuning the length of the cavity – moving a coupling iris and a short circuit piston- it is possible to work in a TE102 or TE103 resonant mode (Figure 12b). In the TE102 mode the sample is placed at the maximum of the magnetic field component of the microwave (H mode), whereas in the TE103 mode the sample is placed at the maximum of the E field (E mode). During the synthesis process the sample temperature can be measured with a pyrometer [37], which consists in a non-contacting device that intercepts and measures thermal radiation.

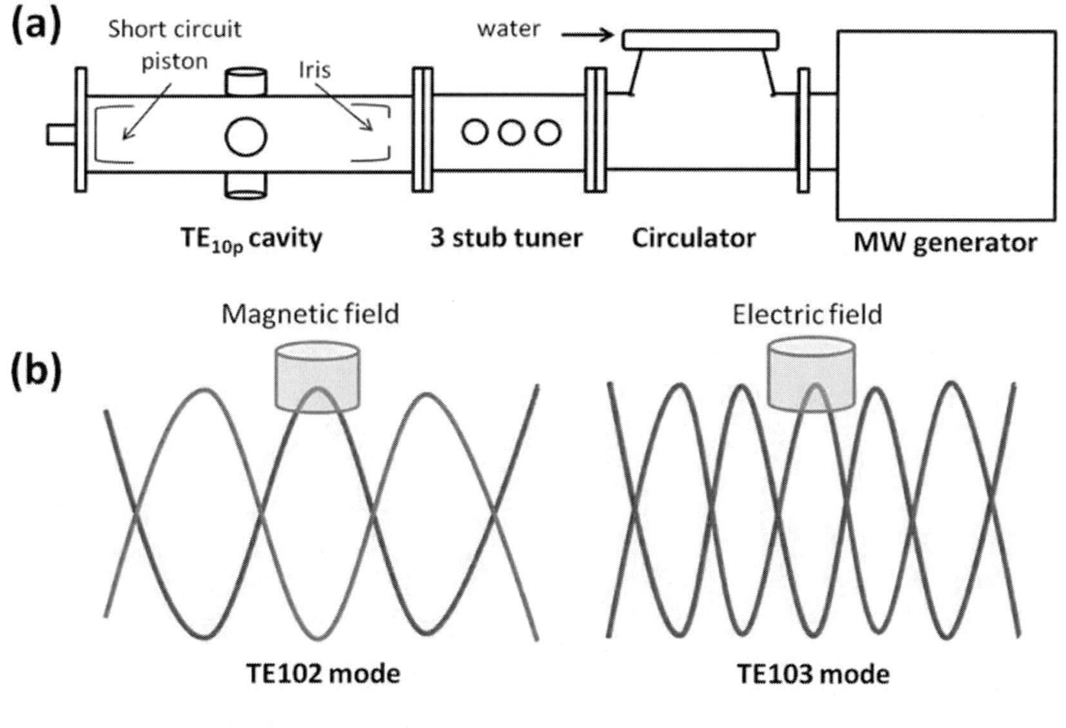

Figure 12. a) Single-mode microwave apparatus. b) Magnetic (red) and perpendicular electric (blue) field amplitude profiles in the cavity depending on the excited mode (TE102 or TE103). The cylinder represents the ideal positioning of the sample for each mode.

M. Gupta et al. [15] argued that the electric loss is dominant for dielectric materials (e.g. pure oxide ceramics) and the material is heated significantly only when placed in the E field, but displays modest or almost no heating when placed in the H field. However, for certain semiconducting materials and metals, the magnetic loss can be significant and the material can be heated in the magnetic field. For mixed systems containing more than one material the heating mode depends on the constituents added. If ceramic and metal are combined, sample heating takes place in both electric and magnetic fields.

4.2.1. Synthesis of Cubic $BaTiO_3$

$BaTiO_3$ is a well known ferroelectric material with high dielectric permittivity ($\varepsilon > 1000$) at room temperature and is one of the most widely used ceramic materials in the electronics industry, especially in multilayer ceramic capacitors.

Nano-sized cubic $BaTiO_3$ powder was prepared rapidly at 90 ºC using TiO_2 and $Ba(OH)_2$ precursors by a single-mode microwave-assisted synthesis system. $BaTiO_3$ particles of 30 to 50 nm were obtained after microwave treatment for 5 min at 90 °C (Figure 13). Conventional heating would consume approximately 7 times more energy than the microwave-assisted process at the same temperature [38].

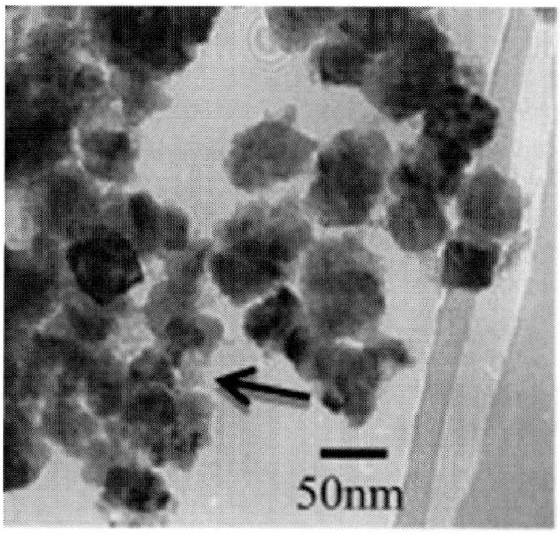

Figure 13. Morphologies of c-$BaTiO_3$ particles prepared at 90 °C for 5 min by single-mode microwave heating. Image reproduced from reference [38] with permission from Elsevier.

4.2.2. Synthesis of Lanthanum Chromite

$LaCrO_3$ perovskite was synthesized in our laboratories by a single mode microwave process (SAIREM) in the TE102 mode (H mode) in just 2 minutes with a microwave power of 150 W, where the temperature was controlled to be about 900 ºC.

As mentioned above, the H mode may be less common for ceramic materials but is feasible here due to the Co^{3+} magnetic structure providing sufficient magnetic losses. Figure 14a shows the temperature profile of the synthesis process. The starting precursors used were $La(NO_3)_3{\cdot}6H_2O$ and $Cr(NO_3)_3{\cdot}9H_2O$ in the appropriate stoichiometric ratio. The space group of the phase pure $LaCrO_3$ obtained was the expected orthorhombic Pbnm (#62) and the cell parameters from Rietveld refinement (Figure 14b) were $\boldsymbol{a}$ = 5.5092(2) Å, $\boldsymbol{b}$ = 5.4731(2) Å and $\boldsymbol{c}$ = 7.7544(3) Å.

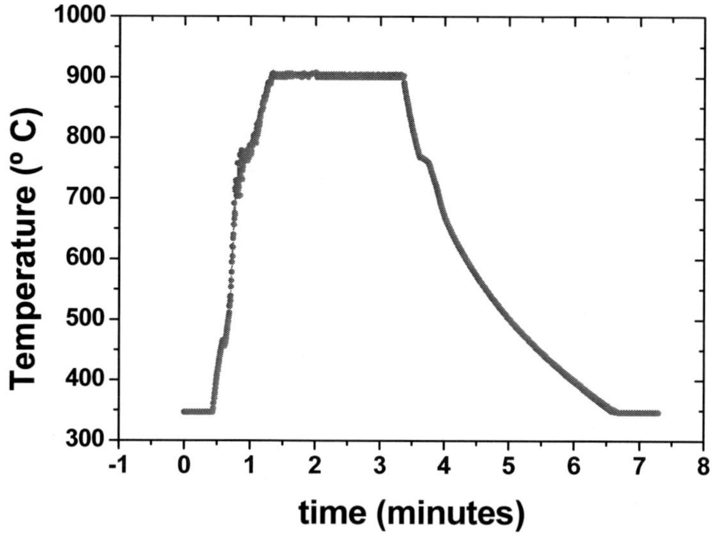

a

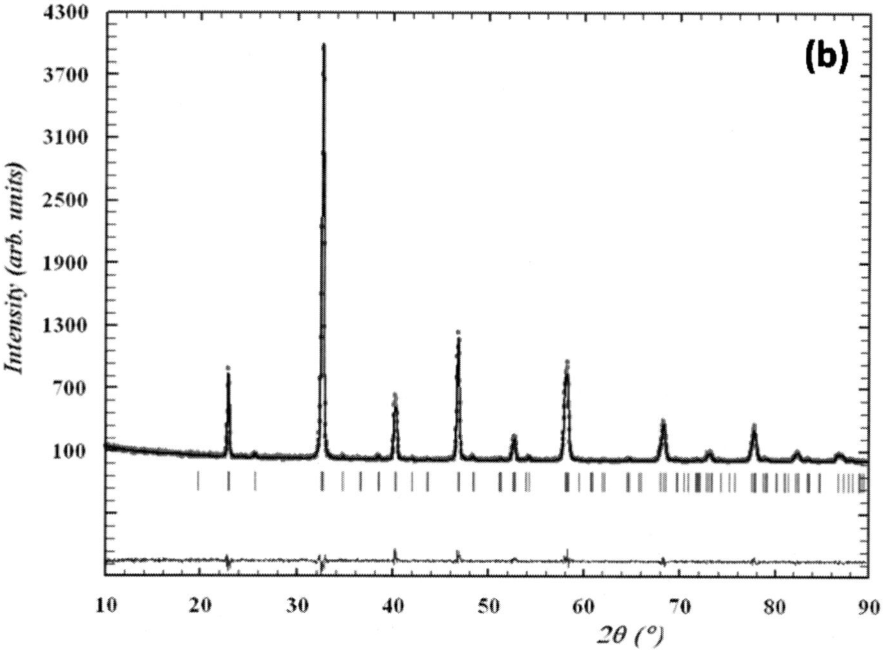

b

Figure 14. a) Temperature control of the $LaCrO_3$ synthesis process using a pyrometer. The pure phase can be obtained in 2 minutes synthesis. b) Rietveld refinement of powder X-ray diffraction patterns of $LaCrO_3$: observed (red dotted lines), refined (black solid lines), and their difference (blue bottom line). Green vertical bars indicate the X-ray reflection positions.

Similarly, D. Grossin et al. [39] demonstrated single mode microwave synthesis of $La_{0.8}Sr_{0.2}MnO_3$ using a microwave TE10p cavity, where SiC was used as a microwave susceptor to facilitate an effective synthesis process within only 10 minutes. The properties of the microwave ceramic were equivalent to the properties obtained conventionally.

4.3. Microwave Assisted / Hydrothermal Synthesis

In the early 1990s, Sridhar Komarneni at the University of Pennsylvania launched a pioneering work by studying and comparing the differences between hydrothermal syntheses performed with conventional means of heating and hydrothermal syntheses performed by heating special autoclaves with microwaves [40]. Since then many functional materials have been produced by microwave assisted hydrothermal synthesis, ranging from binary metallic oxides, oxyhydroxides, ternary oxides to more complex materials and structures such as zeolites or other mesoporous materials.

"Hydrothermal synthesis" is a “wet”, moderate-pressure method of synthesis, which consists in an aqueous solution that is heated above 100ºC in a sealed container (autoclave) and concomitantly the pressure increases. This results in an increase of the dispersion of the system components which react very quickly. It should be further noted that “solvothermal” synthesis has become increasingly popular lately, where non-aqueous solvents are used.

For the hydrothermal synthesis it is important to consider “subcritical” and “supercritical” synthesis conditions, where temperature is below or above the critical temperature of water (T_c = 374 ºC) respectively. Above this critical point the pressure in the synthesis autoclave raises very rapidly with increasing temperature, which makes the reaction process difficult to control. Besides the characteristic microwave parameters of power and time, the method of microwave-assisted hydrothermal synthesis involves additional parameters such as the media reaction, the pH of the solution, the temperature and pressure. Therefore, more advanced synthesis technology is required to control pressure and temperature in the subcritical region of water. Autoclaves are made out of Teflon, in which case the temperature is limited to 250 ºC, or out of quartz.

It should be noted that microwave assisted hydrothermal synthesis is an effective method for the production of nanoparticles, because the particle size can be controlled by tuning the mechanisms of nucleation and growth kinetics through the appropriate choice of the synthesis parameters. In the following, microwave-assisted hydrothermal synthesis of selected perovskite oxides will be presented.

4.3.1. $BiFeO_3$

$BiFeO_3$ exhibits multiferroic properties at room temperature (ferroelectric and magnetic order coexisting in the same phase) and is, therefore, a good candidate for potential multiferroic application in information technology. $BiFeO_3$ can be synthesized in crystalline form by combined microwave hydrothermal synthesis in less than 5 minutes [41-42]. Starting reactants are $Fe(NO_3)_3{\cdot}9H_2O$ and $Bi(NO_3)_3{\cdot}5H_2O$ together with 8M KOH. Microwave assisted hydrothermal synthesis was carried out at a pressure of 15 bars and a microwave power of 500W. The X-ray pattern presented in Figure 15a indicates phase pure $BiFeO_3$ (trigonal, R3c), and successful Rietveld refinement was performed. The unit cell parameters obtained were $\boldsymbol{a}$ = 5. 5799 (1) Å and $\boldsymbol{c}$ = 13. 8692 (4) Å. A detailed micro-structural study of the materials has been performed by means of HRTEM (Figure 15b) confirming the structural information obtained by X-ray diffraction.

In Figure 16 plots of magnetization vs applied magnetic field plots are presented taken at 5 K. A small hysteretic effect may be indicative of the $BiFeO_3$ typical weak ferrimagnetic moment.

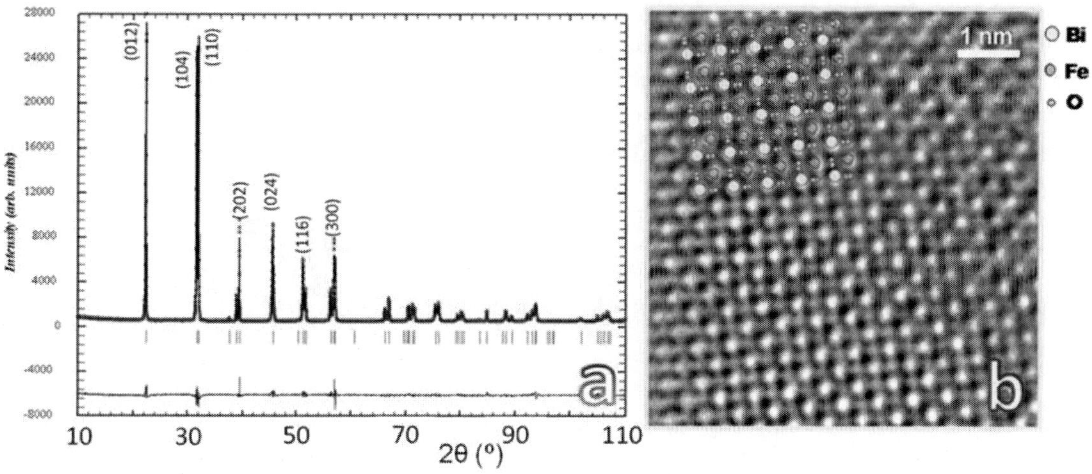

Figure 15. (a) Rietveld refinement of $BiFeO_3$ (b) HRTEM image of $BiFeO_3$.

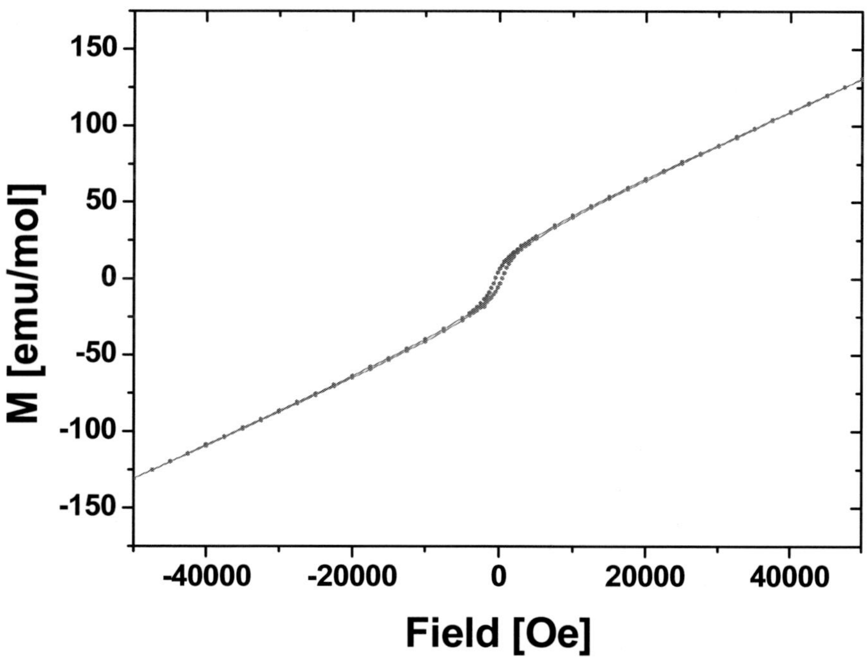

Figure 16. Magnetization vs applied magnetic field hysteresis loop of $BiFeO_3$ at 5 K.

4.3.2. $BaTiO_3$

Novel synthesis techniques have raised considerable interest recently to control the shape and size of $BaTiO_3$ particles to improve ferroelectrics properties for applications. $BaTiO_3$ perovskite was synthesized by the microwave hydrothermal route [43], starting from $Ba(NO_3)_2$ and $TiCl_4$, using KOH as mineralizer at a temperature of 150 °C in a EMC MDS-2000 microwave set-up. Figure 17 shows the X-ray powder diffraction patterns for the microwave hydrothermal $BaTiO_3$ powders prepared for 3, 12, and 20 h. The wide-angle (20°–80° 2θ) diffraction patterns of all samples indicate the single $BaTiO_3$ phase. The insets in

Figure 17 show the reflections in the 2θ range of 44°–46°, where the asymmetrical peak of the sample prepared for 3 h can be well fitted with two distinct Gaussian peaks (dotted lines) with 1:2 intensity ratio. On the other hand, the reflections in the insets for the samples synthesized for 12 and 20 h display enhanced splitting. The splitting or asymmetry of reflections in this region is a result of the distortion of the unit cell, characteristic of tetragonal $BaTiO_3$ [43].

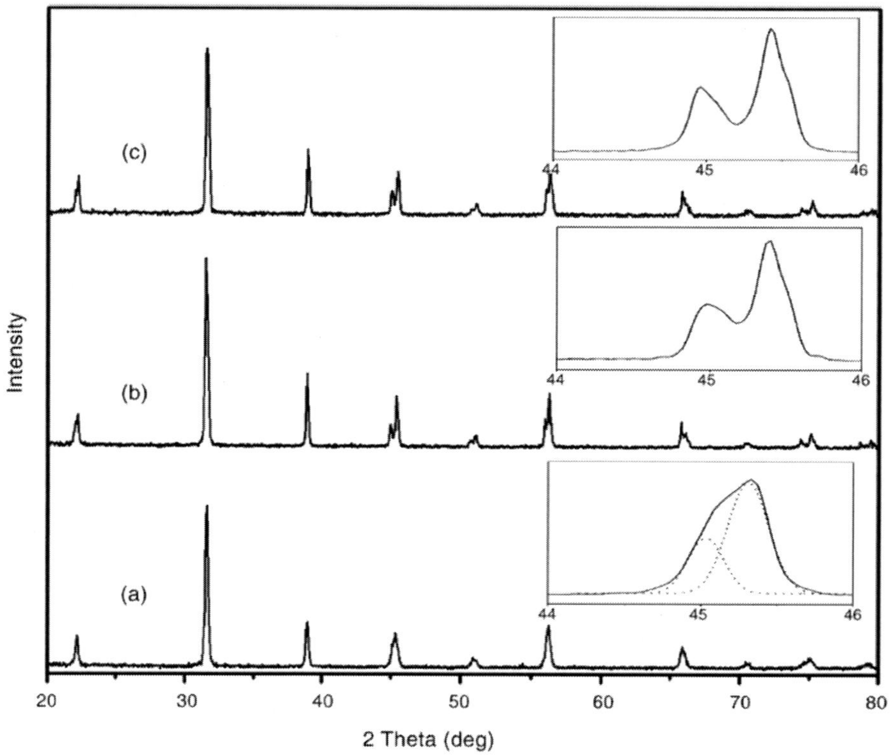

Figure 17. XRD patterns of $BaTiO_3$ powders prepared by microwave hydrothermal methods at 240 °C. The initial Ti concentration was 0.525 mol dm^{-3} with a $Ba(OH)_2$/Ti mole ratio of 3.2. The solutions were heated for different amounts of time: (a) 3h. The reflection around 45° 2θ in the inset is well fitted with two Gaussian peaks (dotted lines) with an intensity ratio of 1:2, (b) 12 h, and (c) 20 h. Image reproduced from reference [43] with permission from Elsevier.

4.3. Microwave Synthesis Combined with Sol-Gel or Combustion

The sol-gel process is a wet-chemical technique widely used in the synthesis of inorganic materials, typically metal oxides. The starting point is a chemical solution (sol) which acts as the precursor for a polymeric network (gel) formed upon hydrolysis and polycondensation reactions. Thermal decomposition of a particular gel (usually containing high amounts of water) leads to the corresponding oxide. This process was discovered already at the end of the 19th century, but was not fully understood until 1931 when Steven S. Kistler demonstrated that the networks in a dry gel are of the same nature than those existing in a moist one. The application of microwave radiation to gels has been demonstrated for the synthesis of cathode materials for solid oxide fuel cells (SOFCs) [20], such as $La_{1-x}Sr_xFeO_{3\pm\delta}$ and $La_{1-x}Sr_xFe_yCo_{1-y}O_{3\pm\delta}$ (Figure 18).

Combustion reactions for the synthesis of oxide materials usually involve the mixing of metal nitrates and a solid fuel (urea, glycine, sacharose) in adequate quantities, where the later provides the necessary energy to ignite the (exothermic) reaction. The ignition can be performed by providing the energy in the form of microwaves. The high speed of the process together with the production of large quantities of gas produce sponge-like porous materials, many times yielding nano-sized particles. For instance, perovskite $LaMnO_3$ powders with an average crystallite size of 12.5 nm can be rapidly synthesized via a microwave-induced autocombustion reaction using glycine as a fuel and nitrate as an oxidant [44].

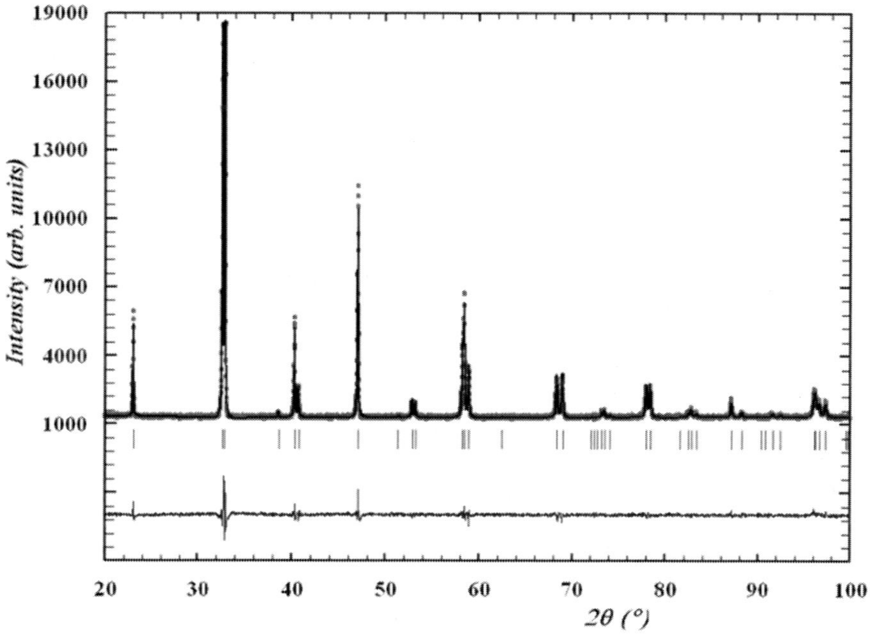

Figure 18. Rietveld refinement of $La_{0.8}Sr_{0.2}Fe_{0.5}Co_{0.5}O_3$; Space group R-3c. Cell parameters: $\boldsymbol{a} = \boldsymbol{b} = 5.4818(1)$ Å and $\boldsymbol{c} = 13.2752(3)$ Å; $R_{exp} = 2.53$.

CONCLUSION

Microwaves are increasingly used as a new synthetic route in Solid State Chemistry. The particular nature of the matter-microwave interactions leads to rapid synthesis of many inorganic materials such as perovskite oxides.

As compared to synthesis techniques where heat is transferred by convection, the reactions in microwave techniques are orders of magnitude faster. Microwave-assisted synthesis of perovskites in the solid state is often limited to “simple” compositions (ternary) but the possibility of combining microwaves with other methods such as hydrothermal synthesis, sol-gel, or combustion allows for better stoichiometric control of complex doped phases. In particular, combining microwave heating with solvothermal synthesis may result in metastable phases and novel morphologies. Single-mode microwave synthesis allows separating the magnetic and electric components and provides an accurate control of the temperature together with much shorter reaction and processing times.

The synthesis of a wide range of perovskite oxide materials was shown to be feasible by microwave techniques, where such materials include superconducting, ferromagnetic, ferroelectric, dielectric and multiferroic perovskite systems.

ACKNOWLEDGMENT

The authors acknowledge funding from the Community of Madrid (Materyener S2009/PPQ-1626) and the Ministry of Science and Innovation (MAT-2007-64006). R.S. acknowledges the Ministerio de Ciencia e Innovación (MICINN) for granting a Ramon y Cajal Fellowship. Many thanks for valuable contributions go to Ian Terry (Durham University, UK), Dra. C. Parada (UCM), Dra. M.E. Villafuerte-Castrejón (UNAM-Mexico), Dr. L. Fuentes (CIMAV-Mexico), Dr. D. Avila (UCM), Dr. S. Marinel (CRISMAT-CNRS, Caen) and I. Herrero-Ansorregui (UCM).

REFERENCES

[1] Mitchell R. H. *Perovskites Modern and Ancient*, Almaz Press Inc.: Canada, 2002; Vol. 57, 317.

[2] Hayes, B. L. *Microwave Synthesis: Chemistry at the Speed of Light*, C. E. M. publishing: 2002; 11-27.

[3] Sutton W. H. *Ceram. Bull.* 1989, 68, 376-386.

[4] Prado-Gonjal, J.; Morán, E. *An. Quím.* 2011, 107(2), 129–136.

[5] Zhao, J.; Yan, W. *Modern Inorganic Synthetic Chemistry*, 2011,8, 173-195.

[6] Clark, D. E.; Folz, D. C.; Folgar, C.; Mahmoud, M. Microwave Solutions for Ceramic Engineers, *The American Ceramics Society*, Inc.: Westerville, O. H., 2005; 5-30

[7] Gupta, M.; Leong, W. W. *Microwaves and Metals, John Wiley and sons*: Asia, 2007, 35-60.

[8] Rao, K. J.; Vaidhyanathan, B.; Ganduli, M.; Ramakrishnan P. A. *Chem. Mater.* 2009, 11, 882-895.

[9] Sahu, R. K.; Rao, M. L.; Manoharan, S. S. *Journal of Materials Science* 2001, 36, 4099-4102.

[10] Lidström, P.; Tierney, J.; Wathey, B.; Westman, *J. Tetrahedron*, 2001, 57, 9225-9283.

[11] Balaji, S.; Mutharasu, D.; Subramanian, N. S.; Ramanathan, K. *Ionics*, 2009, 15, 765-777.

[12] Parada, C.; Morán, E. *Chem. Mat.* 2006, 18, 2719-2725.

[13] Liu, S.-F.; Abothu, I. R.; Komarneni, S. *Mater. Let.* 1999, 38, 344-350.

[14] Komarneni; S.; Katsuki, H. *Pure Appli. Chem.* 2002, 9, 1537-1543.

[15] Das, S.; Mukhopadhyay, A. K.; Datta, S.; Basu, D. Bull. *Mater. Sci.*, 2009, 32, 1–13.

[16] Herrero, M. A.; Kremsner, J. M.; Kappe, C. O.; *J. Org. Chem.* 2008, 73, 36-47.

[17] Wroe, R.; Rowley, A. T. J. *Mater. Sci.* 1996, 31, 2019-2026.

[18] De la Hoz, A.; Díaz-Ortiz, A.; Moreno, A. *Chem. Soc. Rev.* 2005, 34,164-178.

[19] Chandrasekaran S., Basak T., Ramanathan S., *J. Mater. Process. Techn.* 2011, 211, 482-487.

[20] Prado-Gonjal, J.; Arévalo-López, A. M.; Morán, E. *Mat. Res. Bull.* 2011, 46, 222-230.

[21] Backhaus-Ricoult, M. *Solid State Sci.* 2008, 10, 670-688.

[22] Choi, Y.; Mebane, D. S.; Lin, M. C.; Liu, M. *Chem. Mater.* 2007, 19, 1690-1699.

[23] Petrovic, S.; Terlecki-Baricevic, A.; Karanovic, L.; Kirilov-Stefanov, P.; Zdujic, M.; Dondur, V.; Paneva, D.; Mitov, I.; Rakic, V. *Applied Catalysis B:Environmental* 2008, 79, 186-198.

[24] Sahu, J. R.; Serrao, C. R.; Ray, N.; Waghmare, U. V.; Rao, C. N. R., *J. Mater. Chem.* 2007, 17, 42-44.

[25] Serrao, C. R.; Kundu, A. K.; Krupanidhi, S. B.; Waghmare, U. V.; Rao, C. N. R. *Phys. Rev. B.* 2005, 72, 220101–220104.

[26] Prado-Gonjal, J.; Schmidt, R.; Ávila, D.; Amador, U.; Morán, E., *J. Eur. Ceram. Soc.* 2012, 32, 611–618.

[27] Vaidhyanathan, B.; Raizada, P.; Rao, K. J., *J. Mater. Sci. let.* 1997, 16, 2022-2025.

[28] Bednorz, J. G.; Müller, A. *Zeitschrift für Physik B Condensed Matter* 1986, 64, 189-193.

[29] Baghurst, D. R.; Chippindale, A. M.; Mingos, D. M. P. *Nature*, 1988, 322, 24.

[30] Marinel, S.; Bourgault, D.; Belmont, O.; Sotelo, A.; Desgardin, G. *Physica C*, 1999, 315, 205-214.

[31] Herrero-Ansorregui, I.; Prado-Gonjal, J.; Morán, E., U. C. M. master thesis, 2012 (unpublished work).

[32] Chung, S.-Y.; Kim, I.-D.; Kang S.-J. *Nature Materials* 2004, 3, 774-778.

[33] Homes, C.; Vogt, T.; Shapiro, S. M.; Wakimoto, S.; Ramirez, A. P. *Science* 2001, 293,673-676.

[34] Schmidt, R.; Stennett, M. C.; Hyatt, N. C.; Pokorny, J.; Prado-Gonjal, J.; Li, M.; Sinclair, D. C., *J. Eur. Ceram. Soc.* 2012, 32, 3313–3323.

[35] Yu, H.; Liu, H.; Luo, D.; Cao, M., *J. Mater. Proces. Tech.* 2008, 208, 145-148.

[36] Hatagalung, S. D.; Ibrahim, M. I. M.; Ahmad, Z. A. *Ceramics International* 2008, 34, 939-942.

[37] Marinel, S.; Savary, E.; Gomina, M., *J. M. P. E. E.* 2010, 44, 57-63.

[38] Katsuki, H.; Furuta, S.; Komarneni, S. *Mater. Lett.* 2012, 83, 8–10.

[39] Grossin, D.; Marinel, S.; Noudem, J.-G. *Ceramics International*, 2006, 32, 911-915.

[40] Komarneni, S.; Roy, R.; Li, Q. H. *Mat. Res. Bull.* 1992, 27, 1393-1405.

[41] Prado-Gonjal, J.; Villafuerte-Castrejón, M. E.; Fuentes, L.; Morán, E. *Mat. Res. Bull* 2009, 44, 1734-1737.

[42] Prado-Gonjal, J.; Ávila, D.; Villafuerte-Castrejón, M. E.; González-García, F.; Fuentes, L.; Gómez, R. W., Pérez-Mazariego, J. L.; Marquina, V.; Morán, E. *Solid State Sci.* 2011, 13, 2030-2036.

[43] Sun, W.; Li, C.; Li, J; Liu, W. *Mat. Chem. Phys.* 2006, 97, 481–487.

[44] Weifan, C.; Fengsheng, L.; Leili, L.; Yang, L. *J. Rare Earths*, 2006, 24, 782–787.

In: Perovskite
Editors: Jinghua Zhang and Huan Li

ISBN: 978-1-62417-800-9

Chapter 5

DEFECT STRUCTURE AND DEFECT-INDUCED EXPANSION OF PEROVSKITE OXIDES

A. Yu. Zuev and D. S. Tsvetkov*
Ural Federal University, Department of Chemistry,
Ekaterinburg, Russia

ABSTRACT

Perovskites based upon substituted $LaBO_{3-\delta}$, where B is a 3d transition metal, are the state-of-the-art materials for a variety of different high and intermediate temperature devices such as solid oxide fuel cells (SOFCs) and mixed ionic and electronic conducting (MIEC) membranes. The unique feature of the oxides is their ability to undergo both thermal expansion and that induced by the defects of oxygen nonstoichiometry in the oxide lattice. The latter is chemical or defect-induced expansion.

There are different possible mechanisms to explain the isothermal expansion observed. If formation of oxygen vacancies in oxide lattice at low oxygen partial pressures/high temperature is accompanied by the reduction of 3d metal then one of them may be based on a change of mean size of the reducible cation due to substitution of "large" $B^{(n-1)+}$ for smaller B^{n+} as a result of the reduction.

The priority purpose of this chapter is to explain the defect-induced expansion for substituted chromites, manganites and cobaltites on the basis of their defect structure using both the experimental data available in the literature and those recently obtained by the authors.

Theoretical values of the chemical expansion were computed on the basis of a relative change of the mean ionic radius. In order to enable such computation modeling analysis of the defect structure of the perovskites studied was carried out. As a result, most adequate models were found. The experimental data and those calculated on the basis of model proposed were shown to be in good agreement for all oxides investigated. Change of a spin state of cobalt in different oxidation states was found to contribute significantly in an expansion value of cobalt containing perovskites.

* Corresponding author. E-mail: Andrew.Zuev@usu.ru.

1. Introduction

Perovskites based upon substituted oxides $LaBO_{3-\delta}$, where B=*3d*-transition metal, are the state-of-the-art materials for a variety of different devices for moderate high temperature applications such as solid oxide fuel cells (SOFCs) and mixed ionic and electronic conducting (MIEC) membranes. The unique feature of the oxides is their ability to undergo both thermal expansion and that induced by the defects of oxygen nonstoichiometry in the oxide lattice. The latter is chemical or defect-induced expansion. This property is extremely sensitive to the defect structure of the oxide material. Therefore reliable data on the defect structure of oxide materials is of key importance for understanding the origin of their chemically induced lattice strain. On the other hand, chemical expansion of oxide crystal cell precisely measured with ppm-resolution allows better understanding its local electronic and defect structure.

2. Thermal and Isothermal Expansion of Solid Oxides

Thermal expansion is a fundamental property of materials which originates from anharmonicity of the atomic vibrations leading to variation of interatomic distances with temperature. This phenomenon can be quantified using the volumetric (α_V) and linear (α_L) coefficients of thermal expansion (CTEs):

$$\alpha_V = \frac{1}{V_0}\left(\frac{\partial V}{\partial T}\right)_P, \tag{1}$$

$$\alpha_L = \frac{1}{L_0}\left(\frac{\partial L}{\partial T}\right)_P, \tag{2}$$

where V_0 and L_0 are the values of volume and length in a selected initial state, and P is the total pressure. Note that, in many branches of materials science, the CTE is defined phenomenologically as

$$\varepsilon = \frac{\Delta L}{L_0} = \alpha_L \Delta T\ , \tag{3}$$

where ε is so-called uniaxial strain, and ΔL is the length increment corresponding to the temperature change ΔT. In general, CTE can be treated as a linear function of temperature as a first approximation. For engineering materials used at elevated temperatures, however, this effect should be minimized to a possible extent. In the case of complicated devices such as solid oxide fuel cells (SOFCs) consisting of several different components, a close match in their thermal expansion coefficients is the mandatory requirement.

The thermal expansion of crystalline solids depends on their lattice symmetry and may be iso- or anisotropic. For example, graphite has a layered structure and its expansion in the direction perpendicular to the layers is quite different from that along the layers. For isotropic materials, $\alpha_V \approx 3\alpha_L$ and thermal expansion of ceramics is equal to that of single crystal. In anisotropic solid materials the total volume expansion is distributed unequally among the three crystallographic axes and, hence, thermal expansion of ceramics cannot correctly reproduce that of single crystal. Monocrystalline samples should be used for dilatometric measurements in this case or, alternatively, thermal expansion should be studied using *in situ* X-ray diffraction (XRD) to determine temperature dependence of the lattice parameters.

Strelkow [1] was first who found a significant positive deviation from linear trend of ε temperature dependence for several halides such as NaCl, AgCl, and AgBr just below the melting points of these compounds. He called this phenomenon as anomalous expansion. Later Mott and Gurney [2] and also Seitz [3] discussed anomalous thermal expansion in silver halides and proposed that this anomaly in expansion should be attributed to point defects arising in the crystal lattice at elevated temperatures. However Mott, Gurney and Seitz seem to be unacquainted with the data on thermal expansion obtained earlier by Strelkow. Lawson [4] plotted the logarithm of the anomalous length increase found by Strelkow for silver halides [1] versus reciprocal absolute temperature and showed that resultant points fall reasonably well on straight lines suggesting the abnormal expansion arises from an activation process.

Further Lawson [4] compared the activation heat for abnormal expansion to that observed earlier for silver halides ionic conductivity and found good agreement between both heats which suggested that the anomalous expansion arises from the increase in the number of lattice defects which, in turn, are responsible for the ionic conductivity.

Since it was not clear before Lawson work [4] whether these defects are Schottky ones or, in opposite way, Frenkel ones it was of great interest to ascertain whether the data on expansion are capable of discriminating between two kinds of defect species.

Taking into account ionic radii Lawson compared possible values of the volume expansion which could be caused by the different disordering mechanisms and found that only those calculated on the basis of Schottky defects presence are completely consistent with those reported by Strelkow [1].

The use of CTE only is, therefore, insufficient for adequate description of solids when substantial amounts of defects form at elevated temperatures. This is the case for mixed-conducting oxides with perovskite structure the lattice volume of which is a function of both temperature and oxygen vacancies concentration.

What is why strain of their lattice can be quantified in terms of both the standard volumetric CTE (α_V) and volumetric chemical expansivity (α_C) induced by the oxygen vacancy formation [5]

$$\alpha_C = \frac{1}{V_0}\left(\frac{\partial V}{\partial X_{V_O}}\right)_{T,P}, \qquad (4)$$

where X_{Vo} is the oxygen vacancies mole fraction, and V_0 is the specific volume of an oxygen-stoichiometric composition at a given temperature. The former quantity is defined, for example, as $\delta/3$ in $ABO_{3-\delta}$ perovskite oxides where δ is the oxygen nonstoichiometry. α_C defined by Eq.(4) is valid under assumption that the chemical expansion is uniaxial. Using the aforementioned definitions, the total derivative of the uniaxial strain ε in the absence of additional pressure or mechanical forces can be given as

$$d\varepsilon(T, X_{V_O}) = \frac{1}{3}\alpha_V dT + \frac{1}{3}\alpha_C dX_{V_O}. \tag{5}$$

Eq.(5) makes it possible to determine both thermal and chemical constituents of the uniaxial strain by dilatometric measurements as shown, for instance, for Sr-substituted cobaltites $La_{1-x}Sr_xCoO_{3-\delta}$ [6].

To date, partially substituted perovskites $ABO_{3-\delta}$, where A and B are rare–earth and 3d–transition metal cations, respectively, are the most extensively studied oxides with respect to their chemical expansion. However, the thermodynamic approach given by Eq.(5) does not enable to reveal a chemical expansion mechanism.

Despite it is generally recognized that point defects are responsible for isothermal expansion the chemical expansion mechanism is still controversial topic. Different reasons such as changing Coulomb forces, atomic packing, local structure, preferred coordination, association between dopants and vacancies and others are discussed in this respect [7]. If electronic defects have localized nature then another reason may lead to chemical expansivity. Within the framework of such assumption the oxygen vacancy formation is accepted to be accompanied by the reduction of *3d*-metal cations.

As a result, average size of the B-site cations increases due to the apparent substitution of "large" $B^{(z-1)+}$ for smaller B^{z+}. Oxygen vacancy formation may also contribute to chemical expansion observed due to a change of coulomb interaction between ions. If this effect is significant then it may prevail over the effect of cation radius change [8].

One can conclude, therefore, that the question on the nature of chemical expansion in oxides including perovskite structured ones still remains open. However, experimental data on oxygen nonstoichiometry [9, 10, 11, 12, 13] and chemical expansion [14, 15, 16, 17] of several lanthanum chromites, cobaltites, and manganites are available in the literature.

The refinement of the selected lanthanum chromites, manganites, and cobaltites defect structure by means of the quantitative modeling on the basis of the available data on oxygen nonstoichiometry was, therefore, the first aim of the chapter. The second one was to explain the chemical expansion of these perovskite structured oxides using the defect structure models proposed for them.

3. Chemical Expansion Modeling

If oxide lattice chemical expansion is caused by a change of mean ionic radius due to reduction of most reducible cation then a model allowing its calculation can be developed. In order to do it the following assumptions should be accepted. (i) A closely packed lattice of oxide is formed by ions with rigid spheres. (ii) Expansion in each of three space directions is of equal value.

The latter seems to be valid only for oxides with pseudo-cubic structure such as perovskites. Then the isothermal defect-induced expansion can be computed by using the following equation

$$\frac{\Delta L}{L_0} = \frac{\sum_i \left(c_i r_i - c_{i_0} r_i\right)}{\sum_i c_{i_0} r_i}, \tag{6}$$

where c_i and r_i are the concentration and ionic radius, respectively, of an ion i of $ABO_{3-\delta}$ perovskite containing oxygen vacancies. The subscript, 0, in and L_0 denotes in Eq.(6) that these properties belong to the perovskite in a reference state e.g. with certain oxygen nonstoichiometry or without that ($\delta = 0$) at a given temperature.

The sum is taken over all atoms belonging to the perovskite formula $ABO_{3-\delta}$ and c_i refers to the molar concentration per unit formula in this compound. The molar concentration of oxygen was assumed to have the constant value of 3. In other words, oxygen vacancy radius is believed to be identical to that of oxygen ion. Substitution of appropriate concentrations in general expression Eq.(6) yields the model equation of oxide chemical expansion which enables its value calculation on the basis of the defect structure accepted for a given oxide. The crystal radii [18] should be accepted for all ionic species instead of more often used effective ionic radii because the former fit better the physical sizes of ions in solids as compared to the latter [13].

As already mentioned above in order to calculate chemical expansion of a perovskite oxide according to model Eq.(6) a modeling analysis of its defect structure has to be carried out.

4. Defect Structure Modeling

4.1. Doped Lanthanum Chromites $La_{1-x}Me_xCr_{1-y-z}Al_zMg_yO_{3-\delta}$ (Me=Ca, Sr)

The characteristic feature of the defect structure of the chromites containing Ca, Sr or Mg as a dopant is that tetravalent chromium is formed under oxidizing conditions as a result of charge compensation due to acceptor doping. The process of oxygen release from the chromite lattice under reducing conditions is, therefore, accompanied by the transition of Cr^{4+} to Cr^{3+} according to the "small polaron" reaction

$$O_O^\times + 2Cr_{Cr}^\bullet \Leftrightarrow V_O^{\bullet\bullet} + \frac{1}{2}O_2 + 2Cr_{Cr}^\times \quad K_r = \frac{[V_O^{\bullet\bullet}]\cdot[Cr_{Cr}^\times]^2 \cdot p_{O2}^{1/2}}{[O_O^\times]\cdot[Cr_{Cr}^\bullet]^2}, \tag{7}$$

where is an oxygen vacancy, and are tetra- and trivalent chromium, respectively, written in the Kröger-Vink notation for point defects that used here and hereafter.

It should be noted here, that reaction Eq. (7) is common for all chromites and does not depend on their composition, if they are able to release or uptake oxygen. The simplest case, when the equilibration reaction Eq. (7) describes completely the process of oxygen release and uptake by the chromite lattice, has been described in great detail [17]. This behavior is described as a "small polaron" model, because the electron transport (polaron conduction) in chromites can be understood in terms of a charge transfer reaction where an electron hole is transferred from a chromium ion in the high oxidation state Cr^{4+} to a neighbor ion in the low oxidation state Cr^{3+}. Substitution of the species concentrations in Eq.(7) by the use of evident relation $[V_O^{\bullet\bullet}] = \delta$, electroneutrality condition $[Cr_{Cr}^{\bullet}] = x + y - 2\delta$, and mass conservation law $[Cr_{Cr}^{\times}] = 1 - z - y - [Cr_{Cr}^{\bullet}]$, where *z* and *y* are concentration of Al and Mg, respectively, yields finally the modeling expression

$$\log(pO_2 / atm) = \frac{\sqrt{K_r} \cdot (x + y - 2\delta) \cdot \sqrt{3 - \delta}}{(1 - z - 2y - x + 2\delta) \cdot \sqrt{\delta}}, \qquad (8)$$

which allows to verify the model proposed for doped chromites $La_{1-x}Me_xCr_{1-y-z}Al_zMg_yO_{3-\delta}$ using data on their oxygen nonstoichiometry.

4.2. Perovskite structured oxides $La_{1-y}Sr_yB_{1-x}B'_xO_{3-\delta}$ (B=Mn, Fe, and Co; B′=3d–metal)

It is generally recognized that the defect structure of the oxygen deficient $LaBO_{3-\delta}$ (B=Mn, Fe, and Co) is based on the simultaneous presence of B–species in different oxidation states such as B^{3+}, B^{4+}, and B^{2+} cations in its structure. The following defect equilibria may, therefore, be written

$$O_O^{\times} + 2B_B^{\bullet} \Leftrightarrow V_O^{\bullet\bullet} + 1/2O_2 + 2B_B^{\times} \quad K_1 = \frac{[V_O^{\bullet\bullet}][B_B^{\times}]^2 p_{O_2}^{1/2}}{[O_O^{\times}][B_B^{\bullet}]^2} \qquad (9)$$

and

$$2B_B^{\times} \Leftrightarrow B_B^{\bullet} + B_B^{/} \quad K_2 = \frac{[B_B^{\bullet}][B_B^{/}]}{[B_B^{\times}]^2}, \qquad (10)$$

where $[B_B^{\times}], [B_B^{\bullet}],$ and correspond to concentration of B^{3+}, B^{4+}, and B^{2+}, respectively.

Reactions (9) and (10) completely describe the defect structure of oxide compositions without other *3d*-metal. For oxides with partial substitution of other *3d*-metal for a host metal on B–site the defect structure can be more complicated due to an electron exchange between this metal and a host metal. However, different simplifications are acceptable depending on the nature of a dopant and a host metal. For instance, for a lanthanum cobaltite simultaneously doped with strontium on A–site and nickel on B–site we can accept as a first approximation that *Ni* keeps constant oxidation state 2+ over all range of p_{O2} and T investigated as a more electronegative element as compared to cobalt. In contrast to that, we can treat iron as cation Fe^{4+} in a lanthanum cobaltite simultaneously doped with strontium on A–site and iron on B–site since that is less electronegative element as compared to cobalt.

Model of the defect structure can be represented by a set of equations consisting of expressions for equilibrium constants of appropriate defect reactions and necessary mass balance and electroneutrality conditions along with the definition of oxygen nonstoichiometry as $[V_O^{\bullet\bullet}] = \delta$. The following set of independent equations describing the selected defect model can be given, for instance, for the doped lanthanum cobaltite $La_{1-x}Sr_xCo_{1-y}Ni_yO_{3-\delta}$

$$\begin{cases} K_1 = \dfrac{p_{O_2}^{1/2}[V_O^{\bullet\bullet}][Co_{Co}^{\times}]^2}{[O_O^{\times}][Co_{Co}^{\bullet}]^2} = K_1^0 \exp\left(-\dfrac{\Delta H_1}{RT}\right) \\ K_2 = \dfrac{[Co_{Co}^{\bullet}][Co_{Co}']}{[Co_{Co}^{\times}]^2} = K_2^0 \exp\left(-\dfrac{\Delta H_2}{RT}\right) \\ [Sr_{La}'] + [Co_{Co}'] + [Ni_{Co}'] = [Co_{Co}^{\bullet}] + 2[V_O^{\bullet\bullet}] \\ [Co_{Co}^{\bullet}] + [Co_{Co}^{\times}] + [Co_{Co}'] + [Ni_{Co}'] = 1 \\ [Ni_{Co}'] = y \\ [V_O^{\bullet\bullet}] = \delta \\ [O_O^{\times}] = 3 - \delta \\ [Sr_{La}'] = x \end{cases} \tag{11}$$

This set can be definitely solved analytically. Obvious advantage of an analytical solution is an explicit form of the fit expression. For instance, a solution of the set [11] with physical meaning for $La_{1-x}Sr_xCoO_{3-\delta}$ (y=0) is the following

$$[Co_{Co}^{\bullet}] = \frac{4K_2(1-2\delta+x)+2\delta-x-\sqrt{A(\delta)}}{2(4K_2-1)}, \tag{12}$$

where $A(\delta) = 16K_2\delta(x-\delta) + 4\delta(\delta-x) + 4K_2(1-x^2) + x^2$, K_2 is the equilibrium constant for the reaction (10) written for B=Co. The further appropriate substitutions yields the necessary fit expression

$$p_{O_2}^{1/4} = \frac{\sqrt{K_1}\sqrt{3-\delta}(4K_2(1-2\delta+x)+2\delta-x-\sqrt{A(\delta)})}{2\sqrt{\delta}(-1+\sqrt{A(\delta)})}, \tag{13}$$

where K_1 is the equilibrium constant for reaction (9) written for B=Co. Since the changes in oxygen stoichiometry of oxides are usually measured within relatively narrow temperature range defect formation enthalpies can be treated as constants over complete temperature range investigated.

This assumption allows substituting equilibrium constants in appropriate fit equation, for instance, Eq. (13), by their temperature dependences as following

$$K_i = K_i^0 \exp(-\frac{\Delta H_i}{RT}) \tag{14}$$

and to treat the data on oxygen nonstoichiometry obtained at different temperatures simultaneously as $\delta=f(pO_2, T)$ according to the model of the defect structure proposed.

5. Defect Structure Analysis

5.1. Doped Lanthanum Chromites $La_{1-x}Me_xCr_{1-y-z}Al_zMg_yO_{3-\delta}$ (Me=Ca, Sr)

Oxygen nonstoichiometry of the lanthanum chromites selected measured in Ref. [13] by thermogravimetric technique is given, as an example, in Figs. 1-3 as a simultaneous function of p_{O2} and temperature. The results of nonlinear surface fitting for these chromites using the model described by Eq. (8) are also shown in Figures 1-3. As seen the model proposed fits the experimental data on oxygen nonstoichiometry really good over the complete p_{O2} and temperature ranges investigated. The nonlinear surface fitting parameters determined for all chromites investigated according to the models proposed for their defect structure are summarized in Table I.

Table 1. The results of the defect structure analysis according to the models described in the text

Compound	K_i	ΔH_i, kJ/mol	lnK_i^0	R^2
$La_{0.9}Ca_{0.1}CrO_{3-\delta}$	K_r	295.6	9.9	0.998
$LaCr_{0.79}Mg_{0.05}Al_{0.16}O_{3-\delta}$	K_r	242.4	6.8	0.98
$La_{0.95}Ca_{0.05}Cr_{0.84}Al_{0.16}O_{3-\delta}$	K_r	231.8	5.9	0.999
$LaMnO_{3-\delta}$	K_1	384.8	20.7	0.999
	K_2	-7.2	-5.0	
$LaCoO_{3-\delta}$	K_1	694.3	53.0	0.992
	K_2	517.7	45.0	
$(La_{0.7}Sr_{0.3})_{0.99}Co_{0.9}Ni_{0.1}O_{3-\delta}$	K_1	108.392	7.257	0.99
	K_2	93.948	4.706	
$(La_{0.6}Sr_{0.4})_{0.99}Co_{0.9}Ni_{0.1}O_{3-\delta}$	K_1	106.490	8.617	0.97
	K_2	90.375	4.641	
$La_{0.8}Sr_{0.2}CoO_{3-\delta}$	K_1	124.975	9.827	0.99
	K_2	76.484	2.876	
$La_{0.6}Sr_{0.4}CoO_{3-\delta}$	K_1	93.628	8.172	0.99
	K_2	87.354	4.178	
$La_{0.3}Sr_{0.7}CoO_{3-\delta}$	K_1	88.049	8.349	0.99
	K_2	134.023	8.138	

As seen the experimental data coincide completely with the model. The values of equilibrium constant K_r of defect reaction (7) obtained as a result of the successful fitting procedure allows to calculate the concentration of all defect species defined within the framework of the defect structure model as a simultaneous function of oxygen partial pressure (or oxygen nonstoichiometry) and temperature. These concentrations will be employed further in order to calculate chemical expansion of the chromites according to the model equation (6).

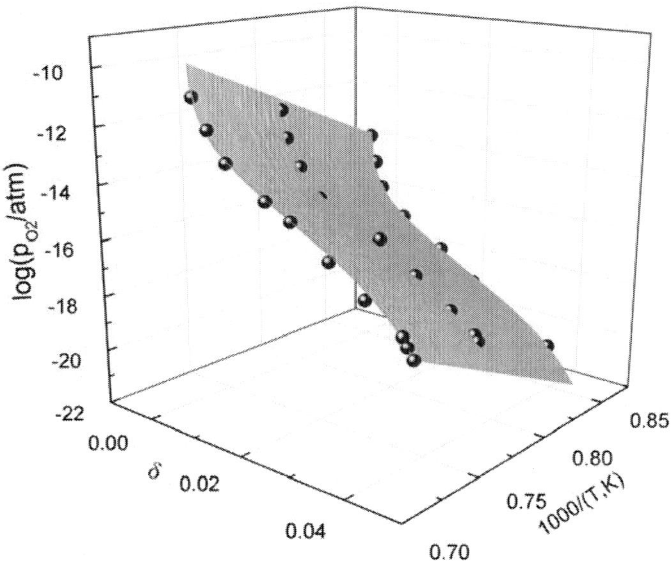

Figure 1. Oxygen nonstoichiometry of $La_{0.9}Ca_{0.1}CrO_{3-\delta}$ vs. p_{O2} and T. Points – experimental data, surface – fitted model.

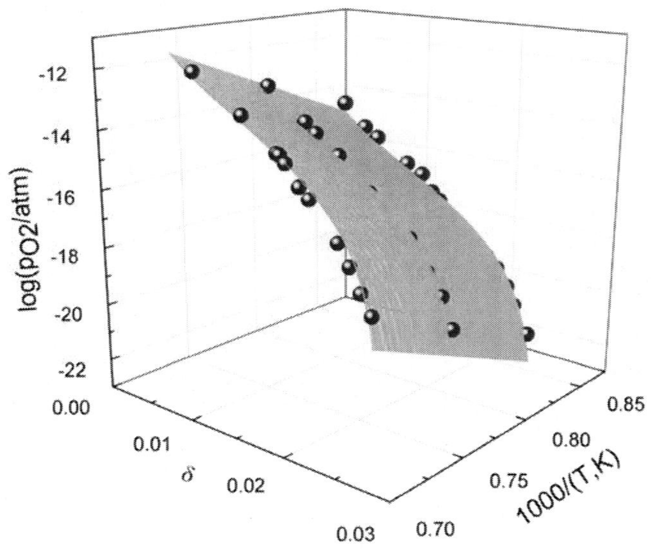

Figure 2. Oxygen nonstoichiometry of $LaCr_{0.79}Mg_{0.05}Al_{0.16}O_{3-\delta}$ vs. p_{O2} and T. Points – experimental data, surface – fitted model.

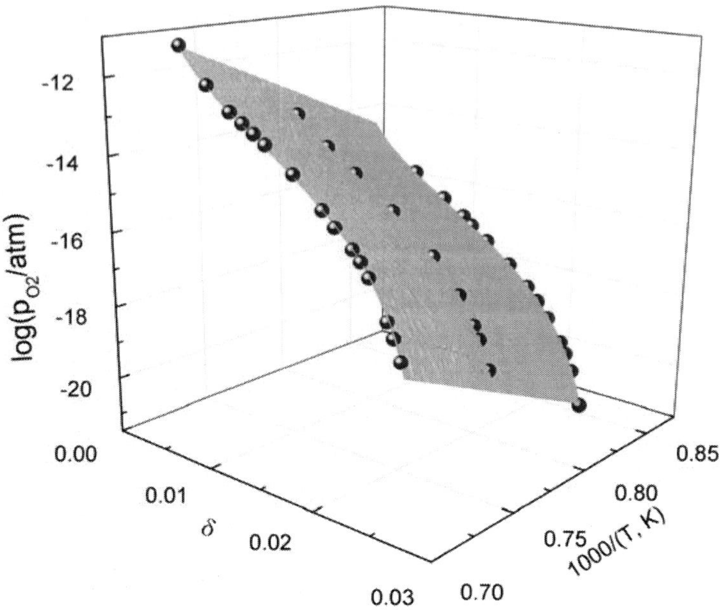

Figure 3. Oxygen nonstoichiometry of $La_{0.95}Ca_{0.05}Cr_{0.84}Al_{0.16}O_{3-\delta}$ vs. p_{O2} and T. Points – experimental data, surface – fitted model.

5.2. Perovskite Structured Oxides $La_{1-y}Sr_yB_{1-x}B'_xO_{3-\delta}$ (B=Mn, Fe, and Co; B$'$=3d–Metal)

Oxygen nonstoichiometry of the undoped $LaMnO_{3\pm\delta}$ and $LaCoO_{3-\delta}$ measured as a function of pO_2 at different temperatures in Ref. [12] and Ref. [13], respectively, by coulometric technique is given in Figures 4 and 5, respectively. The results of nonlinear least square fitting for $LaMnO_{3\pm\delta}$ in oxygen deficit region and $LaCoO_{3-\delta}$ using the model described by Eq. (13) along with Eq. (14) are shown *ibidem*. In addition, the fitted parameters determined for this model along with a goodness of fit criteria, R-Squared, are summarized in Table 1. As seen the model proposed fits the experimental data on oxygen nonstoichiometry for both $LaMnO_{3-\delta}$ and $LaCoO_{3-\delta}$ really excellent over the complete p_{O2} and temperature ranges investigated. The parameters of temperature dependence of appropriate equilibrium constants obtained as a result of the successful fitting procedure enable to calculate the concentration of all defect species defined within the framework of the defect structure model for $LaMnO_{3-\delta}$ $LaCoO_{3-\delta}$ and as a simultaneous function of oxygen partial pressure (or oxygen nonstoichiometry) and temperature. Oxygen nonstoichiometry of the doped lanthanum cobaltites $La_{1-x}Sr_xCoO_{3-\delta}$ ($x = 0.2$, 0.4 and 0.7) and $(La_{1-x}Sr_x)_{0.99}Co_{0.9}Ni_{0.1}O_{3-\delta}$ ($x = 0.3$ and 0.4) measured in Ref. [6] by coulometric technique and in Ref. [11] by TG one, respectively, is given in Figures 6–10 as a simultaneous function of p_{O2} and temperature.

The results of nonlinear surface fitting for these cobaltites using the models proposed are also shown in Figures 6–10. One can see that the model surfaces are in good agreement with the experimental data on oxygen nonstoichiometry over the complete p_{O2} and temperature ranges investigated for all doped lanthanum cobaltites studied like it was for the undoped

$LaCoO_{3-\delta}$. The fitted parameters of these models along with a goodness fit criteria, R-Squared, are also summarized in Table 1.

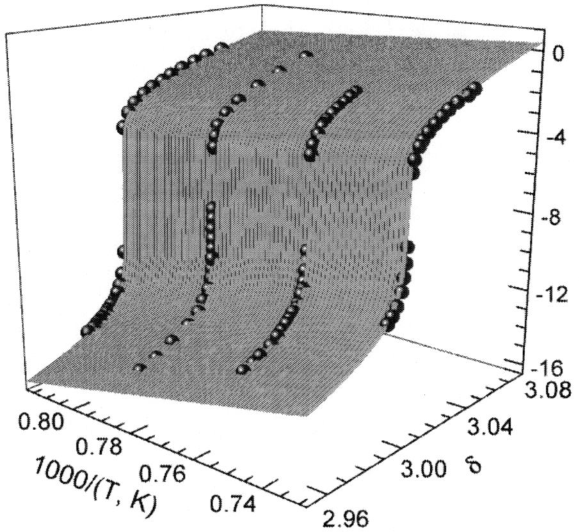

Figure 4. Oxygen nonstoichiometry of $LaMnO_{3\pm\delta}$ vs. p_{O2} and T. Points – experimental data, surface – fitted model.

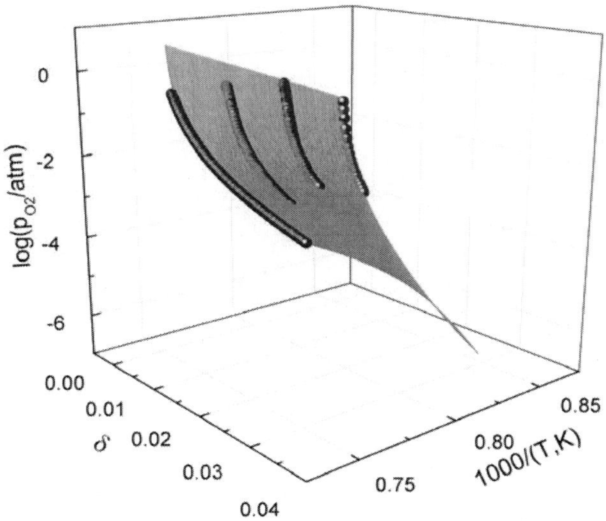

Figure 5. Oxygen nonstoichiometry of $LaCoO_{3-\delta}$ vs. p_{O2} and T. Points – experimental data, surface – fitted model.

The parameters of temperature dependences of equilibrium constants K_1 and K_2 enable calculation of the concentrations of all defect species defined within the framework of the defect structure verified. These concentrations will be employed further in order to calculate chemical expansion of the lanthanum cobaltites selected according to the model equation (6).

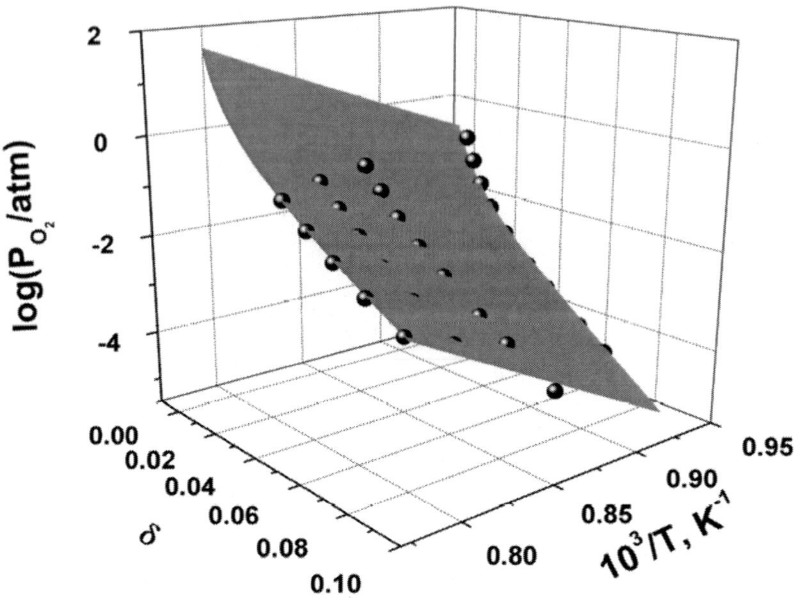

Figure 6. Oxygen nonstoichiometry of $La_{0.8}Sr_{0.2}CoO_{3-\delta}$ vs. p_{O2} and T. Points – experimental data, surface – fitted model. (Reproduced with permission from *J. Electrochem. Soc.*, 159, F594 (2012). Copyright 2012, The Electrochemical Society).

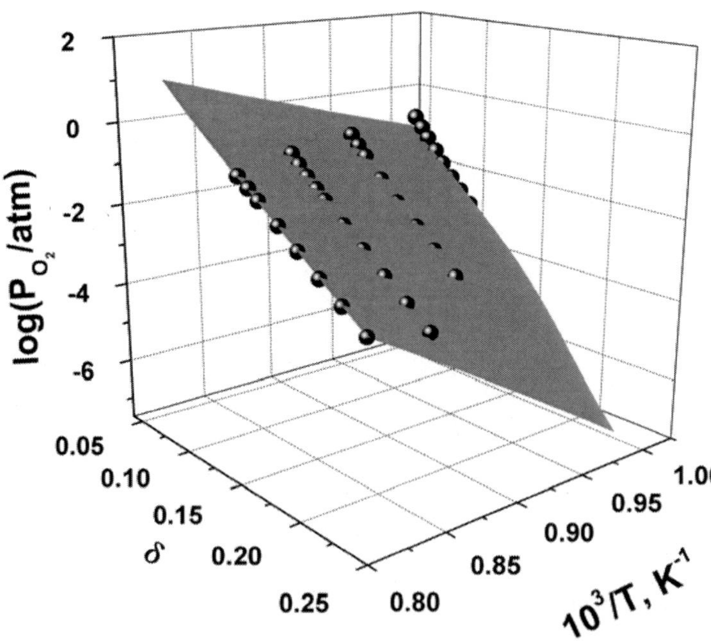

Figure 7. Oxygen nonstoichiometry of $La_{0.6}Sr_{0.4}CoO_{3-\delta}$ vs. p_{O2} and T. Points – experimental data, surface – fitted model. (Reproduced with permission from *J. Electrochem. Soc.*, 159, F594 (2012). Copyright 2012, The Electrochemical Society).

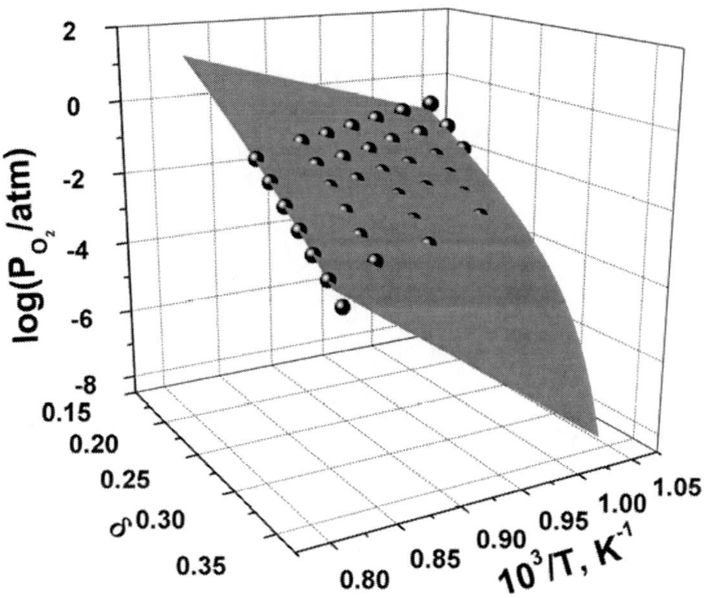

Figure 8. Oxygen nonstoichiometry of $La_{0.3}Sr_{0.7}CoO_{3-\delta}$ vs. p_{O2} and T. Points – experimental data, surface – fitted model. (Reproduced with permission from *J. Electrochem. Soc.*, 159, F594 (2012). Copyright 2012, The Electrochemical Society).

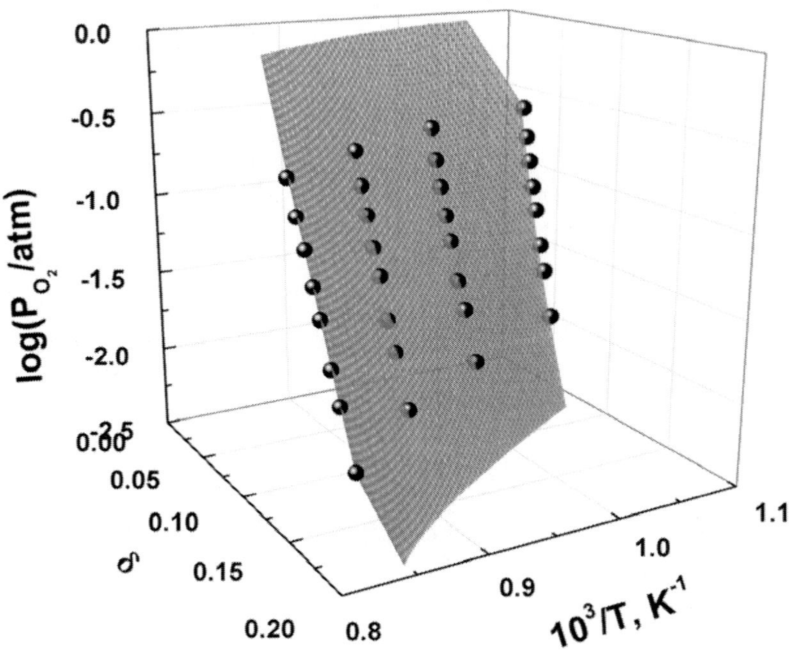

Figure 9. Oxygen nonstoichiometry of $(La_{0.6}Sr_{0.4})_{0.99}Co_{0.9}Ni_{0.1}O_{3-\delta}$ vs. p_{O2} and T. Points – experimental data, surface – fitted model. (Reproduced with permission from *J. Electrochem. Soc.*, 159, F594 (2012). Copyright 2012, The Electrochemical Society).

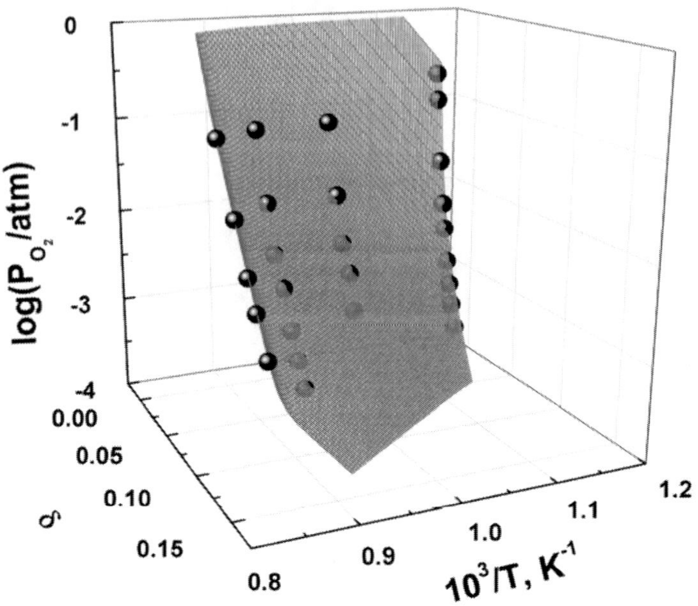

Figure 10. Oxygen nonstoichiometry of $(La_{0.7}Sr_{0.3})_{0.99}Co_{0.9}Ni_{0.1}O_{3-\delta}$ vs. p_{O2} and *T*. Points – experimental data, surface – fitted model. (Reproduced with permission from *J. Electrochem. Soc.*, 159, F594 (2012). Copyright 2012, The Electrochemical Society).

6. Chemical Expansion Analysis

6.1. Doped Lanthanum Chromites $La_{1-x}Me_xCr_{1-y-z}Al_zMg_yO_{3-\delta}$ (Me=Ca, Sr)

The chemical expansion of selected chromites $La_{1-x}Me_xCr_{1-y-z}Al_zMg_yO_{3-\delta}$ (Me=Ca, Sr) measured at 1273 K [15, 16, 17] is given in Figure 11 as a deviation from the reference state normalized to the length of a sample with stoichiometric composition regarding oxygen ($\delta_0 = 0$) plotted versus oxygen nonstoichiometry. This length was accepted as a reference state as well. The remarkable features of the dependences shown in Figure 11 are the following: (i) they are linear in shape and (ii) they all belong to a common line.

Taking into account the defect structure model accepted for $La_{1-x}Me_xCr_{1-y-z}Al_zMg_yO_{3-\delta}$ (M=Sr, Ca) appropriate concentrations can be substituted in the model Eq. (6) and the latter can be transformed, for instance, to the following equation for z = 0 and y=0

$$\frac{\Delta L}{L_0} = \frac{(r_{Cr^{3+}} - r_{Cr^{4+}})2\delta}{(r_{Cr^{4+}} - r_{Cr^{3+}} - r_{La^{3+}} + r_{Me^{2+}})x + r_{Cr^{3+}} + r_{La^{3+}} + 3r_{O^{2-}}}, \quad (15)$$

where *x* is a dopant concentration.

As follows from Eq. (15) predicted chemical expansion of doped lanthanum chromites (i) is really a linear function with respect to their oxygen nonstoichiometry and (ii) it does not depend on temperature. The slope of the linear relationship Eq. (15) depends on the doping level of a lanthanum chromite in strict sense. However, calculation according to Eq. (15)

using crystal radii values r_{Cr}^{3+}=0.755 Å *(CN=6)*, r_{Cr}^{4+}=0.69 Å *(CN=6)*, and r_{Ca}^{2+}=1.48 Å *(CN=12)* [18] yields, for instance, the slope values of 0.02157 and 0.02163 for x=0.1 and 0.3 in $La_{1-x}Ca_xCrO_{3-\delta}$, respectively. Replacement of Sr (r_{Sr}^{2+}=1.58 Å *(CN=12)* [18]) for Ca results in the slope values of 0.02154 and 0.02153 for x=0.1 and 0.3, respectively. By comparing these results, it is evident that the computed oxygen nonstoichiometry coefficient of isothermal expansion can be regarded as a constant irrespective of the dopant nature and level. As seen in Figure 11 the predicted expansion behavior according to Eq. (15) completely coincides with that observed for the doped lanthanum chromites. Calculated values of chemical expansion are seen to be in very good agreement with those experimentally measured.

Thus aforementioned agreement gives rise to draw a conclusion that chemical lattice expansion of the doped lanthanum chromites observed with deviation from their stoichiometric composition with respect to oxygen is solely caused by the mean ionic radius increase.

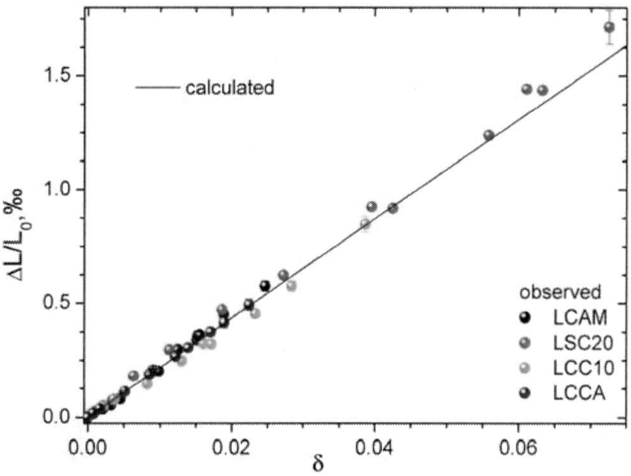

Figure 11. Chemical expansion of $La_{1-x}Me_xCr_{1-y-z}Al_zMg_yO_{3-\delta}$ (Me=Ca, Sr) as a function of the oxygen nonstoichiometry at 1273K, normalized as described in the text. The points represent experimental values and line is given by Eq. (15). LCAM – $LaCr_{0.79}Al_{0.16}Mg_{0.05}O_{3-\delta}$; LSC20 – $La_{0.8}Sr_{0.2}CrO_{3-\delta}$; LCC10 – $La_{0.9}Ca_{0.1}CrO_{3-\delta}$; LCCA – $La_{0.95}Ca_{0.05}Cr_{0.84}Al_{0.16}O_{3-\delta}$.

6.2. Perovskite Structured Oxides $La_{1-y}Sr_yB_{1-x}B'_xO_{3-\delta}$ (B=Mn, Fe, and Co; B′=3d–metal)

The chemical expansion measured at different temperatures and normalized to stoichiometric composition with respect to oxygen (δ=0) for the undoped lanthanum cobaltite [14] and manganite [12] is given as a function of oxygen nonstoichiometry in Figures 12 and 13, respectively. It is noteworthy that dependences shown in Figures 12 and 13 are not linear in shape and chemical expansion values measured at the same value of oxygen nonstoichiometry but at different temperatures differ from each other in strict sense.

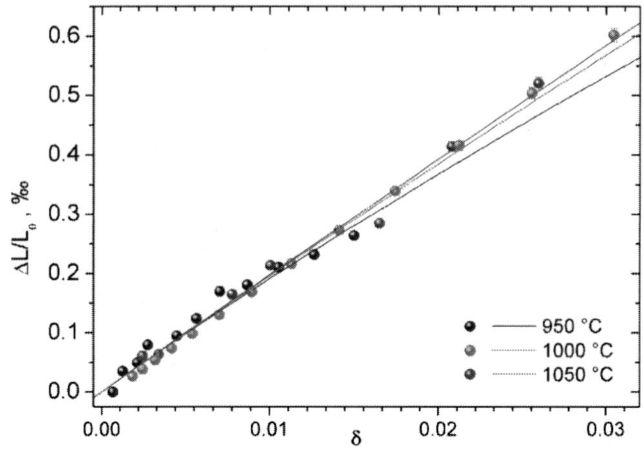

Figure 12. Chemical expansion of $LaCoO_{3-\delta}$ as a function of the oxygen nonstoichiometry at different temperatures, normalized as described in the text. The points represent experimental values and lines are given by Eq. (16). (Reproduced with permission from Solid State Ionics. 179 (2008) 1876. Copyright 2008, Elsevier).

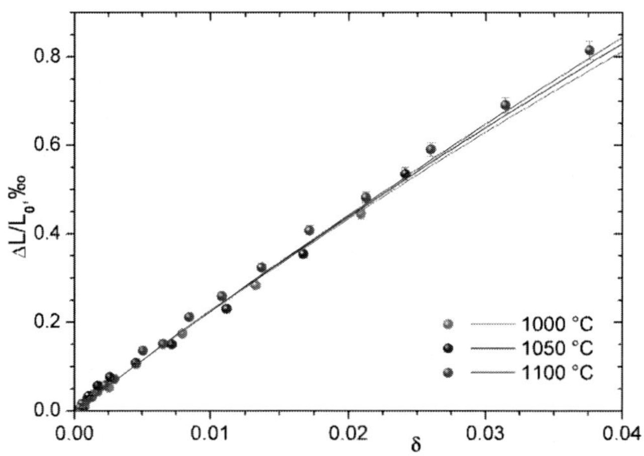

Figure 13. Chemical expansion of $LaMnO_{3-\delta}$ as a function of the oxygen nonstoichiometry, normalized as described in the text. The points represent experimental values and lines are given by Eq. (16). (Reproduced with permission from Solid State Ionics. 181 (2010) 557. Copyright 2010, Elsevier).

Aforementioned particularities obviously indicate that the chemical expansion behavior for the undoped lanthanum cobaltite and manganite does not coincide with that observed for the doped lanthanum chromites and discussed in section 6.1.

Taking into account the defect structure model accepted for undoped $LaBO_{3-\delta}$ (B=Mn, Co) appropriate concentrations can be substituted in the model Eq. (6) and the latter can be transformed to the following equation

$$\frac{\Delta L}{L_0}=\frac{\left(2r_{B^{3+}}-r_{B^{4+}}-r_{B^{2+}}\right)\cdot\left(\sqrt{\delta^2-4K_2\delta^2+K_2}-\sqrt{K_2}\right)+\delta(4K_2-1)\cdot\left(r_{B^{2+}}-r_{B^{4+}}\right)}{\left(2\sqrt{K_2}-1\right)\left(r_{B^{3+}}+\sqrt{K_2}\left(r_{B^{4+}}+r_{B^{2+}}\right)\right)+(4K_2-1)\left(3r_{O^{2-}}+r_{La^{3+}}\right)},\tag{16}$$

where K_2 is the equilibrium constant of the reaction (10).

Two particularities can be highlighted with respect to the model function Eq. (16). The first one consists in that chemical expansion calculated according to Eq. (16) is not a linear function of oxygen nonstoichiometry unlike that for the doped lanthanum chromites. The second particularity is that function may change distinctly with temperature, if equilibrium constant K_2 depends on temperature significantly. It is worth to note that, generally speaking, model equation (16) can be fitted to the experimental data on the chemical expansion of $LaBO_{3-\delta}$ given in Figures 4 and 5 for B=Mn and Co, respectively. However, computation of a chemical expansion using the values of the equilibrium constant K_2 determined independently on the basis of oxygen nonstoichiometry measurements seems to be more preferable, since such approach allows to reveal unambiguously whether the isothermal expansion of $LaBO_{3-\delta}$ is solely caused by the dimensional effect. The parameters of temperature dependences of K_2 given in Table 1 were, therefore, employed for model calculations according to Eq. (16). The crystal ionic radii [18] accepted for all ions along with their coordination numbers are: r_O^{2-} =1.26 Å *(CN=6)*, r_{La}^{3+}=1.50 Å *(CN=12)*, r_{Mn}^{2+}=0.81 Å *(LS, CN=6)*, r_{Mn}^{3+}=0.785 Å *(HS, CN=6)*, r_{Mn}^{4+}=0.67 Å *(CN=6)*, r_{Co}^{2+}=0.79 Å *(LS, CN=6)*, r_{Co}^{3+}=0.75 Å *(HS, CN=6)*, and r_{Co}^{4+}=0.67 Å *(HS, CN=6)* where *LS* and *HS* correspond to the low and high spin states, respectively.

The chemical expansion calculated using Eq. (16) is shown in Figures 12 and 13 for $LaMnO_{3-\delta}$ and $LaCoO_{3-\delta}$, respectively. These figures reveal obviously that there is very good agreement between the experimental data and those predicted on the basis of the model proposed for both perovskites. Such coincidence implies obviously that the isothermal expansion of $LaBO_{3-\delta}$ (B=Mn, Co) observed with small deviation from their stoichiometric composition with respect to oxygen at a given temperature is solely caused by the mean ionic radius increase. The latter occurs due to the partial substitution of "large" $B^{(n-1)+}$ cation for smaller B^{n+} that as a result of the reduction. On the other hand, aforementioned agreement seems to be regarded as an additional evidence of the localized nature of electronic defects or, in other words, small polarons existence in the undoped lanthanum manganite and cobaltite.

The chemical expansion of $La_{1-x}Sr_xCoO_{3-\delta}$ (x = 0.2, 0.4, 0.7) measured at 1164 K and that of $La_{0.6}Sr_{0.4}CoO_{3-\delta}$ measured at different temperatures by Chen et al. [6] is given in Figures 14 and 15 as a deviation from the reference state normalized to the length of a sample with stoichiometric composition regarding oxygen ($\delta_0 = 0$) plotted versus oxygen nonstoichiometry. This length was accepted as a reference state as well.

As seen in Figure 14 the oxygen nonstoichiometry dependences of the chemical expansion are curved in shape and they all belong to a common curve. The former indicates the presence of a significant concentration of divalent cobalt and the latter shows obviously that association between the dopant and oxygen vacancies does not play any role in chemical expansion observed for these cobaltites. It follows from Figure 15 that a chemical expansion value measured for $La_{0.6}Sr_{0.4}CoO_{3-\delta}$ at higher temperature exceeds that measured at lower temperature but at the same value of oxygen nonstoichiometry.

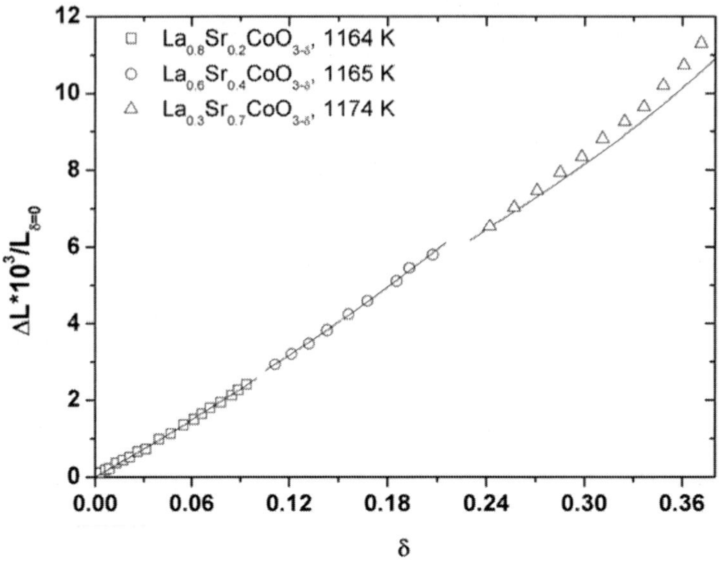

Figure 14. Chemical expansion of $La_{1-x}Sr_xCoO_{3-\delta}$ (x=0.2, 0.4, 0.7) as a function of the oxygen nonstoichiometry at 1164-1174 K, normalized as described in the text. The points represent experimental values and lines are given by Eqs. (6), (17) and (18). (Reproduced with permission from *J. Electrochem. Soc.*, 159, F594 (2012). Copyright 2012, The Electrochemical Society).

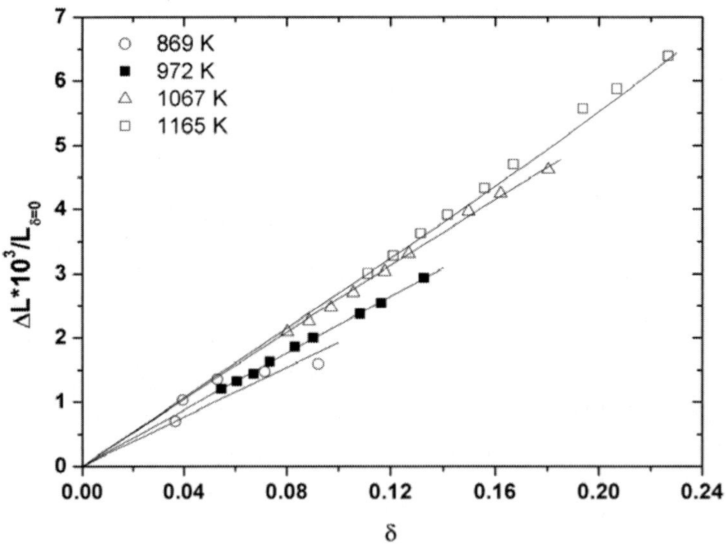

Figure 15. Chemical expansion of $La_{0.6}Sr_{0.4}CoO_{3-\delta}$ as a function of the oxygen nonstoichiometry at different temperatures, normalized as described in the text. The points represent experimental values and lines are given by Eqs. (6), (17) and (18). (Reproduced with permission from *J. Electrochem. Soc.*, 159, F594 (2012). Copyright 2012, The Electrochemical Society).

Such increase of chemical expansion observed with increasing temperature cannot be explained by possible influence of temperature dependence of equilibrium constant, K_2, of cobalt disproportionation reaction (10) since this dependence is really weak for the strontium substituted lanthanum cobaltites in contrast to the undoped $LaCoO_{3-\delta}$ as it follows from appropriate enthalpies comparison (see Table 1). On the other hand, it is known [19] that cobalt may exist not only in different oxidation states but also in different spin states and ionic radius of high spin (HS) cobalt species in certain oxidation state is larger significantly than that of low spin (LS) cobalt species [18].

In order to calculate the chemical expansion of $La_{1-x}Sr_xCoO_{3-\delta}$ (x = 0.2, 0.4, 0.7) according to the model Eq. (6) it is, therefore, necessary to decide which spin state should be accepted for Co^{2+} and Co^{3+} in the lattice of $La_{1-x}Sr_xCoO_{3-\delta}$ since their radii depend on their spins. This problem is not trivial since divalent and trivalent cobalt may exist in the low and high spin state and trivalent cobalt - in the intermediate one as well [6,19, 20]. So far the nature of spin states of cobalt in rare earth cobaltites is the controversial topic [6,19,20]. Nevertheless, a gradual transition from the low spin state via the intermediate state to the high one with increasing temperature or alkali earth metal content is accepted for trivalent cobalt [6,19,20]. On the contrary, aforementioned transition was not discussed in the literature for divalent cobalt so far. However, we can assume that the similar gradual transition occurs for divalent cobalt. Another challenge consists in definition of a crystal radius for cobalt in the intermediate spin state since that was not presented by Shannon [18]. In this respect it is of interest to mention the results of Asai et al. [20] since they showed that a transition from the low spin state to the intermediate one for cobalt in the undoped lanthanum cobaltite is not accompanied by any anomaly during its thermal expansion. That is why we can accept the crystal radius of Co^{3+} (LS) as that of Co^{3+} (IS). The values of chemical expansion of $La_{0.6}Sr_{0.4}CoO_{3-\delta}$ calculated at 969 K under different assumptions upon cobalt spin states are given in Figure 16 as an example. As seen both divalent cobalt and trivalent one seem to occur in mixed spin states. On this reason average radii of trivalent and divalent cobalt can be introduced as

$$r_{Co}{}^{3+} = r_{Co}{}^{3+}(HS,\ CN{=}6)*a_1 + r_{Co}{}^{3+}(LS,\ CN{=}6)*(1-a_1) \qquad (17)$$

and

$$r_{Co}{}^{2+} = r_{Co}{}^{2+}(HS,\ CN{=}6)*a_2 + r_{Co}{}^{2+}(LS,\ CN{=}6)*(1-a_2), \qquad (18)$$

where a_1 and a_2 are fractions of Co^{3+} and Co^{2+}, respectively, in the high spin state, $r_{Co}{}^{3+}(LS,\ CN{=}6)$=0.685 Å and $r_{Co}{}^{2+}(HS,\ CN{=}6)$=0.885 Å [18], are crystal radii of Co^{3+} in low spin and Co^{2+} in high spin, respectively. It is worth to note that a_1 and a_2 are the only parameters that have to be fitted in the model Eq. (6) to the data on chemical expansion while others are defined independently on the basis of results on the defect structure modeling using data on oxygen nonstoichiometry.

The results of model calculation of a chemical expansion according to Eqs. (6,17,18) are given in Figures 14-15, respectively. As seen the model proposed coincides with the experimental data on chemical expansion really good for all $La_{1-x}Sr_xCoO_{3-\delta}$ compositions investigated at 1164 K and those measured for $La_{0.6}Sr_{0.4}CoO_{3-\delta}$ at all temperatures. It is of interest to note that a slope of the calculated oxygen nonstoichiometry dependences of $La_{0.6}Sr_{0.4}CoO_{3-\delta}$ chemical expansion increases gradually with temperature at a given oxygen nonstoichiometry indicating in favor of the spin state transition from low spin to high spin state for the cobalt species with increasing temperature.

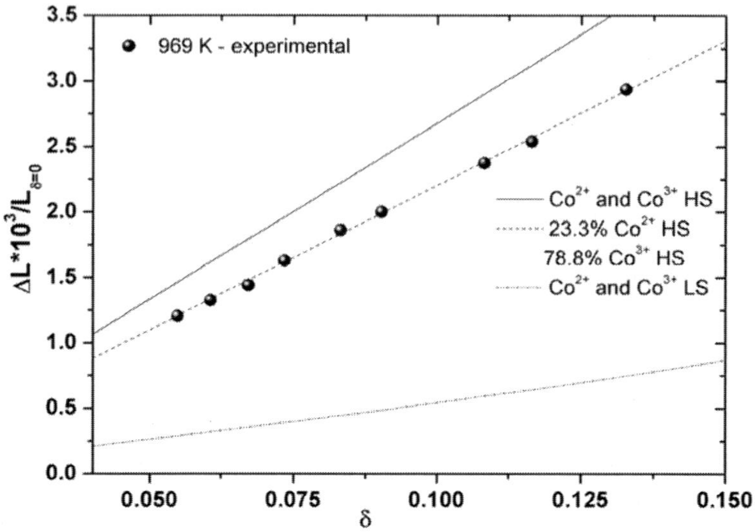

Figure 16. Chemical expansion of $La_{0.6}Sr_{0.4}CoO_{3-\delta}$ as a function of the oxygen nonstoichiometry at 969 K, normalized as described in the text. The points represent experimental values and lines are given by Eq. (6) under different assumptions upon cobalt spin states. (Reproduced with permission from *J. Electrochem. Soc.*, 159, F594 (2012). Copyright 2012, The Electrochemical Society).

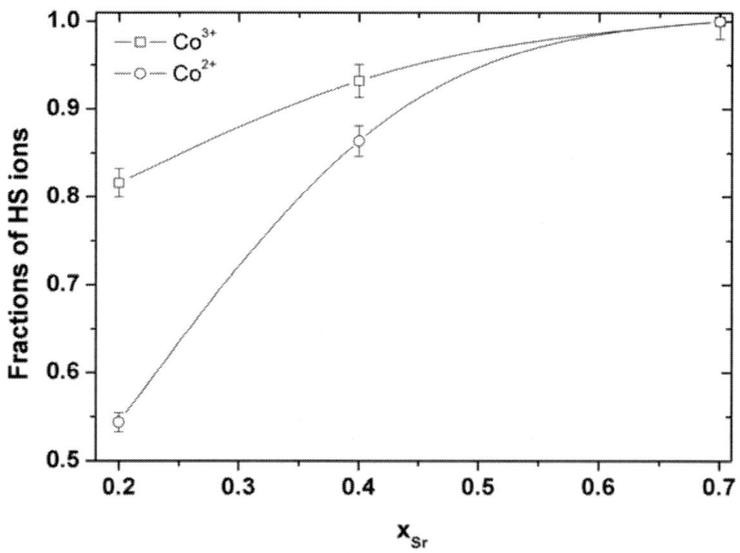

Figure 17. Fractions of Co ions in High Spin (HS) state in $La_{1-x}Sr_xCoO_{3-\delta}$ as a function of strontium content at 1164 K. (Reproduced with permission from *J. Electrochem. Soc.*, 159, F594 (2012). Copyright 2012, The Electrochemical Society).

The fitted values of the high spin fraction of cobalt as function of strontium content, x, and temperature are shown in Figures 17 and 18, respectively. Quite plausible observation follows from these figures. As seen in Figure 17 HS fractions of Co^{2+} and Co^{3+} gradually grow with strontium content increase at the same temperature reaching unity for

$La_{0.3}Sr_{0.7}CoO_{3-\delta}$. Figure 18 shows that aforementioned increase takes place in $La_{0.6}Sr_{0.4}CoO_{3-\delta}$ with increasing temperature as well.

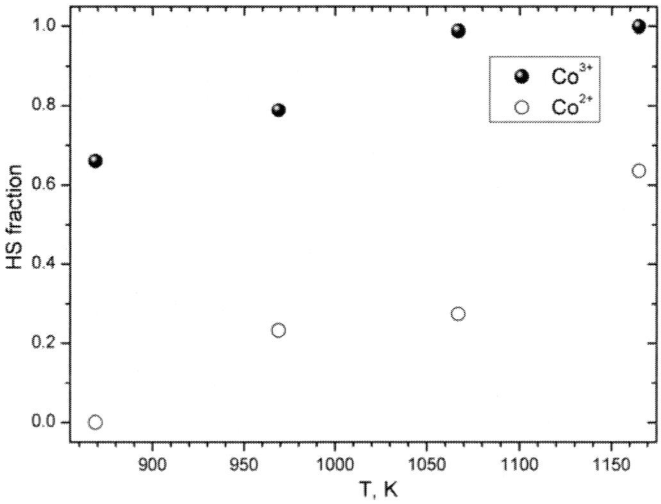

Figure 18. Fractions of Co ions in High Spin (HS) state in $La_{0.6}Sr_{0.4}CoO_{3-\delta}$ as a function of temperature. (Reproduced with permission from *J. Electrochem. Soc.*, 159, F594 (2012). Copyright 2012, The Electrochemical Society).

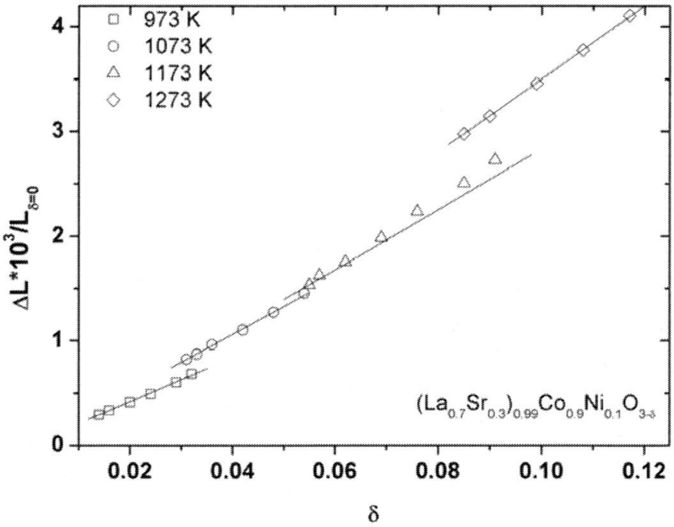

Figure 19. Chemical expansion of $(La_{0.7}Sr_{0.3})_{0.99}Co_{0.9}Ni_{0.1}O_{3-\delta}$ as a function of the oxygen nonstoichiometry, normalized as described in the text. The symbols represent experimental values and lines are given by Eqs. (6, 17, 18). (Reproduced with permission from *J. Electrochem. Soc.*, 159, F594 (2012). Copyright 2012, The Electrochemical Society).

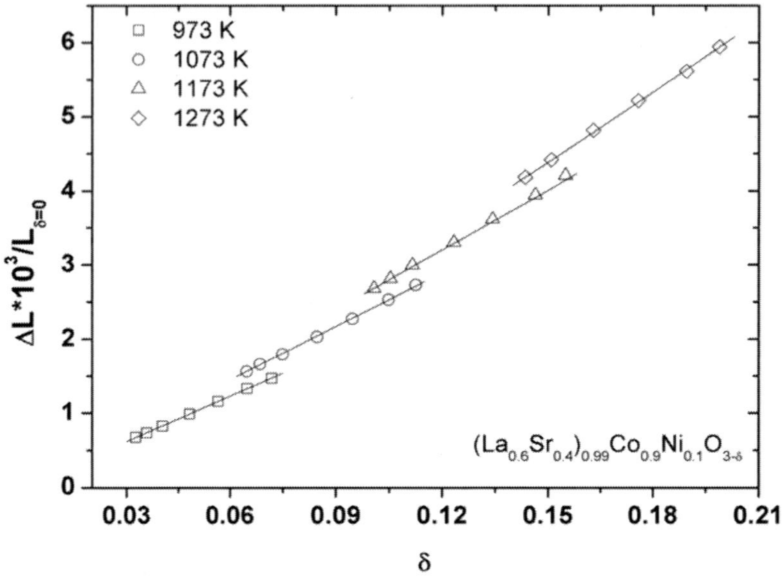

Figure 20. Chemical expansion of $(La_{0.6}Sr_{0.4})_{0.99}Co_{0.9}Ni_{0.1}O_{3-\delta}$ as a function of the oxygen nonstoichiometry, normalized as described in the text. The symbols represent experimental values and lines are given by Eqs. (6, 17, 18). (Reproduced with permission from *J. Electrochem. Soc.*, 159, F594 (2012). Copyright 2012, The Electrochemical Society).

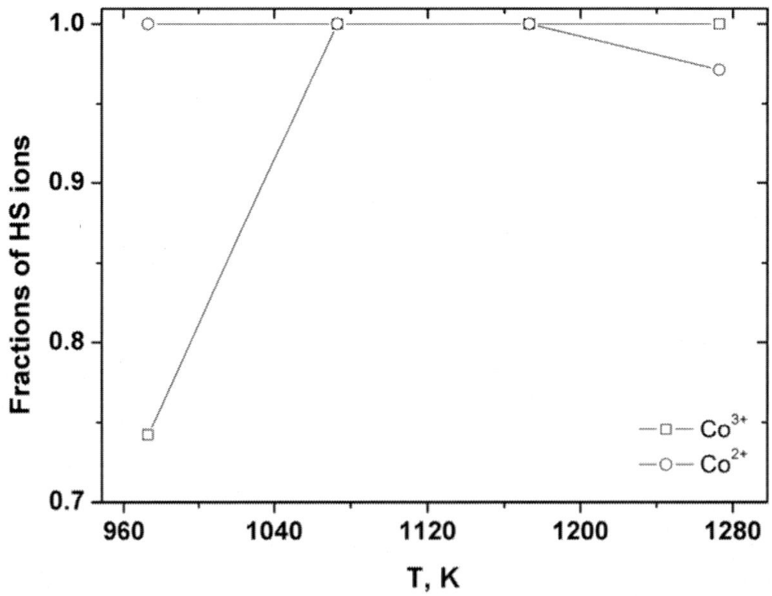

Figure 21. Fractions of Co ions in High Spin (HS) state in $(La_{0.7}Sr_{0.3})_{0.99}Co_{0.9}Ni_{0.1}O_{3-\delta}$ as a function of temperature. (Reproduced with permission from *J. Electrochem. Soc.*, 159, F594 (2012). Copyright 2012, The Electrochemical Society).

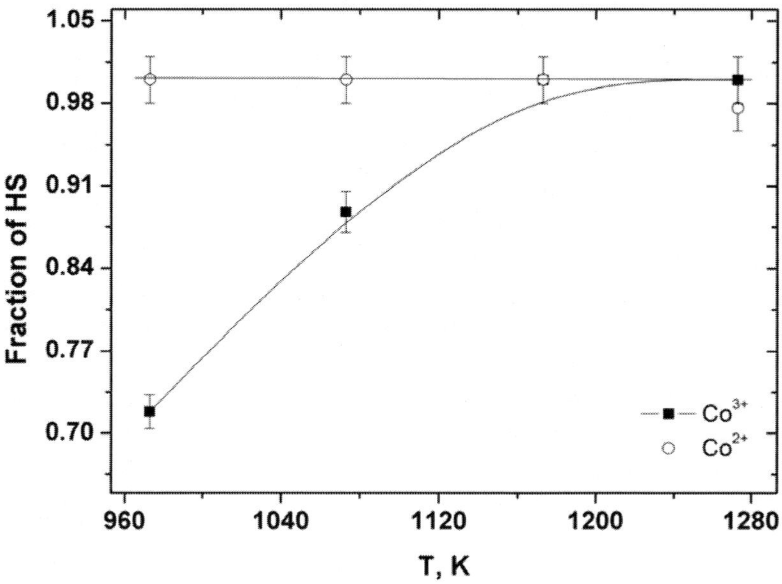

Figure 22. Fractions of Co ions in High Spin (HS) state in $(La_{0.6}Sr_{0.4})_{0.99}Co_{0.9}Ni_{0.1}O_{3-\delta}$ as a function of temperature. (Reproduced with permission from *J. Electrochem. Soc.*, 159, F594 (2012). Copyright 2012, The Electrochemical Society).

Taking into account obtained HS state fractions it is possible to calculate a magnetic moment for the cobaltites studied and compare that to a value calculated earlier by Baskar and Adler [19] from the data on magnetic susceptibility. For instance, magnetic moment of $La_{0.8}Sr_{0.2}CoO_{3-\delta}$ calculated at 1164 K on the basis of the model proposed in the present study comes to value of 4.05 μ_B while that is 3.8 μ_B according to Ref. (19). Moreover, Baskar and Adler [19] showed that magnetic moment of $La_{0.8}Sr_{0.2}CoO_{3-\delta}$ does not depend on oxygen nonstoichiometry in practical term. This is the case for the magnetic moment calculated by us. Such coincidence may indicate in favor of the model proposed by us for chemical expansion of the doped cobaltites as well.

The chemical expansion of $(La_{1-x}Sr_x)_{0.99}Co_{0.9}Ni_{0.1}O_{3-\delta}$ ($x = 0.3$, 0.4) measured by Hjalmarsson et al. [11] is given in Figures 19 and 20 as a deviation from the reference state normalized to the length of a sample with stoichiometric composition regarding oxygen ($\delta_0 = 0$) plotted versus oxygen nonstoichiometry at different temperatures. The results of model calculations for chemical expansion of $(La_{1-x}Sr_x)_{0.99}Co_{0.9}Ni_{0.1}O_{3-\delta}$ ($x = 0.3$, 0.4) according to Eq. [6, 17, 18] are also given in Figures 19 and 20. As seen the model proposed coincides completely with experimental data on chemical expansion for both nickel contained compositions at all temperatures investigated.

The fitted values of the cobalt high spin fractions for $(La_{0.7}Sr_{0.3})_{0.99}Co_{0.9}Ni_{0.1}O_{3-\delta}$ and $(La_{0.6}Sr_{0.4})_{0.99}Co_{0.9}Ni_{0.1}O_{3-\delta}$ are shown in Figures 21 and 22, respectively. As follows from Figure 21 the HS fraction of Co^{3+} gradually grows with temperature for $(La_{0.7}Sr_{0.3})_{0.99}Co_{0.9}Ni_{0.1}O_{3-\delta}$ reaching unity at 1273 K while that of Co^{2+} remains equal to unity over complete temperature range. Comparison of Figure 18, on the one hand, and Figure 22, on the other hand, shows that addition of nickel to strontium contained cobaltite leads to significant increase of HS fraction of Co^{2+}.

CONCLUSION

Defect structure models were developed for different oxides with perovskite structure: the doped lanthanum chromites $La_{1-x}Me_xCr_{1-y-z}Al_z(Mg)_yO_{3-\delta}$ (Me=Ca, Sr), undoped lanthanum manganite and cobaltite $LaBO_{3-\delta}$ (B = Mn, Co) and doped lanthanum cobaltites $La_{1-x}Sr_xCoO_{3-\delta}$, $(La_{1-x}Sr_x)_{0.99}Co_{0.9}Ni_{0.1}O_{3-\delta}$ within the framework of the localized nature of electronic defects. These models were shown to fit well the experimental data on oxygen nonstoichiometry available for these oxides in the literature. Chemical expansion model based on the relative change of the mean ionic radius was proposed for the oxides with pseudo cubic structure. Spin state transition from LS to HS for cobalt ions induced by change of temperature and dopant content was taken into account as well. The model proposed coincides with experimental data on chemical expansion really good for all perovskite structured oxides studied at all temperatures investigated. As a result the spin state distribution of cobalt and magnetic moment was calculated for the doped lanthanum cobaltite studied.

ACKNOWLEDGMENTS

This work was supported by the Russian Fund for Basic Research (Grant No 10-03-01139-a).

REFERENCES

[1] P.G. Strelkow, *Physik. Zeita. Sowjetunion* 12 (1937) 73.
[2] Mott and Gurney, Electronic Processes in Ionic Crystals Clarendon Press, Oxford, 1948.
[3] F. Seitz, *Phys. Rev.* 56 (1939) 1063.
[4] A.W. Lawson, *Phys. Rev.* 78 (1950) 185.
[5] S.B. Adler, *J. Am. Ceram. Soc.* 84 (2001) 2117.
[6] X. Chen, J.Yu, and S.B. Adler, *Chem. Mater.* 17 (2005) 4537.
[7] V.A. Cherepanov, A.N. Petrov and A.Yu. Zuev, in *Solid State Electrochemistry I*: Fundamentals and their Applications, V. Kharton, Editor, p. 43, Wiley-VCH, Weinheim (2009).
[8] S.R. Bishop, K.L. Duncan and E.D. Wachsman, *J. Am. Ceram. Soc.* 93 (2010) 4115.
[9] M.H.R. Lankhorst, H.J.M. Bouwmeester and H. Verweij, *J. Solid State Chem.* 133 (1997) 555.
[10] J. Mizusaki, Y. Mima, S. Yamauchi and K. Fueki, *J. Solid State Chem.* 80 (1989) 102.
[11] P. Hjalmarsson, M. Søgaard and M. Mogensen, *J. Solid State Chem.* 183 (2010) 1853.
[12] Yu. Zuev, D. S. Tsvetkov, *Solid State Ionics* 181 (2010) 557.
[13] A.Yu. Zuev, A.N. Petrov, A.I. Vylkov, D.S. Tsvetkov, J. Mater. Sci. 42 (2007) 1901.
[14] A.Yu. Zuev, A.I. Vylkov, A.N. Petrov and D.S. Tsvetkov, *Solid State Ionics* 179 (2008) 1876.
[15] K. Hilpert, R.W. Steinbrech, F. Boroomand et al., *J. Eur. Ceram. Soc.* 23 (2003) 3009.

[16] F. Boroomand, E. Wessel, H. Bausinger, K. Hilpert, *Solid State Ionics* 129 (2000) 251.
[17] Zuev, L. Singheiser, K. Hilpert, *Solid State Ionics* 147 (2002) 1.
[18] R.D. Shannon, *Acta Crystallogr*. A32 (1976) 751.
[19] D. Baskar and S. B. Adler, *Chem. Mater.* 20 (2008) 2624.
[20] K. Asai, A. Yoneda, O. Yokokura, J.M. Tranquauda, G. Shirane and K. Kohn, *J. Phys. Soc. Japan*. 67 (1998) 290.

In: Perovskite
Editors: Jinghua Zhang and Huan Li
ISBN: 978-1-62417-800-9

Chapter 6

PEROVSKITE AND LEAD BASED CERAMIC MATERIALS: A REVIEW

*Niranjan Sahu**
Department of physics, Purushottam Institute of Engineering and Technology, Rourkela, Odisha, India

ABSTRACT

The history, crystallographic signature of perovskite materials along with physical properties has been discussed. Basically lead-based perovskite material (PT, PZT) has so many technological applications in electronic industries. The property of the lead based material can also be enhanced by substituting different transition materials within the perovskite. The main objectives of the present review articles are to understand the crystal structure and electrical transport properties of PT and PZT based ceramic perovskite.

1.1. FERROELECTRICITY

There has been a continuous interest in the fabrication of new ferroelectric materials since the discovery of ferroelectricity in single crystals of Rochelle salt ($NaKC_4H_4O_6.4H_2O$) (*Valasek*, 1921) [1] and its subsequent extension into the domain of polycrystalline ceramics of barium titanate ($BaTiO_3$) during mid-1940s (*Haertling*, 1999) [2]. Ferroelectricity has also been called Seignette electricity, as Seignette or Rochelle Salt (RS) was the first material found to show ferroelectric properties such as a spontaneous polarization on cooling below the Curie point, ferroelectric domains and a ferroelectric hysteresis loop. A huge leap in the research on ferroelectric materials came in the 1950's, leading to the widespread use of many other ferroelectric ceramics including lead titanate ($PbTiO_3$ or PT), lead zirconate titanate

* Corresponding author: Niranjan Sahu. Sr. Lecturer in the department of physics, Purushottam Institute of Engineering and Technology, Rourkela, India, Odisha. E-mail: drniranjannitr@gmail.com/nsahu76@gmail.com.

($PbZr_{1-x}Ti_xO_3$ or PZT), lead lanthanum zirconate titanate ($Pb_{1-x}La_xZr_{1-y}Ti_yO_3$ or PLZT), and relaxor ferroelectrics like lead magnesium niobate (PMN). These ferroelectric materials have been developed and utilized for a variety of applications. Ferroelectricity is a spontaneous electric polarization of a material that can be reversed by the application of an external electric field. Ferroelectricity is the phenomenon, by virtue of which some materials exhibit spontaneous electric polarization, even in the absence of any externally applied field. Ferroelectric crystals possess regions with uniform polarization called ferroelectric domains. Within a domain, all the electric dipoles are aligned in the same direction. There may be many domains in a crystal separated by interfaces called domain walls. A ferroelectric single crystal, when grown, has multiple ferroelectric domains. A single domain can be obtained by domain wall motion made possible by the application of an appropriate electric field. A very strong field could lead to the reversal of the polarization in the domain, known as domain switching [3]. The name refers to certain magnetic analogies, though it is somewhat misleading as it has no connection with iron (ferrum) at all. The development of new ceramic processing and thin film technology lead the materials to a wide range of industrial and commercial applications. Application of these materials includes high dielectric constant capacitors, piezoelectric sonar, ultrasonic transducers, medical diagnostic transducers, buzzers, gas igniters, positive coefficient sensors and switches, ultrasonic motors, electro optic valves, thin film capacitors and ferroelectric thin film memories (*Lines and Glass*, 1977) [4]. The biggest uses of ferroelectric ceramics is in the areas such as dielectric ceramics for capacitor applications, ferroelectric thin films for non volatile memories, piezoelectric materials for medical ultrasound imaging and actuators, and electro-optic materials for data storage and displays. The historical roots leading to the discovery of ferroelectricity can be traced back to the last century and the work of the famous crystal physicists Weiss, Pasteur, Pockels, Hooke, Growth, Voigt and the brothers Curie. Beginning with the pioneering work on Rochelle salt and potassium dihydrogen phosphate, the study of ferroelectrics accelerated rapidly during World War II with the discovery of barium titanate. Then came a period of rapid proliferation with more than 100 new ferroelectrics identified in the next decade, including lead zirconate titanate, the most widely used piezoelectric transducers.

1.2. CRYSTALLOGRAPHIC SIGNATURE OF FERROELECTRICITY

The lattice structure described by the Bravais unit cell of the crystal governs the crystal symmetry. Though there are thousands of crystals in nature, they all can be grouped together into 230 microscopic symmetry types or space groups based on the symmetry elements [5]. Most of the crystals possess symmetry elements in addition to the repetitions expressed by the crystal lattice. The operation of any single symmetry element of the group leaves the pattern of symmetry unchanged. In studying the physical properties of crystals, only the orientations of the symmetry elements and not their relative positions are important. Hence, if only the orientations of the symmetry elements are taken into account, then the macroscopic symmetry elements in crystals reduce to a center of symmetry, mirror plane, 1, 2, 3, 4 or 6 fold rotation axes and 1, 2, 3, 4 or 6 fold inversion axes. A combination of these symmetry elements gives us the macroscopic symmetry also called as point groups. It can be shown by the inspection of the 230 space groups that there are just 32 point groups. The seven crystal systems can be divided into these point groups according to the point group symmetry it possess. Out of 32 point groups, 21 are non-centrosymmetric; however 20 non-centrosymmetric point groups are piezoelectric [6]. Among these 20 piezoelectric, 10 have a unique polar axis and possess spontaneous polarization (which is in general temperature dependent), are called pyroelectric. Similar to pyroelectric materials, ferroelectrics possess spontaneous polarization but the magnitude and direction of the polarization can be reversed by an external electric field. Thus the conditions necessary to classify materials as ferroelectric are: (i) the existence of spontaneous polarization and (ii) reversible spontaneous polarization. Figure 1.1 shows the inter-relationship of piezoelectric and subgroups on the basis of symmetry.

These point groups do not show polarity. The remaining 21 point groups do not have a center of symmetry (i.e. non-centro symmetric). A crystal having no center of symmetry possesses one or more crystallographically unique directional axis. All non-centro symmetric point groups, except the 32 point groups, show the piezoelectric effect along unique directional axes. The ferroelectric materials can be classified into four subgroups based on the unit cell structure such as: (a) the tungsten-bronze group, (b) oxygen octahedral (ABO_3) group, (c) the pyrochlore group and (d) the bismuth-layer structure group [7]. Out of these four, ABO_3 perovskites have attracted much attention due to various factors, which will be discussed in the following sections.

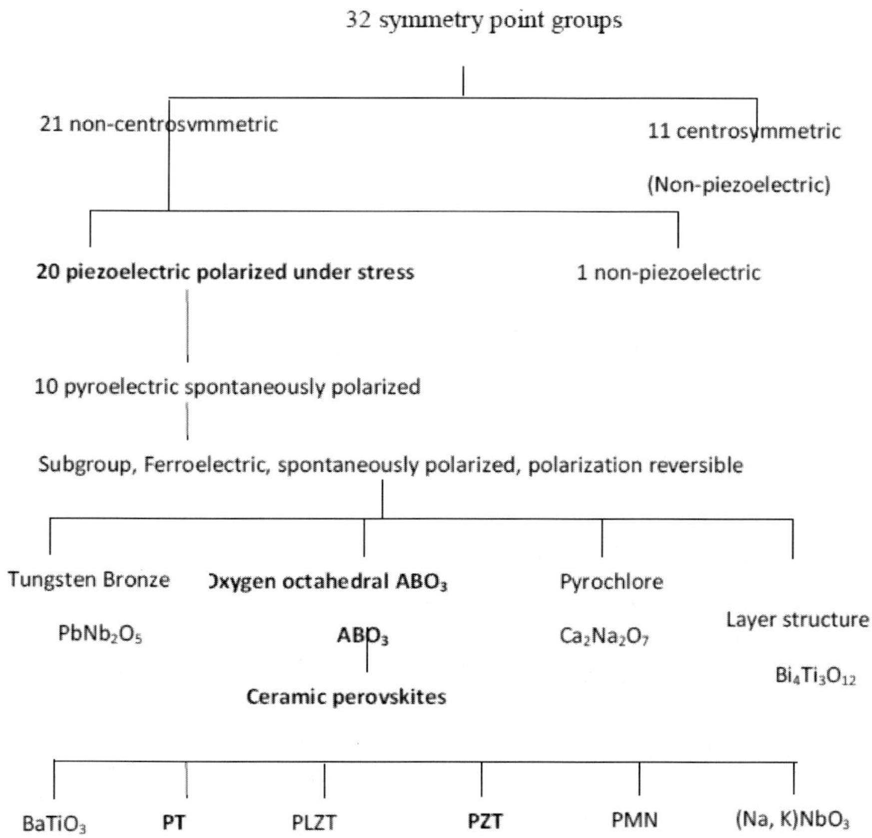

Figure 1.1. Flow chart of inter-relationship of piezoelectric and subgroups on the basis of symmetry.

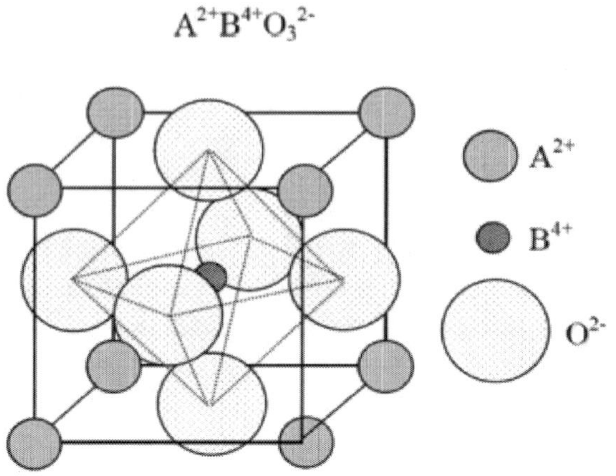

Figure 1.2. An ideal cubic ABO_3 perovskite unit cell.

1.3. Perovskite Structure

In Ural Mountains Gustav Rose discovered CaTiO3 in 1839 and named it Perovskite after an eminent Russian mineralogist, Count Lev Alexevich Von Perovski. Perovskite is a family name of a group of materials and the mineral name of calcium titanate ($CaTiO_3$) having a structure of the type ABO_3.

The ideal perovskite structure is simple cubic, and this is what we generally imply when we refer to the perovskite structure. The mineral perovskite is $CaTiO_3$ and is actually orthorhombic at room temperature, becoming cubic at temperatures above 900°C [8].

The simple cubic structure (space symmetry $Pm\bar{3}m$) consists of corner sharing oxygen octahedra (BO_6) arranged in three dimensions with smaller, highly charged cations (B: Ti^{4+}, Zr^{4+}, Sn^{4+}, Nb^{5+}, Ta^{5+}, W^{6+} etc.) located in the middle of the octahedral, and lower charged, larger cations (A: Na^{+1}, K^{+1}, Ca^{2+}, Ba^{2+}, Pb^{2+} etc.) in between the octahedral. The coordination number of A-site cation is 8 to 12 whereas the coordination number of B-site cation is 6. Its simplest structure is cubic, which is the highest temperature form of many mixed oxides of the ABO_3 type. The ideal ABO_3 structure is shown in Figure 1.2. Most perovskite type ferroelectrics are compounds with either $A^{2+}B^{4+}O_3^{2-}$ or A^{1+} $B^{5+}O_3^{2-}$ type formulation.

The ideal cubic perovskite structure is distorted by cation substitutions (either A site or B site) and/or oxygen off-stoichiometry. It is mainly due to cation size mismatch, John-Teller effect and the electron instability.

The ideal cubic structure is modified into a lower symmetry structure such as orthorhombic, rhombohedral, etc. due to the following three mechanisms, i.e. (i) cooperative tilting of BO_6 octahedra, (ii) displacement of the cations, and (iii) distortions of the BO_6 octahedra. The tilting of the BO_6 octahedra is most relevant in deciding the overall space group symmetry of the particular PT perovskite (*Knight* 1994; *Yang et al.* 1999) [9, 10]. This distortion is characterized by the Goldsmith tolerance factor,

$$t = \frac{d_{A\text{-}O}}{\sqrt{2}(d_{B\text{-}O})} \tag{1.1a}$$

or

$$t = \frac{(R_A + R_O)}{\sqrt{2}(R_B + R_O)} \tag{1.1b}$$

where, $d_{A\text{-}O}$ and $d_{B\text{-}O}$ is the average A cation-oxygen and B cation-oxygen distances respectively. R_A and R_B are the ionic radii of A and B-site cations and, R_O for the O-site anions. In the case of a cubic perovskite t = 1. Stable perovskite structure is found for "t" values in the range of 0.88 to about 1.09. A perovskites structure containing small cations will lead to t < 1, while large cations will give rise to t > 1. The displacement of the cations and the distortion of the octahedra are usually correlated and driven by electronic instability of the metal cations.

The John-Teller distortion in KCuF (*Okazaki et al.*, 1961) [11] and the cation displacement in $PbZrO_3$ (*Corker et al.*, 1997) [12] are the typical examples. Octahedral tilting is a more common distortion mechanism that can be realized by tilting the rigid BO_6 octahedra while maintaining their corner sharing connectivity (Glazer 1972) [13]. This type of distortion is generally observed when the A-site cation is too small for the cubic structure. Very few ion combinations fulfill these strict relative size criteria. The perovskite structure is particularly important for several reasons: i. many perovskites are ferroelectric, ii. many perovskites are piezoelectric and iii. many perovskites have a high dielectric constant.

Many piezoelectric (including ferroelectric) ceramics such as Barium Titanate ($BaTiO_3$), Lead Titanate ($PbTiO_3$), Lead Zirconate Titanate (PZT), Lead Lanthanum Zirconate Titanate (PLZT), Lead Magnesium Niobate (PMN), $SrTiO_3$, Potassium Niobate ($KNbO_3$), Potassium Sodium Niobate ($K_xNa_{1-x}NbO_3$), and Potassium Tantalate Niobate ($K(Ta_xNb_{1-x})O_3$) have a perovskite type structure.

The size of an atom or ion, and how well it fits into its crystallographic site is one of the most important aspects in determining the stability of an ionic structure. The O^{2-} anion and the larger cation (A^{2+}) have similar radii, so that the structure is not just determined by O^{2-}. The larger cation and the anion combine to form a "close-packed" arrangement with the smaller cation, B^{4+}, sitting in the oxygen octahedral interstices.

By the 1950s the solid solution system $Pb(Zr,Ti)O_3$ (PZT), which also possess the perovskite structure, was found to be ferroelectric. PZT compositions are now the most widely exploited of all piezoelectric ceramics both in research and industry. An up to date brief description of PT and PZT systems can be found in *Moulson* (1990) [14]. A great attention has been given to the perovskite oxides due to their excellent physico-chemical properties and wide range of applications.

A few of them are listed in Table 1.1. Nevertheless, the perovskite structure is extremely common. This is because the perovskite structure can distort in a number of ways to accommodate ions with non-ideal sizes. This leads to a rich array of phase transitions, which makes perovskites one of the most technologically and geologically important structure types.

Table 1.1. Properties and applications [15] of perovskite materials

Property	Applications
Dielectric property	Multilayer capacitor, dielectric resonator, thin film resistor
Optical property	Electro optic modulator, laser host, switch
Ferroelectric/piezoelectric property	Piezoelectric transducers, PTC thermistor, electrostrictive actuators.
Magnetic property	Magnetic bubble memory, ferromagnet
Ionic/proton conducting property	Solid electrolyte, SOFC electrolyte, hydrogen sensor

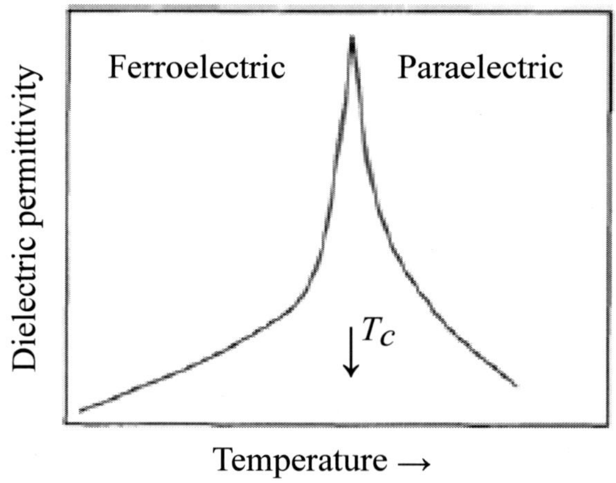

Figure 1.3. An illustration of dielectric permittivity as a function of temperature for normal ferroelectrics [reference 18].

1.4. Curie Point and Phase Transitions

All ferroelectric materials have a transition temperature called the Curie point (T_c). At a temperature $T > T_c$ the crystal does not exhibit ferroelectricity, while for $T < T_c$ it is ferroelectric. On decreasing the temperature through the Curie point, a ferroelectric crystal undergoes a phase transition from a non-ferroelectric phase to a ferroelectric phase.

If there is more than one ferroelectric phases, the temperature at which the crystal transforms from one ferroelectric phase to another is called the transition temperature. Early research works on ferroelectric transitions have been summarized by *Nettleton et al.* [16, 17]. Near the Curie point or transition temperatures, thermodynamic properties including dielectric, elastic, optical and thermal constants show an anomalous behavior. This is due to a distortion in the crystal as the phase structure changes. One distinctive characteristic of ferroelectrics is the hysteresis loop in the polarization versus electric field curve. A net spontaneous polarization exists even at zero electric field and the loop disappears at the Curie temperature Tc, where the material undergoes a phase transition from high temperature paraelectric and low temperature ferroelectric state. The temperature dependence of the dielectric constant above the Curie point ($T > T_c$) in ferroelectric crystals is governed by the Curie-Weiss law [18]:

$$\varepsilon_r = \varepsilon_0 + \frac{C}{T - T_c} \tag{1.2}$$

where ε_r the relative permittivity of the material, ε_0 is the permittivity of vacuum; C is the Curie constant and T_c is the Curie temperature, and it is the actual temperature where the crystal structure changes. The sharp anomaly corresponds to the phase transition at T_c is shown in Figure1.3. A phase transition occurs at the temperature T_c.

Above the transition temperature the substance is in the paraelectric phase, in which the elementary dipoles of the various unit cells in the crystal are oriented randomly and, below the transition temperature, the substance is in the ferroelectric phase.

1.4.1. Diffuse Phase Transition

Many phase transitions in macroscopic homogeneous materials are characterized by the fact that the transition temperature is not sharply defined. Of these, so-called diffuse phase transition temperature (DPT) [19], the transition is smeared out over a certain temperature interval, resulting in a gradual change of physical properties in this temperature region. Though this phenomenon is observed in several types of materials, however, the most remarkable examples of DPT are found in ferroelectric materials. Ferroelectrics diffuse phase transitions (FDPT) are first mentioned in the literature in the early 1950's.

Some characteristics of the DPT are: (a) broadened maxima in the permittivity versus temperature curve, (b) gradual decrease of spontaneous and remnant polarizations with rising temperature, (c) transition temperatures obtained by different techniques which do not coincide, (d) relaxation character of the dielectric properties in transition region and (e) no Curie-Weiss behavior in certain temperature intervals above the transition temperature. The diffuseness of the phase transition is assumed to be due to the occurrence of fluctuations in a relatively large temperature interval around the transition.

Usually two kinds of fluctuations are considered: (a) compositional fluctuation and (b) polarization (structural) fluctuation. From the thermodynamic point of view, it is clear that the compositional fluctuation is present in ferroelectric solids-solutions and polarization fluctuation is due to the small energy difference between high and low temperature phases around the transition.

This small entropy difference between ferroelectric and paraelectric phase will cause a large probability of fluctuation. *Kanzing* (1957) [20] has been observed from X-ray diffraction that in a narrow temperature range around the transition $PbTiO_3$ single crystal splits up into FE (ferroelectric) and paraelectric (PE) micro regions. Complex perovskite type ferroelectrics with distorted cation arrangements show DPT which is characterized by a broad maximum for the temperature dependence of dielectric constant ε_r and dielectric dispersion in the transition region. For DPT, ε_r follows modified temperature dependence [21],

$$\frac{1}{\varepsilon_r} - \frac{1}{\varepsilon_m} = \frac{(T - T_m)^{\gamma}}{C'} \tag{1.3}$$

where T_m is the temperature at which ε_r reaches maximum, ε_m is the value of ε_r at T_m, C′ is the modified Curie Weiss constant and is the critical exponent, explains the diffusivity of the materials, which lies in the range $1< \gamma < 2$ (*Tiwari*, 1994) [22]. The smeared out ε_r versus T response has generally been attributed (*Kirillov et al.*, 1973) [23] to the presence of micro regions with local compositions vary from the average composition over a length scale of 100 to 1000 Å. Different micro regions in a macroscopic sample are assumed to transfer at different temperature, so-called Curie range, leading DPT which is due to compositional fluctuations. The dielectric and mechanical properties of the FE system below their T_C are functions of the state of polarization and stress.

So ferroelectrics have major application today because of their characteristic electro-optic, dielectric and hysteresis properties. For many practical applications, it is desired to use the very large property maxima in the vicinity of the ferroelectric phase transition, to move the transition into the temperature range of interest and to broaden the very large sharp peak values. In DPT the dielectric maxima is now much rounder and polarization persists for a short range of temperature above T_m.

1.4.2. Polarization Mechanism in Ferroelectric Materials

Basically, polarizability is a consequence of the fact that the molecules, which are the building blocks of all substances, are composed of both positive charges (nuclei) and negative charges (electrons). When a field acts on a molecule, the positive charges are displayed in a direction of the field. The effect is therefore to pull the opposite charges apart, i.e., to polarize the molecule [24]. There are different types of polarization processes, depending on the structure of the molecules which constitute the solid. Basically, there are three types polarization mechanisms contribute to the total polarizability. In general, therefore, we may write for the total polarizability:

$$\alpha = \alpha_e + \alpha_i + \alpha_d \tag{1.4}$$

which is the sum of the various contributions; α_e , α_i , and α_d are the electronic, ionic, and dipolar polarizabilities, respectively. The different polarizabilities are active in different frequencies as shown in Figure 1.4. The electronic contribution is present in any type of substances, but the presence of the other two terms depends on the material under consideration.

Thus the term is present in ionic covalent crystals such as Si and Ge, which are non-ionic and non-dipolar, the polarizability is entirely electronic in nature. The relative magnitudes of the various contributions in above equation are such that, in non-dipolar, ionic substances the electronic part is often of the same order as the ionic.

In dipolar substances, however, the greatest contribution comes from the dipolar part. The various polarizabilities may be segregated from each other because each contribution has its own characteristic features which distinguish it from the others. Dipolar polarizability, for instance, exhibits strong dependence on temperature, while the other two contributions are essentially temperature independent. The behaviour of different type polarization is shown in Figure 1.4.

1.5. LEAD BASED PEROVSKITE MATERIALS

1.5.1. Lead Titanate ($PbTiO_3$, PT)

Technological development has been a key to meeting the new demands of the evolving society. Therefore, there is a constant need for new and improved materials. The perovskite structure has and continues to attract interest, because of its suitability for applications in telecommunication. lead titanate ($PbTiO_3$), or simply PT, is one of the most important members in the ferroelectric and piezoelectric families. PT is a classic example of ferroelectric perovskites with a high lattice strain or tetragonality (c/a = 1.064). Its temperature-driven cubic to tetragonal phase transition at T_C = 763 K is taken as an example of a purely displacive transition [25]. PT exhibits a unique negative thermal expansion (NTE) in the perovskite family.

Lead titanate ($PbTiO_3$) has strong ferroelectric behavior because of the Pb-O and Ti-O covalent bonds as compared to $BaTiO_3$ where covalent bonding exists only between Ti and oxygen atoms. Lead titanate is a ferroelectric material having a structure similar to $BaTiO_3$ with a high Curie point 490°C.

On decreasing the temperature through the Curie point a phase transition from the ferroelectric tetragonal phase to paraelectric cubic phase takes place. Lead titanate ceramics are difficult to fabricate in the bulk form as they undergo a large volume change on cooling below the Curie point.

It is the result of a tetragonal (c/a =1.064) to cubic (c/a =1.00) phase transformation leading to a strain of > 6%. Hence, pure $PbTiO_3$ ceramics crack and fracture during fabrication. The spontaneous strain developed during cooling can be reduced by modifying the lead titanate with lanthanum (La) dopants and also other dopants such as Ca, Sr, Ba, Sn, and W to obtain a crack free ceramic.

During the selection of doping element one has to consider the ionic radius of ions (listed in Table 1.2) to get a tolerance factor of stable structure.

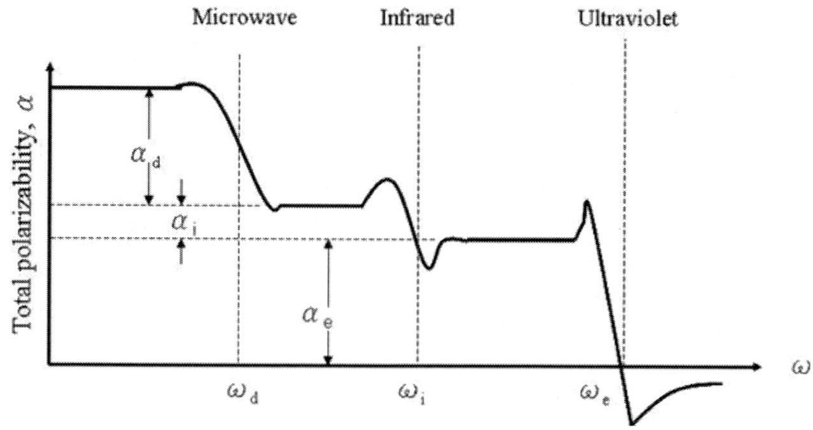

Figure 1.4. The behavior of different type of polarization [24].

One representative modified lead titanate composition that has been extensively investigated recently is $(Pb_{0.76} Ca_{0.24})((Co_{0.50} W_{0.50})_{0.04} Ti_{0.96})O_3$ with 2 mol. % MnO added to it.

This composition has a decreased c/a ratio and Curie point of 255°C [23, 24]. As an example, PLT unit cell consists of a corner linked network of oxygen octahedral with Ti^{4+} ions occupying the B sites and Pb^{2+} and La^{3+} ions situated at the A site created by linking octahedra. As a result of different oxidation states between Pb^{2+} and La^{3+}, some of the A sites and B sites are vacant to maintain electrical neutrality in the structure [26]. The atomic radii of pure and doped PT, PZT are given in Table 1.2.

Ferroelectricity is a spontaneous electric polarization of a material that can be reversed by the application of an external electric field. Ferroelectricity is the phenomenon, by virtue of which some materials exhibit spontaneous electric polarization, even in the absence of any externally applied field.

Ferroelectric crystals possess regions with uniform polarization called ferroelectric domains. Within a domain, all the electric dipoles are aligned in the same direction. There may be many domains in a crystal separated by interfaces called domain walls. A ferroelectric single crystal, when grown, has multiple ferroelectric domains.

A single domain can be obtained by domain wall motion made possible by the application of an appropriate electric field. A very strong field could lead to the reversal of the polarization in the domain, known as domain switching. The spontaneous polarisation depends on temperature in PT. A typical curve of polarzation versus temperature is shown Figure 1.5.

1.5.2. Lead Zirconate Titanate [$Pb(Zr_{1-X}Ti_x)O_3$, PZT]

Lead zirconate titanate (PZT) is a binary solid solution of $PbZrO_3$ an antiferroelectric (orthorhombic structure) and $PbTiO_3$ a ferroelectric (tetragonal perovskite structure). It may exist in either phase; cubic, orthorhombic, rhombohedral, or tetragonal, depending on composition (Zr/Ti ratio) and temperature. PZT has a perovskite type structure with the Ti^{4+} and Zr^{4+} ions occupying the B site at random, Pb ions the corners and oxygen the face centers. The ratio of $PbZrO_3$ to $PbTiO_3$ affects the properties. At high temperatures PZT has the cubic perovskite structure which is paraelectric.

Table 1.2. The atomic radii of component ions in PT and modified PT, PZT and some doped PZT [reference 25, 26]

Ions	Ionic radius (Å)
Pb^{2+}	1.20
Zr^{4+}	0.79
Ti^{4+}	0.68
O^{2-}	1.32
La^{3+}	1.01
Al^{3+}	0.51
Mn^{3+}, Mn^{4+}	0.66; 0.60
Fe^{2+}, Fe^{3+}	0.74; 0.64

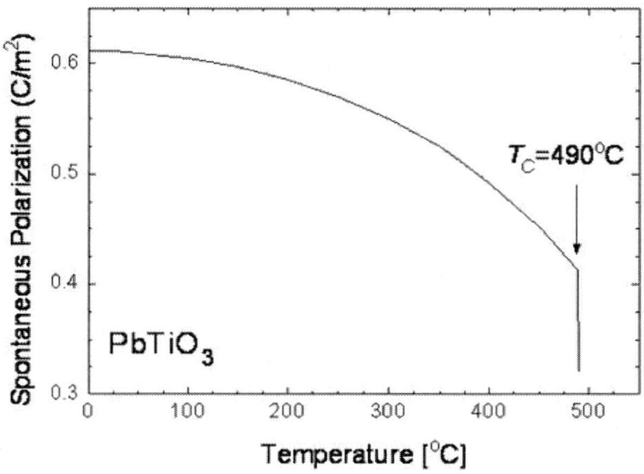

Figure 1.5. The spontaneous polarization versus temperature plot of the "$PbTiO_3$" [reference 26].

On cooling below the Curie point line, the structure undergoes a phase transition to form a ferroelectric tetragonal or rhombohedral phase. In the tetragonal phase, the spontaneous polarization is along the <100> set of directions while in the rhombohedral phase the polarization is along the <111> set of directions.

Most physical properties such as dielectric and piezoelectric constants show an anomalous behavior at the morphotropic phase boundary (MPB). The MPB separating the two ferroelectric tetragonal and orthorhombic phases has a room temperature composition with a Zr/Ti ratio of 52/48. PZT ceramics with the MPB composition show excellent piezoelectric properties.

The poling of the PZT ceramic is also easy at this composition because the spontaneous polarization within each grain can be switched to one of the 14 possible orientations (eight [111] directions for the rhombohedral phase and six [100] directions for the tetragonal phase). Below the Zr/Ti ratio of 95/5 the solid solution is antiferroelectric (AFE) with an orthorhombic phase. On the application of an electric field to this composition a double hysteresis loop is obtained. This is because of the strong influence of the antiferroelectric $PbZrO_3$ phase [27].

In order to suit some specific requirements for certain applications, piezoelectric ceramics can be modified by doping them with ions which have a valence different than the ions in the lattice. Piezoelectric PZT ceramics having the composition at the MPB can be doped with ions to form "hard" and "soft" PZT's. Hard PZT's are doped with acceptor ions such as K^+, Na^+ (for A site) and Fe^{3+}, Al^{3+}, Mn^{3+} (for B site), creating oxygen vacancies in the lattice [28]. Hard PZT's usually have lower permittivities, smaller electrical losses and lower piezoelectric coefficients.

On the other hand, soft PZT's are doped with donor ions such as La^{3+} (for A site) and Nb^{5+}, Sb^{5+} (for B site) leading to the creation of A site vacancies in the lattice. The soft PZTs have a higher permittivity, larger losses, higher piezoelectric coefficient and are easy to pole and dipole. It can be used for applications requiring very high piezoelectric properties. The direction in PZT structure is shown in Figure 1.6 (a-c).

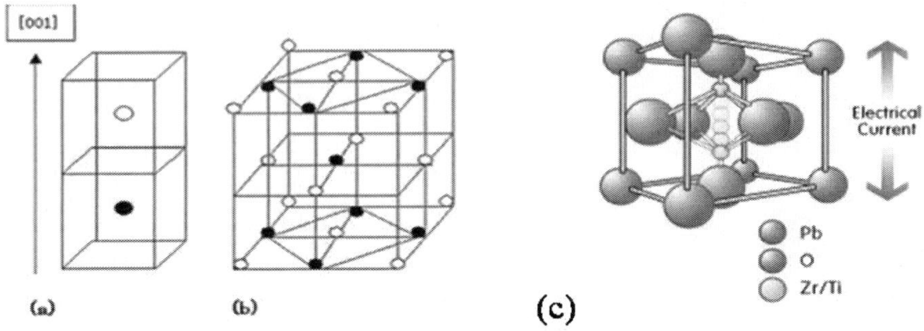

Figure 1.6. (a-c) At high temperatures PZT has the cubic perovskite structure which is paraelectric [reference 27].

Table 1.3. Applications and devices of piezoelectric ceramics [29]

Classification	Applications
High voltage generators	Gas appliances, cigarette lighter, fuse (igniters) for explosive, flash bulbs
High power ultrasonic generators	Ultrasonic cleaner, sonar, echo sounding, ultrasonic machining, atomization, polarization
Transducers for sound and ultrasound in air	Microphones (e.g. for telephone), burglar alarm system, remote control, loud speaker, buzzers, medical ultrasonic equipment
Resonators and filters	Ratio and televisions, remote control, electronic instrumentation
Delay lines	Color televisions, computers, electronic instrumentation, telephones
Keyboards	Computers, printers, desk calculators, slot machines, telephones

On cooling below the Curie point line, the structure undergoes a phase transition to form a ferroelectric tetragonal or rhombohedral phase. In the tetragonal phase, the spontaneous polarization is along the <100> set of directions while in the rhombohedral phase the polarization is along the <111> set of directions.

Uses of PZT ceramics in electronic appliances are manifold, depending on whether the device employs the direct piezoelectric effect, the converse piezoelectric effect or both. This phenomenon is benefited in numerous device applications as given in Table 1.3.

CONCLUSION

The discovery of ferroelectric perovskite ceramic material has been discussed briefly. The different lead based perovskite material has been studied. The effect of different type of polarization mechanism in ferroelectric material is introduced. Perovskite structure and application of this material in electronic industry were revisited.

REFERENCES

[1] J. Valasek, "Piezoelectric and allied phenomena in Rochelle salt", *Phys. Rev.*17 (1921) 475.

[2] G. H. Haertling, *J. Am. Ceram. Soc.* 82 (1999) 797.

[3] M. Deri, Ferroelectric Ceramics (Gordon and Breach, New York, 1969).

[4] M. E. Lines and A. M. Glass, Principles and Applications of Ferroelectric and Related Materials (Clarendon Press, Oxford, 1977).

[5] H. D. Megaw, Ferroelectricity in Crystals (Methuen, London, 1957).

[6] Y. Xu, Ferroelectric Materials and their Applications (North Holland, Amsterdam, 1991).

[7] G. Shirane, R. Pepinsky, B. Frazer, *Acta Cryst.* 9 (1956) 131.

[8] K. K. Deb, *Ferroelectrics,* 82 (1998) 45.

[9] K. S. Knight, *Solid. Stat. Ion*, 74 (1994)109.

[10] J. H. Yang, W. K. Choo, J. H. Lee, C. H. Lee, Acta *Cryst. B*, 55 (1999) 348.

[11] A. Okazaki, Y. Suemune, *J. Phys. Soc. Jpn.,* 16 (1961) 176.

[12] D. L. Corker, A. M. Glazer, J. Dec, K. Roleder, R. W. Whatmore, *Acta Cryst. B*, 53 (1997) 135.

[13] A. M. Glazer, *Acta Cryst. B*, 28 (1972) 3384.

[14] A. J. Moulson, J. M. Herbert, “Electroceramics”, Chapman and Hall Ltd, (1990) London.

[15] K. Ishikawa, T. Uemori, *Phys. Rev. B*, 60 (1999) 11841.

[16] R. E. Nettleton, *Ferroelectrics*, 1, 3, 87, 93, 111, 121,127, 207, 221 (1970)

[17] R. E. Nettleton, *Ferroelectrics*, 2, 5, 77, 93 (1971)

[18] S. Neirman, *J. of Mater. Sci*, 23 (1988) 3973.

[19] D. Hennings, A. Schnell, G. Simon, *J. Am. Ceram. Soc.* 65 (1982) 539.

[20] W. Kanzig, “Ferroelectrics and Antiferroelectrics”, Academic Press, (1957) New York.

[21] R. Ramesh, D. G. Schlom, *Science* 296 (2002) 1975.

[22] V. S. Tiwari, D. Pandey, *J. Am. Ceram. Soc. II*, 77 (1994) 1819.

[23] V. V. Kirillov, V. A. Isupov, “*Ferroelectrics*, 5 (1973) 3.

[24] K. M. Rittenmyer, R. Y. Ting, *Ferroelectrics*, 171 (1990)110.

[25] J. F. Scott, L. Kammerdiner, M. Paris, S. Traynor, V. Ottenbacher, A. Shawabkeh, W. F. Oliver, *J. Appl. Phys*, 64 (1988) 787.

[26] A. von. Hippel, *Rev. Modern Phys.*, 22 (1950) 221.

[27] S. G. Porter, *Ferroelectrics*, 33 (1981) 193.

[28] R. C. Buchanan, ed, *Ceramic Materials for Electronics - Processing, Properties and Applications* (Marcel Dekker, New York, 1986).

[29] J. M. Herbert, *Ferroelectric Transducers and Sensors* (Gordon and Breach, London, 1982).

In: Perovskite
Editors: Jinghua Zhang and Huan Li
ISBN: 978-1-62417-800-9

Chapter 7

PHOTOCATALYTIC PROPERTIES OF PEROVSKITE-TYPE LAYERED OXIDES

Irina Zvereva* and Ivan Rodionov
Saint-Petersburg State University, Faculty of Chemistry,
Department of Chemical Thermodynamics and Kinetics, Russia

ABSTRACT

The chapter presents a comparative study of photocatalytic activity of layered perovskite-type oxides based on original results. Series of Ruddlesden-Popper ($A_2Ln_2Ti_3O_{10}$ and $ANdTiO_4$, where A = H, Li, Na, K; Ln = La, Nd) and Dion-Jacobson ($ANdTa_2O_7$, where A = H, Li, Na, K, Rb, Cs and $ANdNb_2O_7$, where A = Rb, Cs) phases where synthesized and characterized by XRD, SEM, UV-VIS diffuse-reflectance spectroscopy, BET. Photocatalytic activity was determined from the hydrogen evolution rate in aqueous suspensions under UV-irradiation in the presence of isopropyl alcohol as a sacrificial electron donor. Dependence of photocatalytic activity of complex oxides on their characteristics, such as BET surface area, particle morphology, band gap energy and water intercalation ability is discussed.

INTRODUCTION

Photocatalysts for decomposition of water into hydrogen and oxygen under light irradiation have been the subject of extensive investigation since the discovery of this process [1] till the present day. This is due to the obvious practical interest in the using of solar energy to produce hydrogen as an ecological clean fuel. Besides this attractive opportunity, photocatalysts have a number of other important applications. They are used for the design of air- and water purification systems, self-clean coatings, solar cells, etc [2]. However, the efficiency of discovered photocatalysts is still low for many commercial purposes, so the design of new photocatalytic systems is an actual problem. Besides a wealth of "modern"

* E-mail: irinazvereva@yandex.ru, Tel: +7-9043305019.

physical phenomena, such as high-T_c superconductivity, colossal magnetoresistance, ferroelectricity and quantum critical behavior, layered perovskite-type oxides have been shown to demonstrate high photocatalytic performance due to their unique structural features. However, the photocatalytic process is sensitive to a large number of experimental factors, and in literature there is still lack in data on photocatalytic activity of isostructural layered oxides, measured under identical conditions. At the same time, these data are the key for understanding how the composition and structure of a photocatalyst affects its activity.

In this chapter we present results of the comparative study of photocatalytic activity of layered perovskite-type oxides belonging to the series of Ruddlesden-Popper ($A_2Ln_2Ti_3O_{10}$ and $ANdTiO_4$, where A = H, Li, Na, K; Ln = La, Nd) and Dion-Jacobson ($ANdTa_2O_7$, where A = H, Li, Na, K, Rb, Cs and $ANdNb_2O_7$, where A = Rb, Cs) phases in the model reaction of hydrogen evolution from suspensions in 0.1% (mol.) isopropyl alcohol aqueous solution under UV-irradiation.

1. Photocatalysis: Main Concepts and Historical Background

In general, photocatalysis means the acceleration or initiation of chemical reactions induced by light in the presence of special substances called photocatalysts. Like in the case of conventional catalysis, the photocatalyst is not consumed during the reaction. For heterogeneous photocatalysis semiconductor compounds are usually used, the mechanism of catalysis being as follows. The absorption of photons with energy above the band gap leads to formation of electron-hole pairs (Figure 1.a). These charge carriers migrate to the photocatalyst particle surface, where they take part, respectively, in reduction and oxidation reactions with adsorbed species. The possibility of a given reaction is determined by the relative position of its reduction potential Ox/Red and the band edges (E_c and E_v) in the semiconductor. The overall process, which consists of two half reaction is possible only if both potentials lie within the band gap E_g (Figure 1.b).

Nowadays, the most common used photocatalyst is titanium dioxide because of its ability to oxidize almost any organic compound under ultraviolet light, environmental cleanliness and low cost [3] [4]. However, under visible light, which accounts for the maximum sunlight intensity, pure TiO_2 shows no photocatalytic activity, which greatly narrows the field of its application. Modifications such as nitrogen-doping partly solve this problem, making TiO_2 visible light-responsible [5]. However, the activity of such modified samples remains low because of increased electron-hole recombination degree.

The process of water splitting was observed for the first time on a TiO_2 photoelectrode [1]. Pure TiO_2 is inapplicable for water splitting, because its conduction band level is too close to the hydrogen reduction potential. Its activity can be increased by modification with metal particles, such as Ni or Pt [6]. However, to drive an uphill reaction such as overall water splitting, there is need in more effective separation of oxidation and reduction sites to prevent undesirable back reactions. Such separation is possible in photocatalysts with more complex structure, e.g. layered oxides. $K_4Nb_6O_{17}$ was the first layered oxide to be shown an efficient photocatalyst for water splitting owing to its structure, which contains two different types of interlayer space [7].

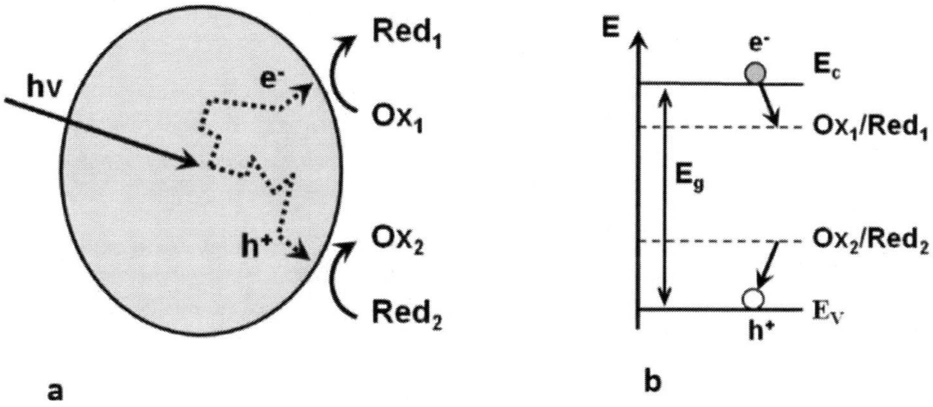

Figure 1. General mechanism of photocatalysis on semiconductor materials.

Both of them can undergo water intercalation. When Ni particles are introduced into the first-type interlayer space, the whole oxide particle becomes an efficient reactor for water splitting. Hydrogen evolves on Ni particles, while oxygen is formed in the second-type interlayer, separated from the first-type interlayer by a niobate sheet.

Layered perovskite-type oxides represent a promising class of photocatalysts. Some members of this class are capable of reversible intercalation of water molecules into the interlayer space [8]-[10]. The composition of these compounds can be widely varied by topochemical reactions (e.g. ion-exchange) due to the high mobility of interlayer cations, which is an effective tool to modify their optical properties and photocatalytic activity [11]-[13]. Many layered perovskite-type oxides have proved to be effective photocatalysts for the decomposition of water, such as $K_2La_2Ti_3O_{10}$ [14], $RbNdTa_2O_7$ [15], $PbBi_2Nb_2O_9$ [16]. For some of them synthesis conditions have been optimized [17],[18], and the influence of different modification methods on the photocatalytic activity has been investigated [19][24]. However, it is difficult to compare the results of photocatalytic property investigation, obtained by different research groups, because the experimental conditions are never exactly the same. The main parameters usually being varied are light intensity, light spectra, composition of the reaction suspension and photocatalyst pretreatment. All these parameters affect the photocatalytic reaction rate, prohibit us to make strict comparison. In this work several classes of layered perovskite-type oxides were investigated for photocatalytic properties under identical experimental conditions. Results of our previous studies can be found in [25], [26].

2. Structure of Perovskite-Type Layered Oxides

Layered perovskite-type oxides are compounds with structure built on block principle. Two-dimensional layers with perovskite structure, composed of corner-shared MO_6 octahedrons, with thickness of n such octahedrons, are alternating with layers having different structure. Depending on the thickness of the perovskite layer (n) one-, two-, three (or more) -

layered perovskite-type oxides are distinguished (Figure 2). When n = ∞, the layered structure is lost, which corresponds to the conventional perovskite.

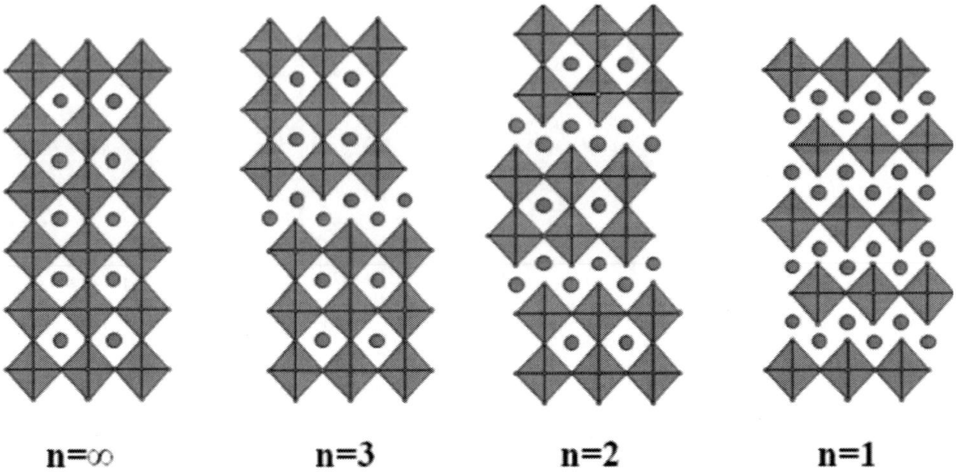

Figure 2. Structures of layered perovskite-type oxides with different layer thickness (n).

Depending on the structure of the interlayer space three major classes of layered perovskite-type oxides are distinguished: Ruddlesden-Popper, Dion-Jacobson and Aurivilius phases (Figure 3). Compounds belonging to the Ruddlesden-Popper phases [27][28] can be represented by the general formula $A_{n+1}B_nO_{3n+1}$. Their structure is characterized by the alternation of perovskite structure layers with layers, having the structure of rock salt. The value n=1 corresponds to the structure type of K_2NiF_4, n=2 - to the structure type $Sr_3Ti_2O_7$ (Figure 4).

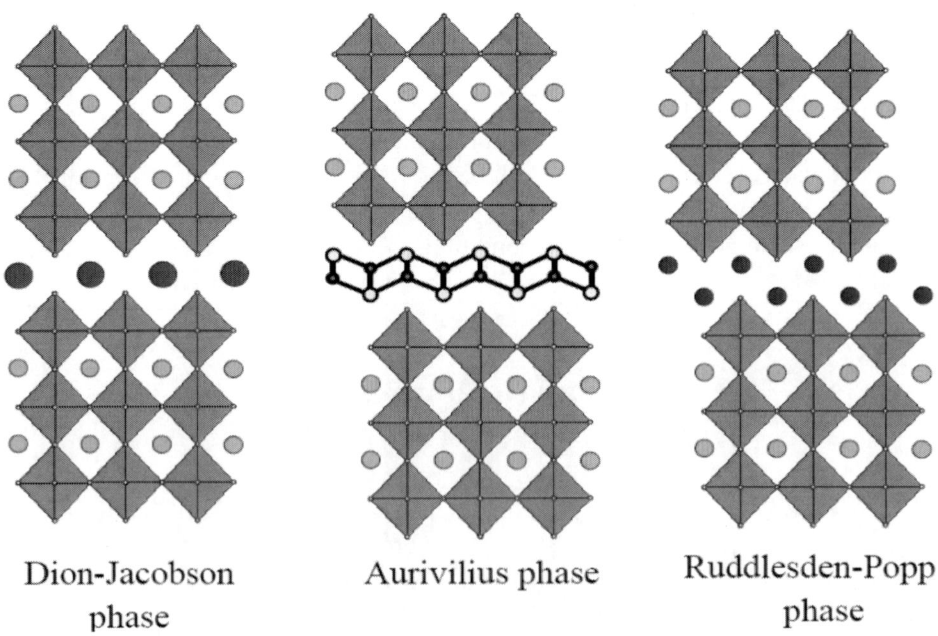

Figure 3. Main structure types of perovskite-type layered oxides.

An important feature of Ruddlesden-Popper phases with number of layers n≥2 is the existence of two non-equivalent structural positions for the A-cation with coordination numbers nine and twelve. If the positions of A are occupied by two different types of cations, their distribution between nonequivalent positions in the crystal lattice is of importance. As shown in the case of aluminates $Ln_2SrAl_2O_7$, [29][30] the change in the distribution from near-statistical for lanthanum to almost completely ordered for lanthanides of the second half results in a significant increase of the structure stability.

Among Ruddlesden-Popper phases the most studied photocatalyst is the triple-layered (n=3) titanate $K_2La_2Ti_3O_{10}$. In contrast with its lithium- and sodium-containing analogs, this compound possesses the ability of water intercalation in the interlayer space, accompanied with an increase of the cell parameter *c* (Figure 5).

Data regarding the total amount of water intercalating into this structure are controversial. In [30] a formula $K_2La_2Ti_3O_{10}{\cdot}H_2O$ is given, while in [32] - $K_2La_2Ti_3O_{10}{\cdot}2H_2O$. It should be noted that the intercalation process of water does not require the layered oxide to contact with liquid water, it occurs even under exposure to humid air. Probably, the amount of intercalated water depends on conditions (humidity, temperature), hence the contradictions in data. Dion-Jacobson phases were first synthesized in 1980 by M. Dion [33] and subsequently extensively studied by A. Jacobson. The general formula for these compounds can be given as $A'[A_{n-1}B_nO_{3n+1}]$ (A'-alkali metal, A - alkaline earth or rare earth element, B - usually pentavalent atom e.g. tantalum, niobium, titanium is possible). In the structure of these compounds perovskite layers are separated by layers composed only of metal cations, usually alkali or alkaline earth. The tantalate series $ANdTa_2O_7$ (A - alkali metal) belong to double-layered (n = 2) Dion-Jacobson phases, with structure shown on Figure 6.

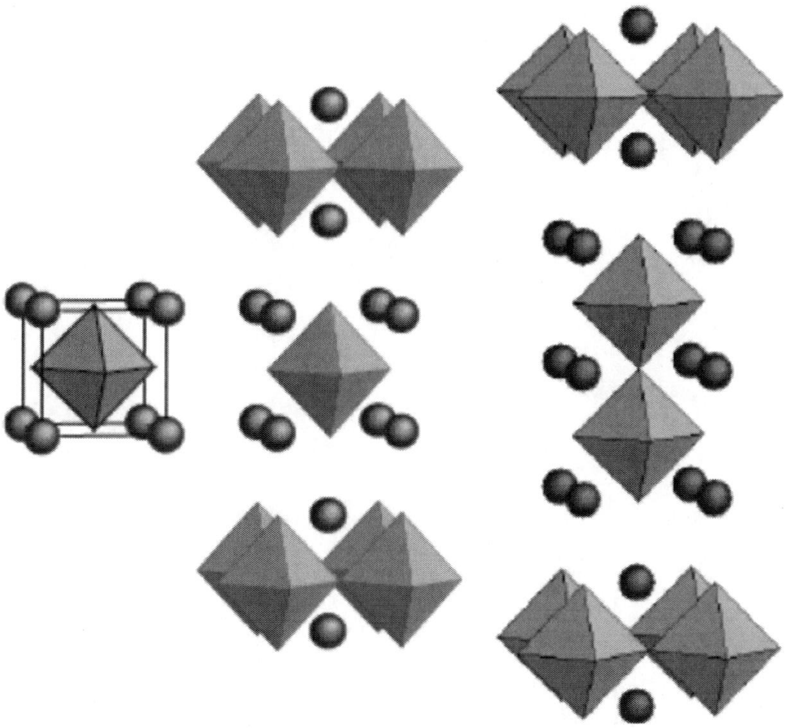

Figure 4. Structure of Ruddlesden-Popper phases (n=1, 2).

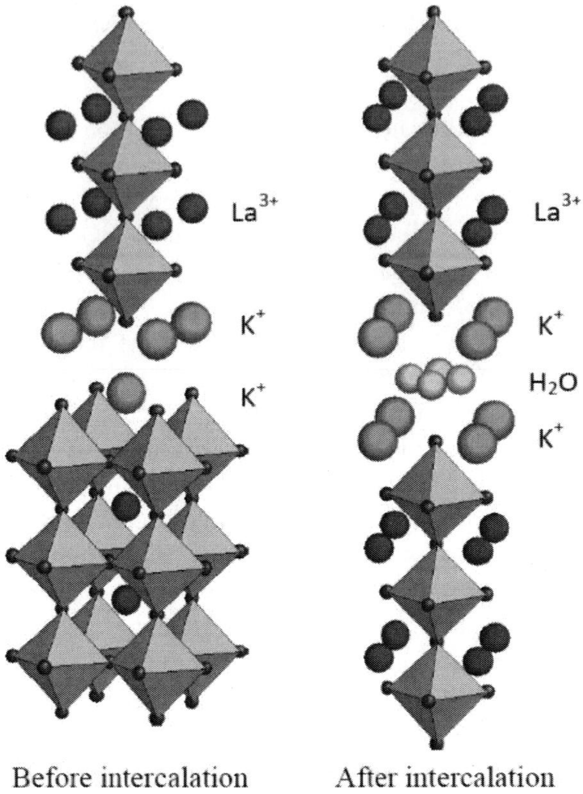

Figure 5. Structure of $K_2La_2Ti_3O_{10}$ before and after water intercalation.

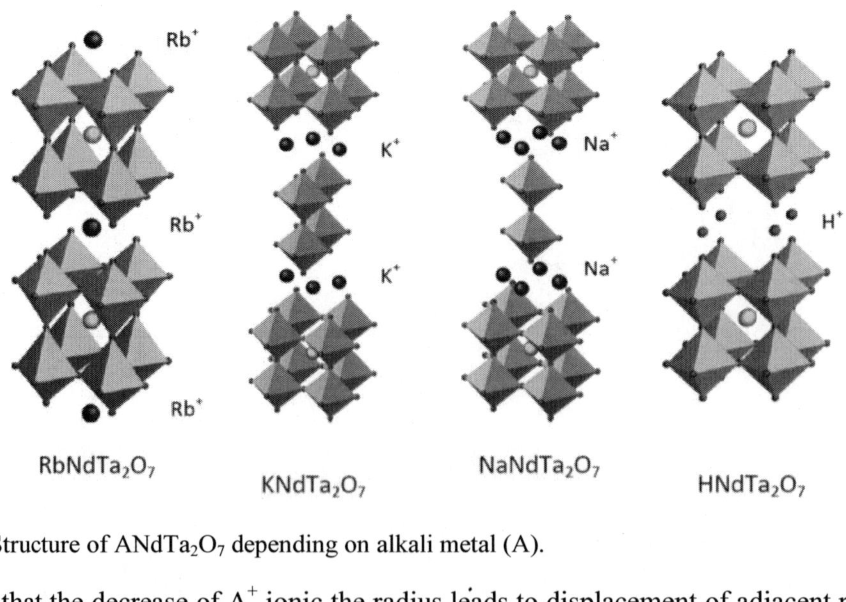

Figure 6. Structure of $ANdTa_2O_7$ depending on alkali metal (A).

Note that the decrease of A^+ ionic the radius leads to displacement of adjacent perovskite layers along the *a* and *b* axes for ½. A special situation is observed in the case of protonated form, where hydrogen forms a bridge between oxygen atoms of adjacent layers and has a coordination number of 2. The change of the A^+ ionic radius and structural differences lead to the ability of only two compounds of this series to intercalate water molecules in the interlayer space, namely $NaNdTa_2O_7$ and $HNdTa_2O_7$, accompanied with an increase in the *c* cell parameter [34]. Aurivillius phases are compounds of general formula $A_{n-1}Bi_2B_nO_{3n+3}$, first described by B. Aurivillius [35]. In their crystal structure perovskite layers alternate with fluorite structure formed by bismuth ions and oxygen. Thus composition and structure of Aurivillius phases is better described by general formula $(Bi_2O_2)A_{n-1}B_nO_{3n+1}$, mainly used in actual papers. Aurivillius phases mainly attract attention because of their ferroelectric properties, [36] but recently also due to photocatalytic activity [37].

3. EXPERIMENTAL

Synthesis

Layered perovskite-type oxides $A_2Ln_2Ti_3O_{10}$, $ALnTiO_4$ (A = Li, Na, K; Ln = La, Nd) and $ANdTa_2O_7$, $ANdNb_2O_7$ (A = Rb, Cs) were synthesized by solid-state method from the corresponding oxides and carbonates. Synthesis temperature was selected on the basis of published data and previously developed techniques and varied in the range of 900-1100°C depending on substance. Samples of $ANdTa_2O_7$ (A = Li, Na, K) were synthesized from $RbNdTa_2O_7$ by ion exchange reaction in melt of corresponding nitrate at 360°C, 10 hours. Protonated forms of layered oxides $H_2Ln_2Ti_3O_{10}$, $HLnTiO_4$ (Ln = La, Nd) and $HNdTa_2O_7$ were synthesized respectively from $K_2Ln_2Ti_3O_{10}$, $NaLnTiO_4$ and $RbNdTa_2O_7$ by ion exchange reaction in 0.1 M hydrochloric acid (five-fold excess).

Characterisation

Phase composition of synthesized samples was controlled by XRD using a powder X-ray diffractometer ARL X'TRA (CuKα). All samples were obtained without noticeable impurities. Diffuse reflectance spectra were recorded on a Shimadzu UV-2550 spectrophotometer, equipped with an ISR-2200 integrating sphere. The optical band gap was determined from the crosspoint of linear sections of the Kubelka-Munk function plot in the coordinates $(K \cdot h\nu)^{1/2} = f(h\nu)$. The morphology of samples was investigated by scanning electron microscopy by Carl Zeiss EVO 40EP and Zeiss Supra 40VP microscopes in the Resource Center "Nanotechnology" of Saint-Petersburg State University. The specific surface area of the samples was determined by argon adsorption (BET) at the temperature of liquid nitrogen.

Photocatalytic Experiments

Photoinduced hydrogen evolution was investigated on a specially designed set-up (Figure 7). To prepare the reaction suspension, 70 mg of photocatalyst were added to 70 mL of 0.1% (mol.) isopropyl alcohol aqueous solution, shaken and left for a day to establish equilibrium between the photocatalyst and solution.

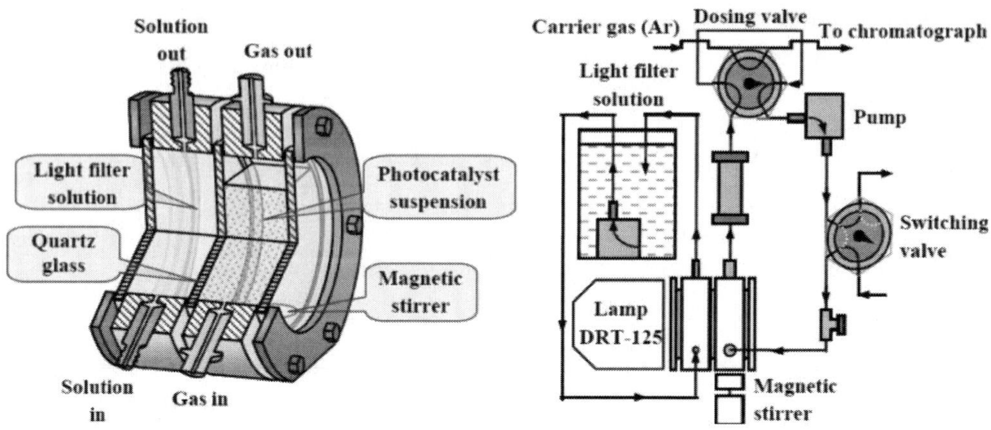

Figure 7. Reactor and set-up for investigation of photoinduced hydrogen evolution.

For experiments without isopropyl alcohol pure distilled water was taken. Straight before the experiment each suspension was sonicated for 10 min ("Sapphire" ultrasound bath, 55W) to disaggregate the photocatalyst particles. The photocatalyst suspension was placed in an external-irradiation reaction cell, equipped with a powerful magnetic stirrer and connected to a closed gas circulation system.

A medium-pressure mercury lamp DRT-125 (125W) was used as a radiation source. Light reaches the reaction cell only after passing through a thermostated at 15°C light filter solution (KCl + NaBr, each 6 g/L), which cuts off radiation of unwanted wavelengths (<220nm). The total irradiation time was 240 min for each experiment. The intensity of incident light was measured by ferrioxalate actinometry. Hydrogen, emitted during the photocatalytic reaction, accumulates in the gas phase, its composition was analyzed by gas chromatography after certain time intervals. A TSVET-800 chromatograph with thermal conductivity detector and 2 m molecular sieve column was used. At the beginning of each experiment, Ar gas was bubbled through the suspension for 15 min to remove dissolved air. This procedure was repeated after 120 min of irradiation to remove evolved hydrogen from the system.

For the analysis of liquid phase during the experiment, the reaction cell was equipped with a special vent, allowing to take samples with chromatographic syringe without any leakage. Analysis of the liquid phase was performed on TSVET-500 chromatograph with flame ionization detector and a 2 m Porapak R column.

2. Hydrogen Evolution from Layered Oxide Suspensions Under UV-Irradiation

Kinetic curves of hydrogen evolution from suspensions of layered oxides in 0.1% (mol.) isopropyl alcohol aqueous solution are presented on Figure 8-10. Hydrogen evolution rates, characterizing the photocatalytic activity of samples, were determined from the slope of these curves at the second stage of experiment, where they are almost linear.

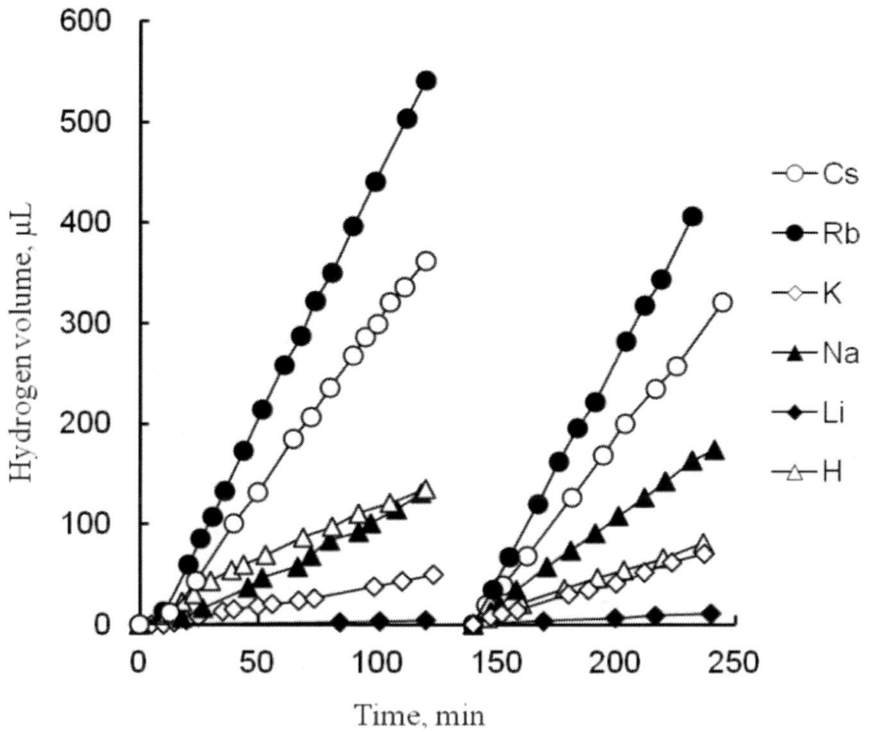

Figure 8. Hydrogen evolution over $ANdTa_2O_7$ (A=H, Li, Na, K, Rb, Cs).

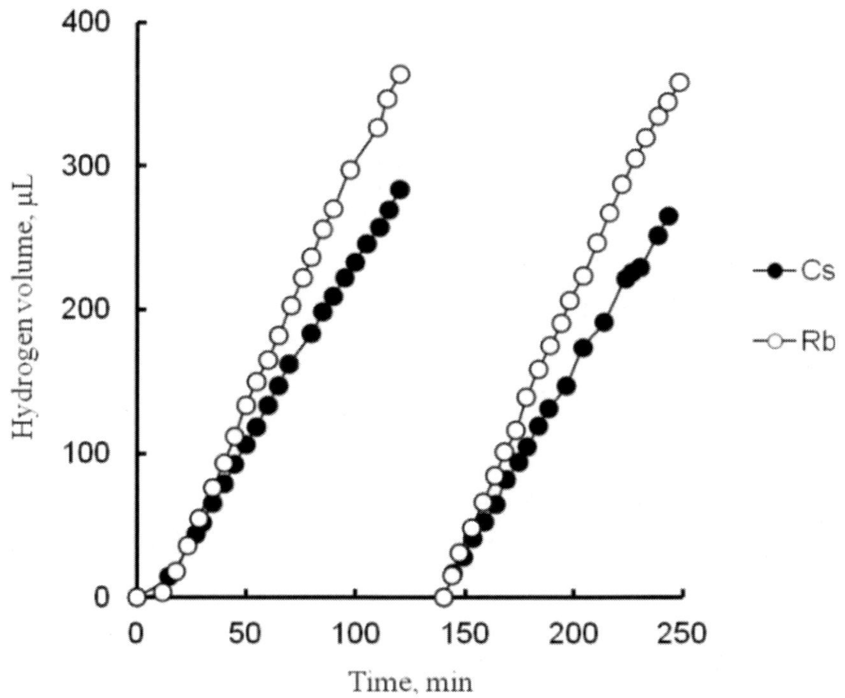

Figure 9. Hydrogen evolution over $ANdNb_2O_7$ (A= Rb, Cs).

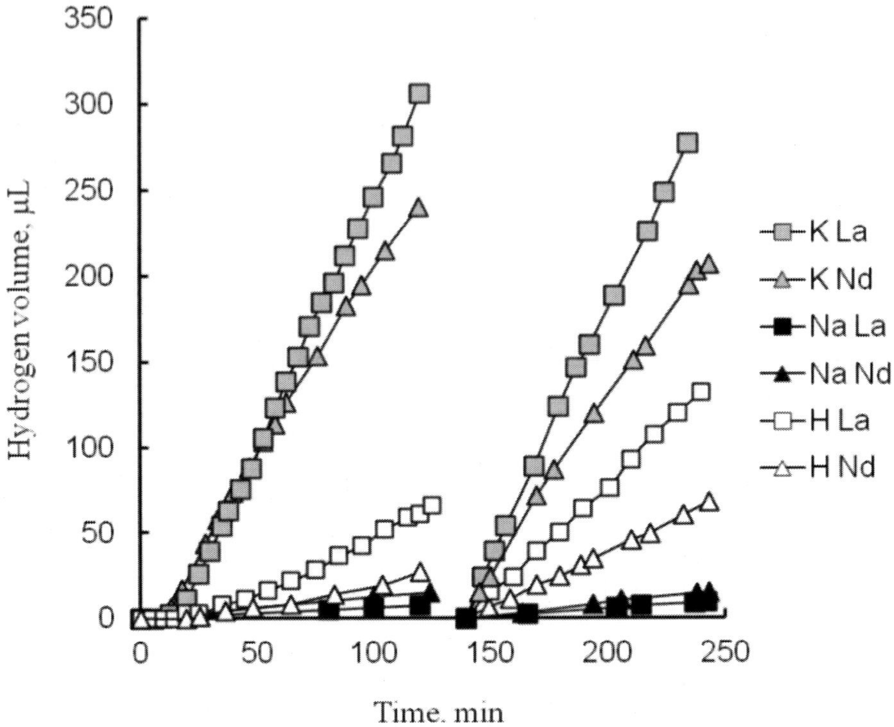

Figure 10. Hydrogen evolution over $A_2Ln_2Ti_3O_{10}$ (A = H, Li, Na, K, Rb, Cs; Ln = La, Nd).

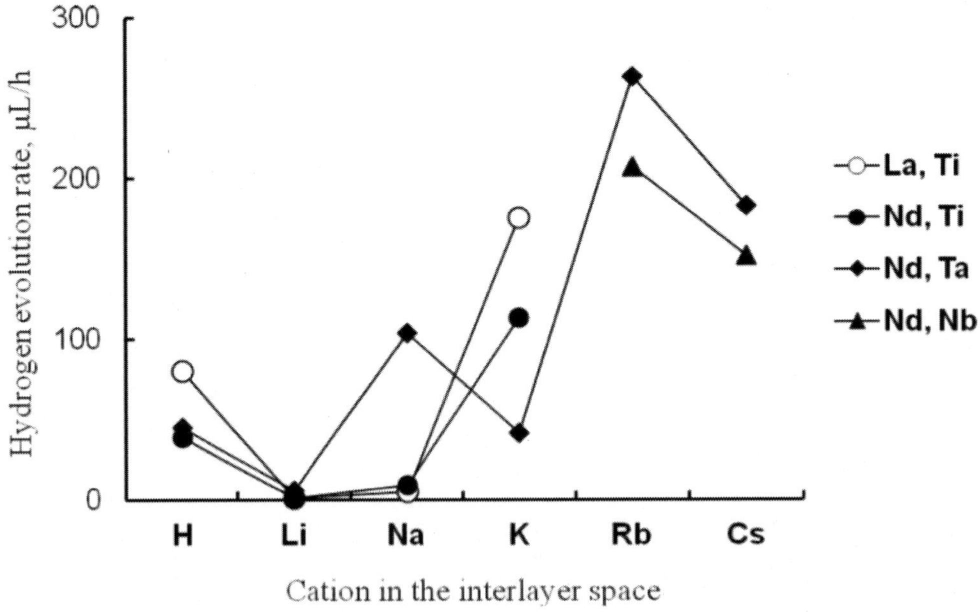

Figure 11. Dependance of photocatalytic activity on cationic composition of complex oxides.

These values are given on Figure 11. and also on Table 1, along with other measured characteristics. For comparison, data obtained for commercial TiO_2 P25 Evonic (Degussa) are also presented. The maximum hydrogen evolution rate (263 μL/h) was achieved over $RbNdTa_2O_7$, belonging to the Dion-Jacobson series. Niobates were slightly less active than tantalates. Among investigated titanates, potassium-containing substances showed the highest activity. Mono-layered $ANdTiO_4$ oxides did not show any notable activity. In general, from Figure 11. there can be seen no obvious dependence between the activity of layered oxide and its composition or structure type. However, some correlations can be found when considering specific surface areas of investigated photocatalysts and their ability to intercalate water in the interlayer space.

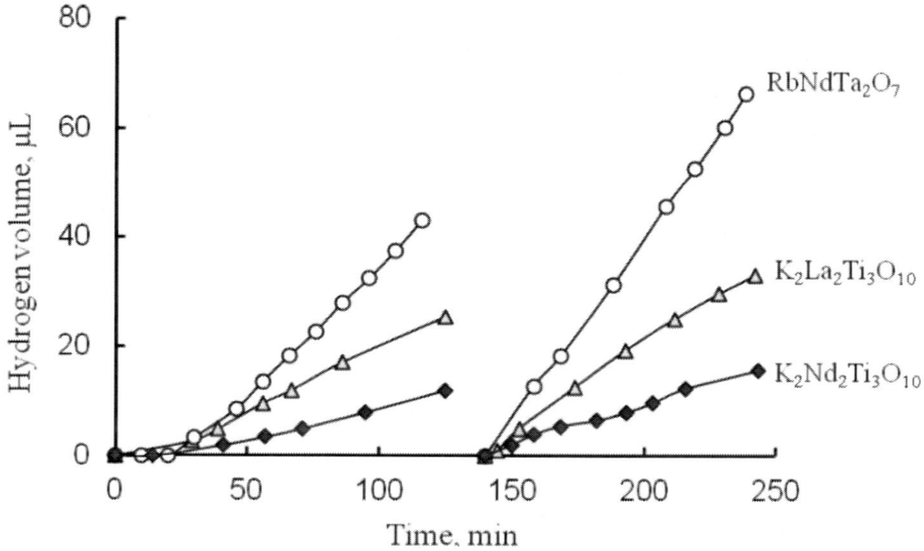

Figure 12. Hydrogen evolution over complex oxides in pure water.

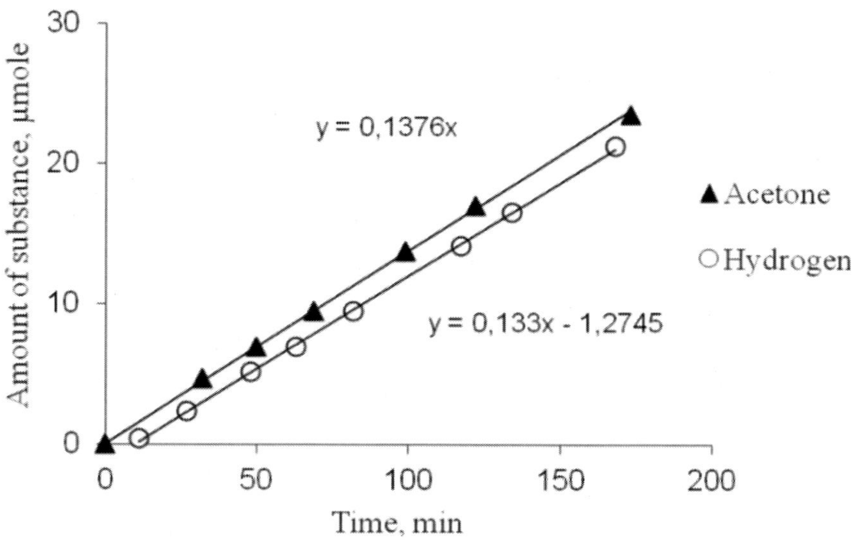

Figure 13. Comparison of hydrogen and acetone evolution rates over $RbNdTa_2O_7$ in a single experiment.

In the absence of isopropyl alcohol the rates of hydrogen evolution were found to be considerably less (Figure 12). Note that the photocatalytic activity decreases in the order $RbNdTa_2O_7$> $K_2La_2Ti_3O_{10}$> $K_2Nd_2Ti_3O_{10}$ both in presence and absence of isopropyl alcohol. This suggests that the obtained values of photocatalytic activities for different complex oxides can be compared correctly.

For $RbNdTa_2O_7$, the sample with the highest activity, it was additionally shown that isopropyl alcohol is oxidized to acetone during the photocatalytic reaction, acetone and hydrogen being produced in a molar ratio of 1:1 (Figure 13).

5. Dependance of Photocatalytic Activity on Layered Oxides Characteristics

Band Gap Energy

The measured band gap energy of tantalates $ANdTa_2O_7$ vary within 4.36-4.54 eV, depending on alkali metal. This is significantly higher than the values found for niobates and titanates - 3.50-3.70 eV. Obviously, there are less photons available for fundamental absorption on tantalates than on titanates and niobates due to the higher value of E_g.

Therefore, at equal observed photocatalytic activity, the quantum efficiency for tantalates will be higher. Values of observed quantum efficiency calculated on the base of light intensity measurements and lamp emission spectra are presented in Table 1. The maximum quantum efficiency is achieved in the case of $RbNdTa_2O_7$ tantalate. Taking into account its small surface area and inability to intercalate water into the interlayer space, the high efficiency can be attributed to relatively low electron-hole recombination rate compared to other tantalates.

Surface Area and Morphology

The minimum BET surface area is observed for samples synthesized by high-temperature solid-state method. Samples obtained by ion exchange in molten nitrate, have higher values of surface area, which may be due to the partial destruction of the precursor oxide particles during the ion exchange process followed by formation of defects such as cracks, pores, etc. The maximum surface area is observed for titanates obtained by ion exchange in hydrochloric acid. Correlation between photocatalytic activity and surface area of layered titanate samples containing identical alkali metals can be seen from Figure 14. Thus, we can propose that the effect of lanthanide on the photocatalytic activity is reduced to its effect on the specific surface area of synthesized sample. This explains why $Na_2Nd_2Ti_3O_{10}$ is more active than its lanthanum-containing analog, while in other cases the neodymium-containing samples showed less activity. Effect of alkali metal on photocatalytic activity can not be explained in terms of specific surface area. The lines corresponding to the different alkali metals on Figure 14. have different slopes due to a combination of factors that are independent of specific surface area, and are related only to the intrinsic characteristics of the substance. Differences in the specific surface area and photocatalytic activity of the samples are related to their particle morphology (Figure 15).

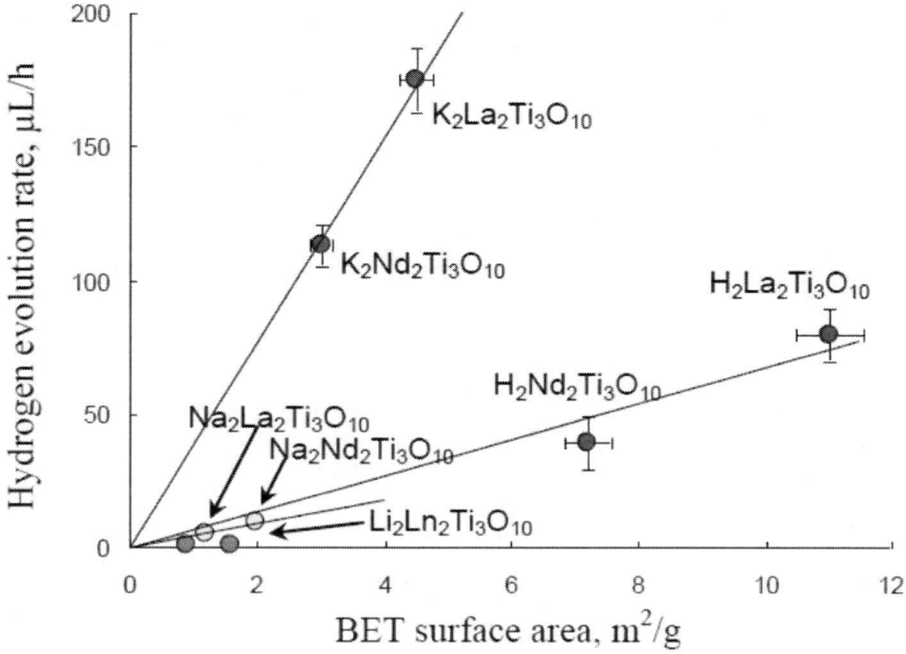

Figure 14. Correlation between photocatalytic activity and specific surface area.

Thus, the layered oxide $Li_2La_2Ti_3O_{10}$ consists of large particles ≈1μm thick with a smooth surface. This oxide has a BET surface area of 0.9 m^2/g, and show miserable photocatalytic activity.

At the same time, the hydrogen-containing sample $H_2La_2Ti_3O_{10}$ consists of lamellar particles, 200 nm thick, which, on closer examination, turn out to be split into even thinner sheets of 10-20 nm thickness. This effect is associated with the synthesis method of this sample, including treatment by hydrochloric acid.

The specific surface area is 11 m^2/g, which leads to a hydrogen evolution rate of 80 μL/h. The morphology of $K_2La_2Ti_3O_{10}$ particles is almost similar to $H_2La_2Ti_3O_{10}$, but the splitting into nanolayers is not observed, the particles have a thickness of 50-100 nm. However, the oxide has the ability to intercalate water in the interlayer space, which distinguishes it from other investigated titanates containing no potassium. This may explain why the rate of hydrogen evolution is 175 μL/h, the surface area being only 4.5 m^2/g.

$Li_2La_2Ti_3O_{10}$ $H_2La_2Ti_3O_{10}$ $K_2La_2Ti_3O_{10}$

Figure 15. SEM micrographs showing morphology of layered titanate particles.

Table 1. Photocatalytic activity of layered oxides compared with other characteristics

Sample	Hydrogen evolution rate ω_V, μL/h	Band gap energy E_g, eV	BET surface area S, m^2	Apparent quantum efficiency, %	Water intercalation
$HNdTa_2O_7$	45	4.36	3.9	0,27	yes
$LiNdTa_2O_7$	6	4.49	4.3	0,043	-
$NaNdTa_2O_7$	104	4.38	5.0	0,61	yes
$KNdTa_2O_7$	42	4.38	1.2	0,25	-
$RbNdTa_2O_7$	263 (40*)	4.54	0.8	2,02	-
$CsNdTa_2O_7$	183	4.40	2.0	1,08	-
$RbNdNb_2O_7$	207	3.70	1.5	0,28	-
$CsNdNb_2O_7$	152	3.64	0.5	0,21	-
$H_2La_2Ti_3O_{10}$	80	3.54	11	0,11	-
$Li_2La_2Ti_3O_{10}$	1	3.64	0,9	0,001	-
$Na_2La_2Ti_3O_{10}$	5	3.50	1.2	0,007	-
$K_2La_2Ti_3O_{10}$	175 (18*)	3.58	4.5	0,24	yes
$H_2Nd_2Ti_3O_{10}$	39	3.51	7.2	0,053	-
$Li_2Nd_2Ti_3O_{10}$	1	3.65	1.6	0,001	-
$Na_2Nd_2Ti_3O_{10}$	9	3.50	2.0	0,012	-
$K_2Nd_2Ti_3O_{10}$	113 (9*)	3.54	3.0	0,15	yes
TiO_2 P25	96	3.10	50	0,08	-

* - measured in the absence of isopropyl alcohol.

Water Intercalation

Among investigated layered oxides only four possess the ability of water intercalation into the interlayer space: $K_2La_2Ti_3O_{10}$, $K_2Nd_2Ti_3O_{10}$, $NaNdTa_2O_7$ and $HNdTa_2O_7$. All of them show considerable photocatalytic activity. The possibility of water intercalation not only increases the effective surface of the photocatalyst. It leads to the formation of new reaction centers located within the crystal structure having different properties from those that are located on the surface. This can lead to selectivity and make the oxidation and reduction reactions proceed at different centers which are spatially separated. However, the intercalation ability is not the only feature of layered perovskite-type oxides, leading to high photocatalytic activity, as demonstrated by such compounds as $RbNdTa_2O_7$, $CsNdTa_2O_7$, $RbNdNb_2O_7$, $CsNdNb_2O_7$. They all contain heavy alkali metals, which results in a large distance between the perovskite slabs in the structure. Further studies are needed to understand how this promotes high efficiency of the photocatralytic process.

Conclusion

Our comparative study of photoinduced hydrogen generation over layered perovskite-type oxides showed that the maximum photocatalytic activity (263 μL/h) and the maximum apparent quantum efficiency (2%) are shown by $RbNdTa_2O_7$. On the example of this complex oxide it was shown, that during the photocatalytic reaction in the presence of isopropyl alcohol acetone and hydrogen are formed at equal rate. In the absence of isopropyl alcohol hydrogen generation rates over $RbNdTa_2O_7$, $K_2La_2Ti_3O_{10}$, $K_2Nd_2Ti_3O_{10}$ increase in the same order as they do in its presence.

Correlation between photocatalytic activity and BET surface area for samples containing the same alkali metal was demonstrated by the triple-layered Ruddlesden-Popper titanate series $A_2Ln_2Ti_3O_{10}$ (A = H, Li, Na, K; Ln = La, Nd). The ability of water intercalation of a layered oxide was found to be related to high photocatalytic activity, but not to be the only feature determining it.

ACKNOWLEDGMENT

This work was supported by Russian Foundation for Basic Research (grant 12-03-00761).

Reviewed by Professor Jhy-Chem Liu, Department of Chemical Engineering, National Taiwan University of Science and Technology, Taiwan

REFERENCES

[1] Fujishima, A.; Honda, K. Electrochemical photolysis of water at a semiconductor electrode // *Nature.* 1972, *238*, 37-38.

[2] Zvereva, I. A.; Kalinkina, L. M.; Rodionov, I. A.; Sankovich, A. M.; Kolesnik, I. V.; Gudilin, E. A. Production of Nanocrystalline Titanium Dioxide Photoactive Coatings for Decomposition of Water Organic Pollutants in a Flow Reactor. *Glass Phys. Chem.* 2012, *38*, 504–510.

[3] Ivanov, V. K.; Maksimov, V. D.; Shaporev, A. S.; Baranchikov, A. E.; Churagulov, B. P.; Zvereva, I. A.; Tret'yakov, Yu. D. Hydrothermal Synthesis of Efficient TiO_2-Based Photocatalysts . *Russ. J. Inorg. Chem.* 2010, *55*, 150-154.

[4] Gavrilov, A. I.; Rodionov, I. A.; Gavrilova, L. Yu.; Zvereva, I. A.; Churagulov, B. P.; Tret'yakov, Yu. D. Hydrothermal Synthesis of nanostructures on the base of TiO_2 for water splitting. *Docl. Chem.* 2012, *444*, 510-513.

[5] Asahi, R.; Morikawa, T.; Ohwaki, T.; Aoki, K.; Taga, Y. Visible-light photocatalysis in nitrogen-doped titanium oxides. *Science*. 2001, *293*, 269-271.

[6] Kudo, A.; Domen, K.; Maruya, K.; Onishi, T. Photocatalytic Activities of TiO_2 Loaded with NiO. *Chem. Phys. Lett.* 1987, *133*, 517–519.

[7] Kudo, A.; Sayama, K.; Tanaka, A.; Asakura, K.; Domen, K.; Maruya, K.; Onishi, T. Nickel-loaded $K_4Nb_6O_{17}$ photocatalyst in the decomposition of H_2O into H_2 and O_2: Structure and reaction mechanism. *J. Catal.* 1989, *120*, 337-352.

[8] Takata, T.; Tanaka, A.; Hara, M.; Kondo, J. N.; Domen, K. Recent progress of photocatalysts for overall water splitting. *Catal. Today*. 1998, *44*, 17-26.

[9] Nishimoto, S.; Matsuda, M.; Miyake, M. Novel protonated and hydrated n=1 Ruddlesden–Popper phases, $H_xNa_{1-x}LaTiO_4 \cdot yH_2O$, formed by ion-exchange/intercalation reaction. *J. Solid State Chem.* 2005, *178*, 811.
[10] Silyukov, O.; Chislov, M.; Burovikhina, A.; Utkina, T.; Zvereva, I. Thermogravimetry study of ion exchange and hydration in layered oxide materials. *J. Therm. Anal. Calorim.* 2012, *110*, 187–192.
[11] Zvereva, I. A.; Silyukov, O. I.; Chislov, M. V.; Ion-exchange reactions in the structure of perovskite-like layered oxides: I. Protonation of $NaNdTiO_4$ complex oxide. *Russ. J. Gen. Chem.* 2011, *81*, 1434-1441.
[12] Kim, H. G.; Becker, O. S.; Jang, J. S.; Ji, S. M.; Borse, P. H.; Lee, J. S. Formation of crystalline $TiO_{2-x}N_x$ and its photocatalytic activity. *J. Solid State Chem.* 2006, *179*, 1214–1218.
[13] Uma, S.; Raju, A. R.; Gopalakrishnan, J. Bridging the Ruddlesden–Popper and the Dion–Jacobson series of layered perovskites: synthesis of layered oxides, $A_{2-x}La_2Ti_{3-x}Nb_xO_{10}$ (A = K, Rb), exhibiting ion exchange. *Mater. Chem.* 1993, *3*, 709-713.
[14] Takata, T.; Shinohara, K.; Tanaka, A.; Hara, M.; Kondo, J. N.; Domen, K. A highly active photocatalyst for overall water splitting with a hydrated layered perovskite structure. *J. Photochem. Photobiol., A.* 1997, *106*, 45–49.
[15] Machida, M.; Yabunaka, J.; Kijima, T. Efficient photocatalytic decomposition of water with the novel layered tantalate $RbNdTa_2O_7$. *Chem. Commun.* 1999, 1939–1940
[16] Kim, H. G.; Hwang, D. W.; Lee, J. S. An Undoped, Single-Phase Oxide Photocatalyst Working under Visible Light. *J. Am. Chem. Soc.* 2004, *126*, 8912-8913
[17] Yang, Y.; Qiu, G.; Chen, Q.; Feng, Q.; Yin, Z. Influence of calcination atmosphere on photocatalytic reactivity of $K_2La_2Ti_3O_{10}$ for water splitting. *Trans. Nonferrous Met. Soc. China.* 2007, *17*, 836-840.
[18] Huang, Y.; Wu, J.; Wei, Y.; Lin, J.; Huang, M. Hydrothermal synthesis of $K_2La_2Ti_3O_{10}$ and photocatalytic splitting of water. *J. Alloy. Compd.* 2008, *456*, 364–367.
[19] Yang, Y.; Chen, Q.; Yin, Z.; Li, J. Study on the photocatalytic activity of $K_2La_2Ti_3O_{10}$ doped with zinc (Zn). *Appl. Surf. Sci.* 2009, *255*, 8419–8424.
[20] Yang, Y.; Chen, Q.; Yin, Z.; Li, J. Study on the photocatalytic activity of $K_2La_2Ti_3O_{10}$ doped with vanadium (V). *J. Alloy. Compd.* 2009, *488*, 364–369.
[21] Thaminimulla, C. T. K.; Takata, T.; Hara, M.; Kondo, J. N.; Domen, K. Effect of Chromium Addition for Photocatalytic Overall Water Splitting on Ni–$K_2La_2Ti_3O_{10}$. *J. Catal.* 2000, *196*, 362–365.
[22] Huang, Y.; Wei, Y.; Cheng, S.; Fan, L.; Li, Y.; Lin, J.; Wu, J. Photocatalytic property of nitrogen-doped layered perovskite $K_2La_2Ti_3O_{10}$. *Sol. Energ. Mat. Sol. C.* 2010, *94*, 761–766.
[23] Tai Y.; Chen J.; Yang C.; Wan B. Preparation of nano-gold on $K_2La_2Ti_3O_{10}$ for producing hydrogen from photo-catalytic water splitting. *Catal. Today.* 2004, *97*, 95–101.
[24] Wu, J.; Huang, Y.; Li, T.; Lin, J.; Huang, M.; Wei, Y. Synthesis and photocatalytic properties of layered nanocomposite $H_2La_2Ti_3O_{10}/Fe_2O_3$. *Scripta Mater.* 2006, *54*, 1357–1362.
[25] Rodionov, I. A.; Silyukov, O. I.; Zvereva, I. A. Study of Photocatalytic Activity of Layered Oxides: $NaNdTiO_4$, $LiNdTiO_4$, and $HNdTiO_4$ Titanates. *Russ. J. Gen. Chem.* 2012, *82*, 635–638.

[26] Rodionov, I. A.; Silyukov, O. I.; Utkina, T. D.; Chislov, M. V.; Sokolova, Yu. P.; Zvereva, I. A. Photocatalytic properties and hydration of perovskite-type layered titanates $A_2Ln_2Ti_3O_{10}$ (A = Li, Na, K; Ln = La, Nd). *Russ. J. Gen. Chem.* 2012, *82*, 1191-1196.

[27] Ruddlesden, S. N.; Popper, P. New compounds of the K_2NiF_4 type. *Acta Crystallogr.* 1957, *10*, 538-539.

[28] Ruddlesden, S. N.; Popper, P. The compound $Sr_3Ti_2O_7$ and its structure. *Acta Crystallogr.* 1958, *11*, 54-55.

[29] Zvereva, I.; Smirnov, Yu.; Gusarov, V.; Popova, V.; Choisnet, J. Complex aluminates $Re_2SrAl_2O_7$ (Re = La, Nd, Sm-Ho): Cation ordering and stability of the double perovskite slab–rocksalt layer P2/RS intergrowth. *Solid State Sci.* 2003, *5*, 343-349.

[30] Zvereva, I. A.; Skorobogatov, G. A. *Synthetic perovskite type layered oxides*. BBM: Saint-Petersburg, RU, 2010; pp 1-224.

[31] Gopalakrishnan, J.; Bhat, V. $A_2Ln_2Ti_3O_{10}$ (A = K or Rb; Ln = La or Rare Earth): A New Series of Layered Perovskites Exhibiting Ion Exchange. *Inorg. Chem.* 1987, *26*, 4299-4301.

[32] Toda, K.; Watanabe, J.; Sato, M. Crystal structure determination of ion-exchangeable layered perovskite compounds, $K_2La_2Ti_3O_{10}$ and $Li_2La_2Ti_3O_{10}$. *Mater. Res. Bull.* 1996, *31*, 1427-1435.

[33] Dion, M.; Ganne, M.; Tournoux, M. Nouvelles familles de phases $MIMII_2Nb_3O_{10}$ a feullets 'perovskites'. *Mater. Res. Bull.* 1981, *16*, 1429-1435.

[34] Machida, M.; Miyazaki, K.; Matsushima, S.; Arai, M. Photocatalytic properties of layered perovskite tantalates, $MLnTa_2O_7$ (M = Cs, Rb, Na, and H; Ln = La, Pr, Nd, and Sm). *J. Mater. Chem.* 2003, *13*, 1433–1437.

[35] Aurivillius, B. Mixed bismuth oxides with layer lattices. I, II, and III. *Arki Kemi.* 1949, *1*, 463; 1949, *1*, 499; 1950, *2*, 519.

[36] Smolenski, G. A.; Isupov, V. A.; Agranovskaya, A. I. Ferroelectrics of the Oxygen-Octahedral Type With Layered Structure. *Sov. Phys. Solid State.* 1961, *3*, 651-655.

[37] Zhang, H.; Chen, G.; Li, X. Synthesis and visible light photocatalysis water splitting property of chromium-doped $Bi_4Ti_3O_{12}$. *Solid State Ionics.* 2009, *180*, 1599-1603.

In: Perovskite
Editors: Jinghua Zhang and Huan Li

ISBN: 978-1-62417-800-9

Chapter 8

ELECTRON STRUCTURE OF PEROVSKITE ELECTRON-IONIC CONDUCTORS

N. Chezhina[*] *and D. Korolev*
Sankt-Petersburg State University, St. Petersburg, Russia

ABSTRACT

Nowadays one of the most urgent problems facing the researchers in the field of solid state chemistry is the problem of obtaining materials for energy-saving technologies. The so called solid oxide fuel cells require the materials for cathodes and electrolytes. Lanthanum gallate with perovskite structure doped with strontium and magnesium is a popular subject to use as an electrolyte, since it is conducting oxygen through the vacancies in the oxygen sublattice resulting from heterovalent doping. However this doped lanthanum gallate appears to be nonstable at high temperatures, at which these systems operate. On doping lanthanum gallate with strontium and magnesium and also with a transition element such systems become stable enough, even though some electron conductivity appears, this is good for cathodes but seems to be bad for electrolytes.

Several problems arise in this context, which led us to study the electron structure of these systems. First: what is the reason for the stabilization of the whole system on doping with transition elements in addition to an alkaline-earth or magnesium. The second: which element is the best for electrolytes and which for cathodes. The third: what must be the quantitative relationships between transition element and an element of the second group of Periodic Table. Up to now these problems were solved empirically, the most commonly used ratio M:Sr or M:Mg was 5:1. There are other problems, such as, why strontium but not calcium or barium is used for SOFC.

The explanations for these questions seem to lie in the electron structure of such systems, which means the state of transition element atoms and interatomic interactions in these systems. Transition elements are paramagnetic, and the whole systems are solid solutions. Therefore their electron structure can be studied by our method of magnetic dilution. This means a systematic study of magnetic susceptibility of a wide series of concentrations in the systems $La_{1-y}Sr_yM_xGa_{1-x}O_{3-\delta}$ or $LaM_xGa_{1-x-y}Mg_yO_{3-\delta}$ with fixed ratio

[*] E-mail: Chezhina@nc2490.spb.edu.

$x{:}y$, starting with solid solutions $LaMO_3$ in $LaGaO_3$ and then passing to $x{:}y$ = 5:1, $x{:}y$ = 2:1 and doping with both Sr and Mg – M:Sr:Mg = 5:1:1.

The examination of magnetic characteristics resulted in the following conclusions: 1) no oxidation of paramagnetic elements is observed in the solid solution upon doping with a bivalent element; 2) in the solid solutions of lanthanum gallate doped with a transition and a diamagnetic elements we see the formation of large (about 1-2 nm in size) clusters; 3) these clusters are not destroyed by the dilution and necessarily include also strontium or magnesium and accompanying vacancies in the oxygen sublattice.

Therefore the reason for the structure stabilization upon doping lanthanum gallate with a diamagnetic and paramagnetic elements is the formation of nanosize clusters, containing both elements and vacancies in the oxygen sublattice. Clustering increases as the content of bivalent element increases.

The study of the impedance spectra of these systems showed that an increase in the bivalent element concentration results in a decrease in conductivity in doped lanthanum gallate. A comparison between several d-elements and calculation of the bond structure allowed nickel to be recommended as a doping element for cathodes whereas cobalt and chromium – for electrolytes. A scheme is suggested to account for for the fact that the ratio M:Sr(Mg)=5:1 appears to be the best for optimizing the conductivity of doped lanthanum gallate.

The general conclusion is that clustering of paramagnetic elements in lanthanum gallate is an important factor for their conductivity. Clustering is necessary for maintaining the stability of the structure, but an increase in clustering following an increase in the bivalent atom concentration results in a decrease in conductivity owing to the fact that the field of negatively charged cluster hampers the mobility of vacancies.

Complex oxides with perovskite structure used to be widely studied in 60-ties of XX century then gave way to the more promising for the time layered systems – Ruddlesden-Popper phases, perovskite-like oxides containing barium and bismuth, etc. However, nowadays standard perovskites, ABO_3 experience a renewed interest from the researchers owing to their demonstrating new and promising properties very important for growing requirements for the materials of electronic industry.

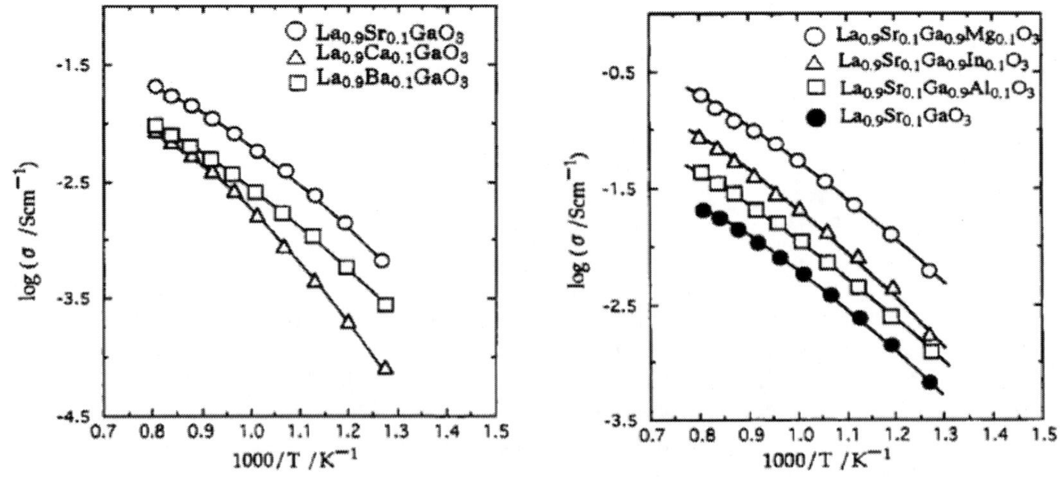

Figure 1. Plots of lg σ vs inverse temperature for doped lanthanum gallate [1].

These are manganites with colossal magnetoresistivity, multiferroics, and electron-ionic conductors for SOFC technologies. Those latter are based on lanthanum gallate doped with various elements.

In 1994 Ishihara [1] was the first to find the ionic conductivity in lanthanum gallate doped with bivalent elements, an increase in the conductivity being observed in the $La_{0.9}Ae_{0.1}GaO_3$ systems (Ae = Ca, Sr, Ba) in the Ca < Ba < Sr series (Figure 1). In the case, when magnesium, aluminum, or indium substituted for a fraction of gallium atoms - $La_{0.9}Sr_{0.1}Ga_{0.9}B_{0.1}O_3$ (B = Al, In, Mg) the conductivity increased in the Al < In < Mg series and was the greatest for lanthanum gallate containing Sr and Mg simultaneously. The thermodynamic examination shows a lesser endothermic effect of the solid solution formation on introduction of Sr (138 ± 12 kJ/mol for $La_{1-x}Sr_xGa_{1-y}Mg_yO_{3-\delta}$ and 166±12 kJ/mol for $La_{1-x}Ba_xGa_{1-y}Mg_yO_{3-\delta}$; $x > 0$, $y \leq 0.2$) or magnesium (275 ± 37 kJ/mol for $LaGa_{1-y}Mg_yO_{3-\delta}$, $y \leq 0.2$) [2]. This suggests that Sr and Mg are the most suitable doping elements in lanthanum gallate from the elements of the II group of Periodic table. However, along with the main perovskite phase the formation of non-conducting phases $SrGaO_3$, $LaSrGa_3O_7$, $LaSrGaO_4$ and La_4SrO_7 was observed [2-7]. Introduction of a transition element to gallium sites stabilizes the system [8, 9], but results in electronic conductivity, which is good for cathodes, but rather a drawback for electrolytes. Lanthanum gallate doped with dia- and paramagnetic elements is a promising material for SOFS due to the fact that both cathode and electrolyte may be composed of the systems close in their coefficient of thermal expansion, which is very important for SOFC. The problem consists of selecting the optimal doping elements and the ratios between dia- and paramagnetic elements.

Potential candidates for substituting in A and B sublattices comprise no less than a half of Periodic system of elements. Some of them are given in Figure 2 [10].

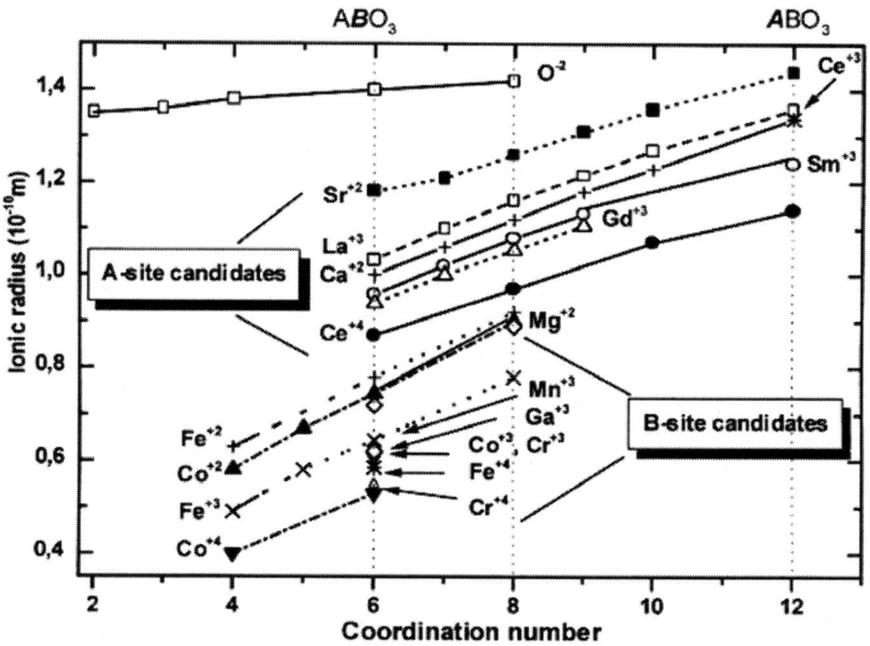

Figure 2. Potential candidates for doping lanthanum gallate [10].

This is the reason for a lot of works devoted to the study of various systems based on lanthanum gallate [11-24], but the search of optimal compositions is carried out empirically without thorough control over the composition resulting from high temperature synthesis. The data of various works contradict each other [19, 23].

For elucidating the qualitative and quantitative regularities in the systems in question the most important seems to be the knowledge of their electron structure. This includes not only the valence and spin states of the atoms, but also their mutual influence and the special features of interatomic interactions. The changes in the type of conductivity on varying the qualitative or quantitative composition of conductors are directly associated with interatomic interactions and atom states in their structure. As for now the question about the electron structure of doped lanthanum gallate remains open, being partially solved on using the quantum chemistry calculations [25-27].

Several problems arise on surveying the applications of such systems. The first: what is the reason for the stabilization of lanthanum gallate structure on doping it with diamagnetic bivalent elements and paramagnetic transition elements simultaneously? The second: what must be the ratio *transition element* : *element of II group* to result in the greatest oxygen conductivity? The third: why in all the current literature we found lanthanum gallate doped with strontium and/or magnesium, but not with calcium and barium? These rather urgent problems require unambiguous answers. In spite of a large body of data on this topic in their great majority the studies are empirical and random. The absence of systematic approach to the study of electron-ionic conductors within the context of "composition – structure – property" hampers the interpretation of experimental results and does not permit their generalization. The data about the electron structure would allow an insight into the problem of electronic and ionic conductivity of doped lanthanum gallates.

We emphasize that lanthanum gallates doped with transition elements are more or less diluted solid solutions. This makes it possible to use the method of magnetic dilution to study their electron structure. Magnetic dilution method [28] means the study of magnetic properties, first of all magnetic susceptibility, of a wide concentration range of solid solutions of paramagnetic oxide in an isomorphous diamagnetic matrix. Extrapolation of magnetic characteristics to the infinite dilution of a solid solution usually gives the susceptibility and effective magnetic moment for a single paramagnetic atom, and the dependencies of magnetic characteristics on the content of paramagnetic atoms allows the exchange interactions and clustering in the solid solutions to be simulated.

We studied magnetic and electrophysical properties of the solid solutions of lanthanum gallate $La_{1-y}Sr_yM_xGa_{1-x}O_{3-\delta}$ or $LaM_xGa_{1-x-y}Mg_yO_{3-\delta}$ with fixed ratio x:y, starting with solid solutions $LaMO_3$ in $LaGaO_3$ and then passing to $x:y = 5:1$, $x:y = 2:1$ and doping with both Sr and Mg – M:Sr:Mg = 5:1:1 (M = Cr, Co, Ni) [29-37] and also calcium and barium doped lanthanum gallate containing Cr with the ratio Cr:Ca(Ba) = 5:1.

EXPERIMENTAL

Solid solutions with $0.01 \leq x \leq 0.10$ were obtained by ceramic or sol-gel procedure [31-33]. The X-ray analysis (The DRON-3 and URS-50 I, CuK_α emission) showed that all the samples are single phase and have the structure of orthorhombic lanthanum gallate.

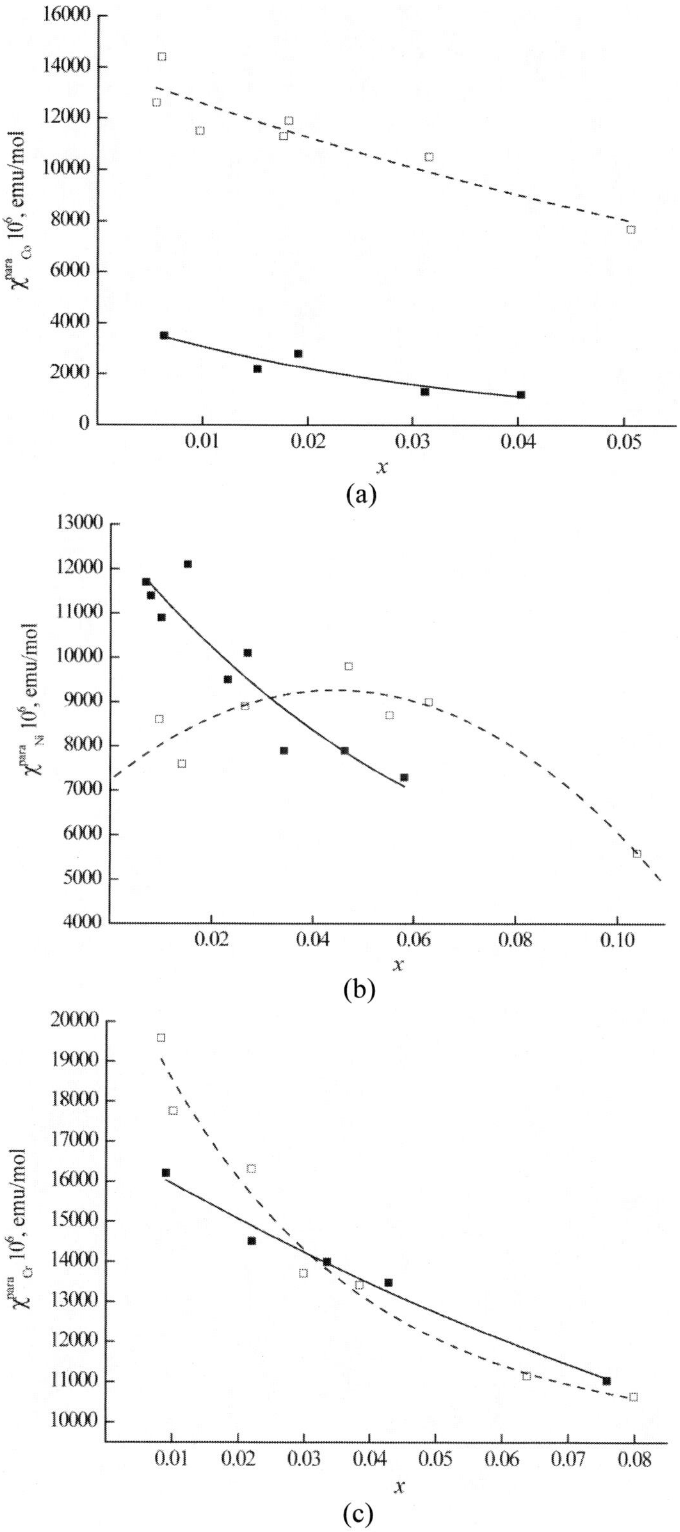

Figure 3. Isotherms of magnetic susceptibility for $LaM_xGa_{1-x}O_3$ (solid line) and $La_{1-0.2x}Sr_{0.2x}M_xGa_{1-x}O_{3-\delta}$ (dash line): (*a*) Co-containing systems; (*b*) Ni-containing systems; (*c*) Cr-containing systems.

High-temperature X-ray analysis in the temperature range 300-1000 K showed that the samples undergo a reversible phase transition orthorhombic↔rhombohedral lattice in the temperature range 320-360 K. The chemical analysis for the contents of paramagnetic and diamagnetic doping element was carried out by the method of atom-emission spectroscopy (the ISR type SPECTRO CIROS device). The accuracy of the chemical analysis was about 3% from x and y in the solid solution formula. Magnetic susceptibility was measured by Faraday method in the temperature range 77-400 K, the accuracy of relative measurements was 1%. The EPR spectra were recorded at room temperature on an SEPR-2+KMYa spectrometer. The magnetization of the samples was measured on an original installation of Sankt-Petersburg Institute of Nuclear Physics, Russian Acad. Sci. The impedance spectra were recorded on an Impedancemeter-Z3000 in the temperature range 298 – 1000 K. The impedance hodographs were treated with the help of specialized program package ZView [38].

RESULTS

Magnetic Data

The magnetic characteristics (magnetic susceptibility and effective magnetic moments) of the solid solutions containing only paramagnetic dopants $LaGa_{1-x}M_xO_3$ and of the solutions containing, for example, strontium with the ratio M:Sr = 5:1 behave totally different (Figure 3) even though for Co and Ni containing systems we can describe μ_{eff} at $x \to 0$ as belonging to single atoms in different spin states, whereas for chromium we have a deviation of μ_{eff} from the single atom value, which cannot be ascribed to single chromium atoms no matter in which oxidation state it is. If we increase the content of strontium or introduce magnesium into lanthanum gallate the difference in the isotherms becomes even more drastic. (Figure 4 *a*, *b*). The effective magnetic moments at infinite dilution go up, moreover, the susceptibility becomes field dependent for cobalt and chromium containing systems. This means that even at infinite dilution the total disaggregation of paramagnetic atoms is not observed (Figure 5 *a*, *b*, *c*). We have high-nuclearity clusters of paramagnetic elements with predominantly ferromagnetic exchange interactions for chromium and cobalt and with antiferromagnetic interactions for nickel.

We made an attempt to estimate the distribution of paramagnetic atoms in the solid solutions using the method of magnetic dilution, which implies that the susceptibility of a solution with a particular transition metal content, x, is a sum of the susceptibilities of single atoms and clusters of paramagnetic atoms multiplied by their fraction in the solid solution [28, 39]:

$$\chi = \sum_i a_i \chi_i .$$

The susceptibilities of simple paramagnetic clusters (dimers or trimers) were calculated by Heizenberg-Dirack-van-Vleck model (HDVV) as a function of exchange parameter J [39]. The problem is that HDVV model can be correctly applied to the atoms with singlet or doublet ground states. For these reasons we made our calculations for the systems containing chromium, the more so that nickel(III) and cobalt(III) can have various spin states even in the systems containing no bivalent dopants [30, 34, 35].

Calculations of χ_M were first made for the $LaCr_xGa_{1-x}O_3$ system. Having 16 temperature points for each concentration of the solid solutions and only four variables – a_2, a_3, J_2 and J_3 in the concentration range from 0 to 10 mol% we were able to obtain J_2 of a dimer -12 cm^{-1}, J_3 for a cluster of three chromium atoms {Cr(III)–O–Cr(III)–O–Cr(III)} and a pattern of the distribution of dimers and trimers shown in Figure 6 *a*, *b*, *c*, *d*. For the solid solutions containing strontium and/or magnesium we were to advance a hypothesis that for the ratio Cr:Sr = 2:1 in the solid solution at the infinite dilution only some clusters X are present and for Cr:Sr:Mg = 5:1:1 – clusters X and clusters Y containing both doping elements (Sr and Cr or Mg and Cr).

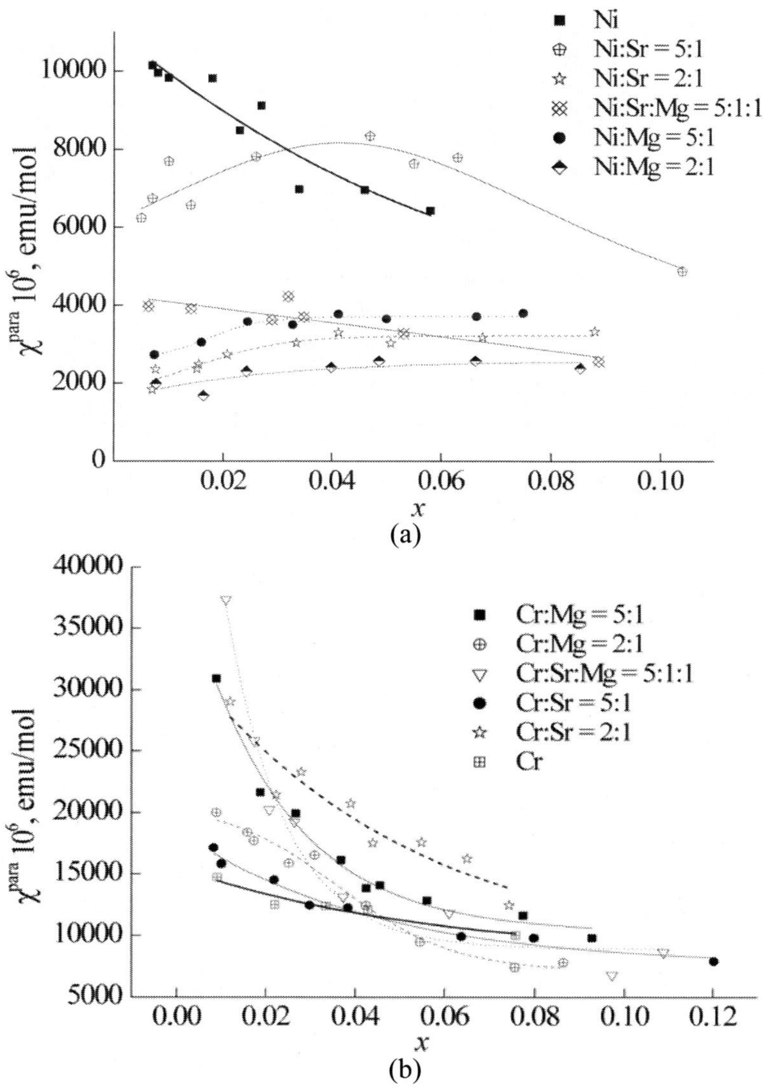

Figure 4. Isotherms of magnetic susceptibility for various systems at 140 K: (*a*) Ni; (*b*) Cr.

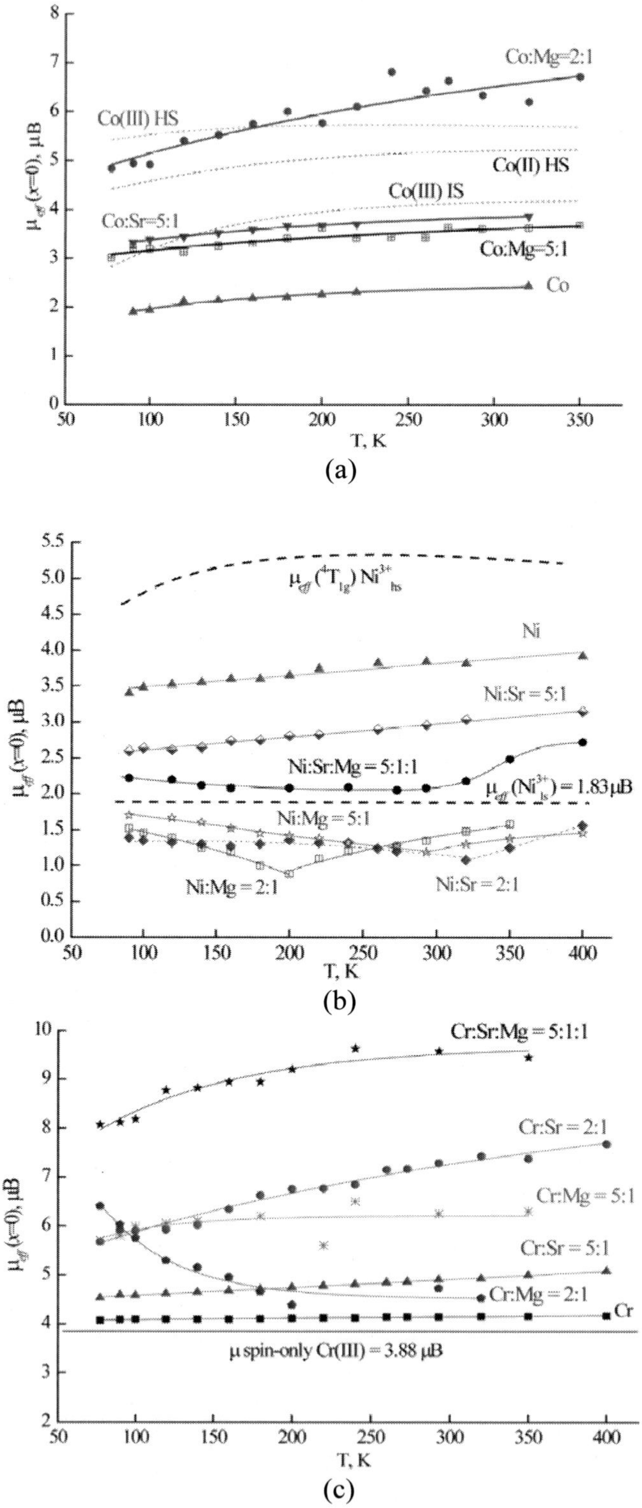

Figure 5. Temperature dependencies of effective magnetic moment at the infinite dilution: (*a*) Co-containing systems; (*b*) Ni-containing systems; (*c*) Cr-containing systems.

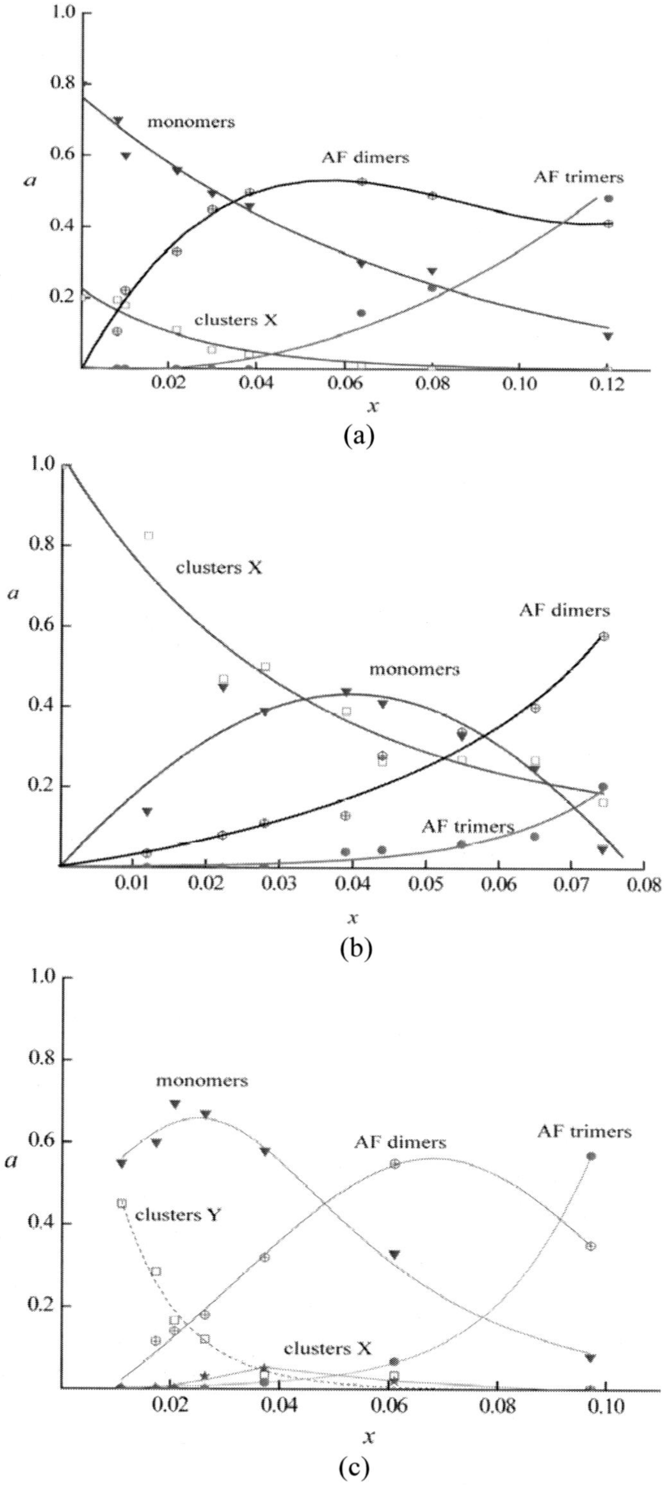

Figure 6. Distribution of single paramagnetic atoms (monomers) and clusters vs concentration of chromium for the systems: (*a*) Cr:Sr = 5:1; (*b*) Cr:Sr = 2:1; (*c*) Cr:Sr:Mg = 5:1:1.

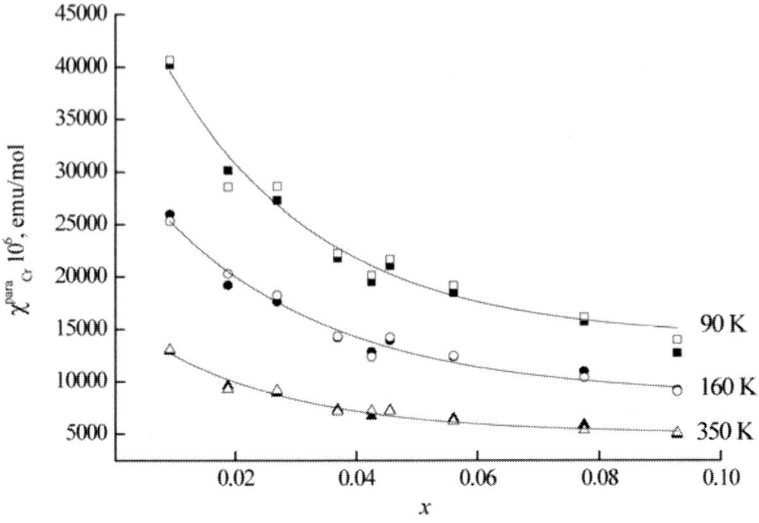

Figure 7. Theoretical and experimental values of paramagnetic component of magnetic susceptibility for the system Cr:Mg = 2:1 (x = 0.0083).

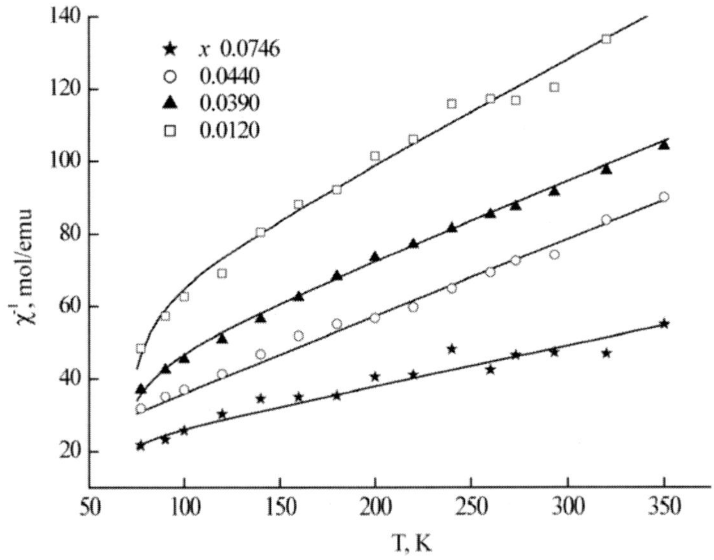

Figure 8. Inverse susceptibility vs temperature for the system Cr:Sr = 2:1 and various concentrations of solid solutions.

Having neither the data about the structure of clusters X, not the possibility to describe the $\mu^{x\to0} = f(T)$ dependence (Figure 5, *c*) with the help of HDVV model, when determining effective magnetic moment of clusters X (μ_{clX}) we postulated that the effective magnetic moment at infinite dilution for the system with the ratio Cr:Sr = 2:1 is determined only by clusters X, i.e. their fraction is equal to 1. Then, as the calculation showed, the effective magnetic moments for the system with Cr:Sr = 5:1 at infinite dilution are a superposition of 20% of clusters X and 80% of monomers, which quite agree with the difference in the Cr:Sr ratio for the systems under study.

By the results of the calculations the difference between theoretical and experimental magnetic susceptibilities for both systems is no more than 2% (Figure 7). Such calculations were carried out also for magnesium containing systems. The main conclusion from the obtained distributions of clusters and single atoms is that as the concentration increases, monomers and small clusters appear in the solid solution owing to paramagnetic atoms appearing near the vacancies of clusters, thus giving rise to monomers dimers and trimers.

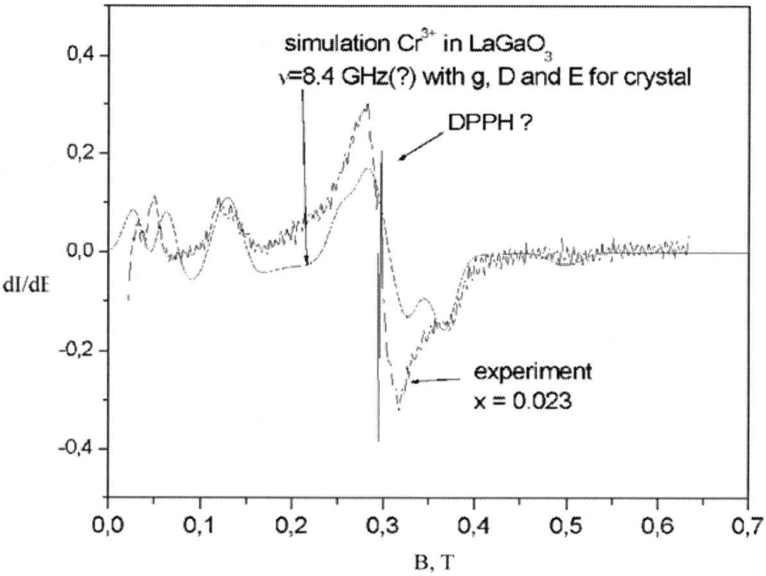

Figure 9. EPR spectra for the system Cr:Sr = 2:1 ($x = 0.0223$): theoretical calculation and experimental data.

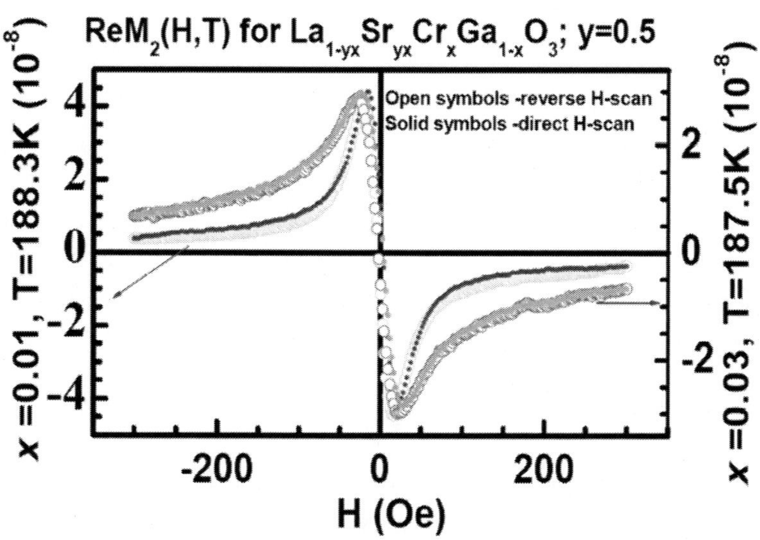

Figure 10. Hysteresis curves for the system Cr:Sr = 2:1 ($x = 0.0223$).

We emphasize that the run of inverse paramagnetic susceptibility plot vs temperature is typical for ferrimagnetic ordering [40] (Figure 8) even for diluted solid solutions at low temperatures. This also testifies for the presence of clusters with predominantly ferromagnetic exchange, the overall exchange in the solid solutions being antiferromagnetic, as is proved by the decrease in χ_M with the increase in concentration.

The EPR spectra of chromium containing solid solutions (Figure 9) have a complex structure and consist of several signals. The theoretical simulation shows a good agreement with the experimental data, each line belonging to trivalent chromium in distorted octahedral fields. In the low field region there is a signal ($g \sim$ 8-10), which can be ascribed to high nuclear clusters of chromium atoms. For solid solutions containing cobalt only a weak signal from high spin Co(III) is observed ($g = 4.3$), whereas for nickel containing – the signal from low spin nickel(III) ($g = 2.15$)

The dependence of susceptibility on the strength of the field made us to consider the dependencies of magnetization on the applied field for the systems Cr:Sr = 2:1, Cr:Sr:Mg = 5:1:1, Cr:Mg = 5:1 and Cr:Mg = 2:1. The examination of the data on magnetic hysteresis (Figure 10) points to the absence of hysteresis loops for the diluted solutions. In addition the magnetization measured at various temperatures and fields coincides (in the M – H/T coordinates, Figure 11). The dependencies in the H/M – M^2 coordinates (Figure 12) gives positive thermodynamic coefficients α in agreement with the theory of paraprocess in ferromagnets [41]. This points the absence of spontaneous magnetization in the samples. The obtained plots testify for the existence of superparamagnetic clusters in the solid solutions. The estimation of the cluster sizes by the procedure of [42] gives $r \sim$ 1-1.8 nm. From structural data we can judge that such clusters must contain no less than 20 atoms.

Now let us try to change the diamagnetic substituent in lanthanum gallate. That means we substitute lanthanum for calcium or barium instead of strontium and study chromium containing solid solutions with the ratio Cr:A = 5:1 – $La_{1-0.2x}(Ca,(Ba))_{0.2x}Ga_{1-x}Cr_xO_3$.

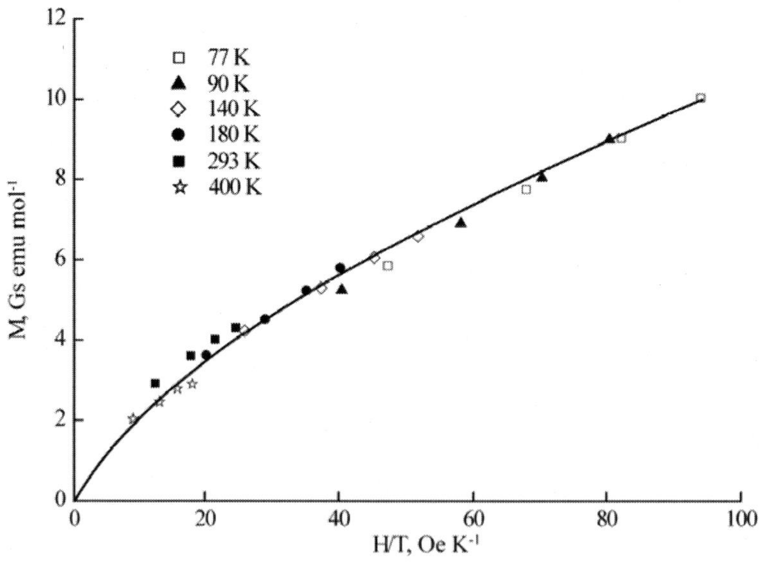

Figure 11. Plot of M – H/T.

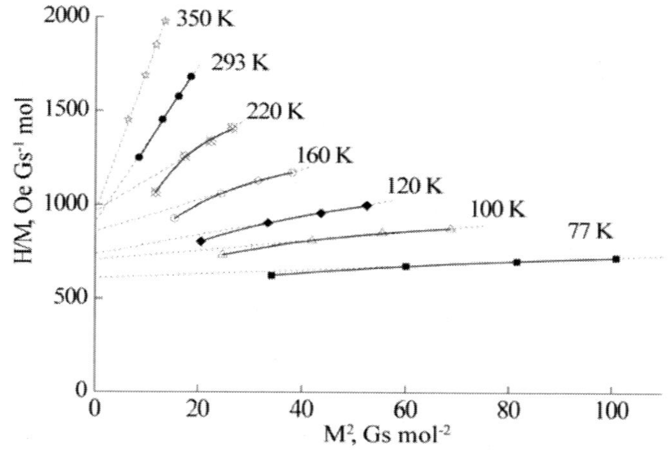

Figure 12. Plot of H/M – M^2.

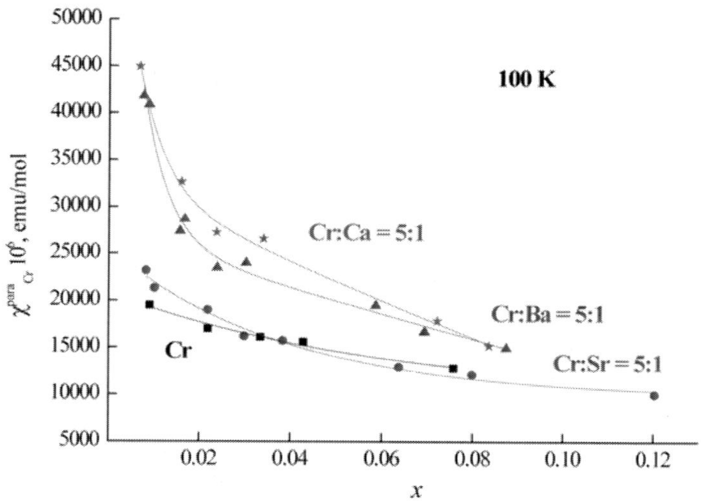

Figure 13. Isotherms of magnetic susceptibility for $LaCr_xGa_{1-x}O_3$ and $La_{1-0.2x}(Ca, Sr, Ba)_{0.2x}Cr_xGa_{1-x}O_{3-\delta}$ at 140 K.

The isotherms of paramagnetic component of magnetic susceptibility for calcium and barium containing gallate lie much higher than for those doped with strontium (Figure 13). This points to an increase in clustering for both system in question. An interesting fact is that clustering decreases on passing from calcium to strontium and again increases on passing to barium (Figure 14).

Electrical Properties

The conductivity of the solid solutions under study increases as the temperature increases, which points their semiconducting properties. The conductivities are small and depend on the nature and content of doping elements (Figure 15 *a*, *b*, *c*, *d*). According to the data of high-temperature X-ray analysis the break in the linear dependencies of lg σ on 1/T is not associated with any polymorphic transformations in the systems under study. Therefore, different activation energies for two parts of straight lines point to different characters (mechanisms) of conductivity. As the temperature increases the ionic transfer becomes more pronounced. For the systems containing nickel the conductivity appears sufficiently high (lg $\sigma > 0$) at $T > 500$ K.

On the basis of obtained dependencies the isotherms of conductivity were constructed (Figure 16 *a*, *b*). Their run testifies for the fact that systems with the ratio M:Sr(Mg) = 5:1 have maximal conductivity, strontium containing systems having higher conductivity compared to magnesium containing analogs.

DISCUSSION

Both the data on EPR and on the calculation of magnetic characteristics show that no oxidation of transition elements occur in our systems on doping with bivalent elements. In such a case we must suggest that the stability of the defect structure is somehow associated with the formation of clusters of paramagnetic elements.

These clusters are not destroyed by the infinite dilution, which means a large, more than 100 kJ per mole interaction energy between atoms [43]. They necessarily include also strontium or magnesium and accompanying vacancies in the oxygen sublattice – otherwise we would never observe such difference in the magnetic behavior of the systems doped with bivalent elements and doped with *d*-element only. We must remember that in the most concentrated solid solution containing 10 mol% of *d*-element and therefore 2 mol% of Sr, for example, only two atoms of lanthanum from 100 are replaced by strontium. They cannot be far away from a paramagnetic element to result in such drastic difference in magnetic characteristics.

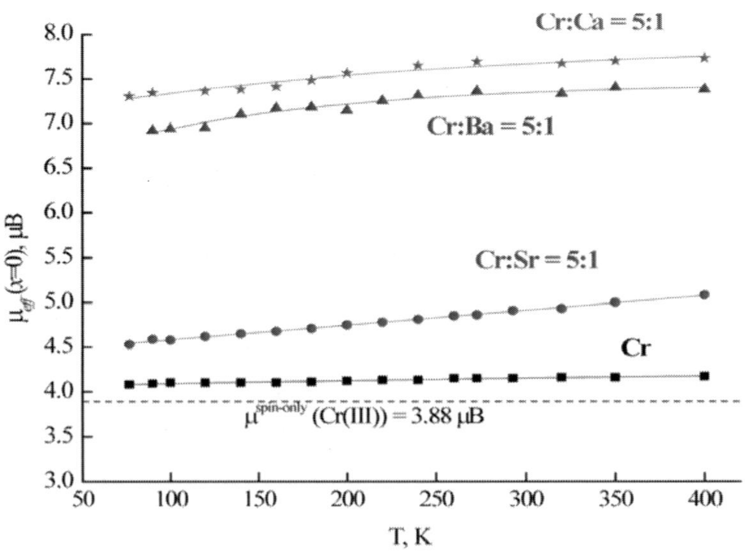

Figure 14. Temperature dependencies of effective magnetic moment at the infinite dilution for the systems: Cr, Cr:Sr = 5:1, Cr:Ca = 5:1, Cr:Ba = 5:1.

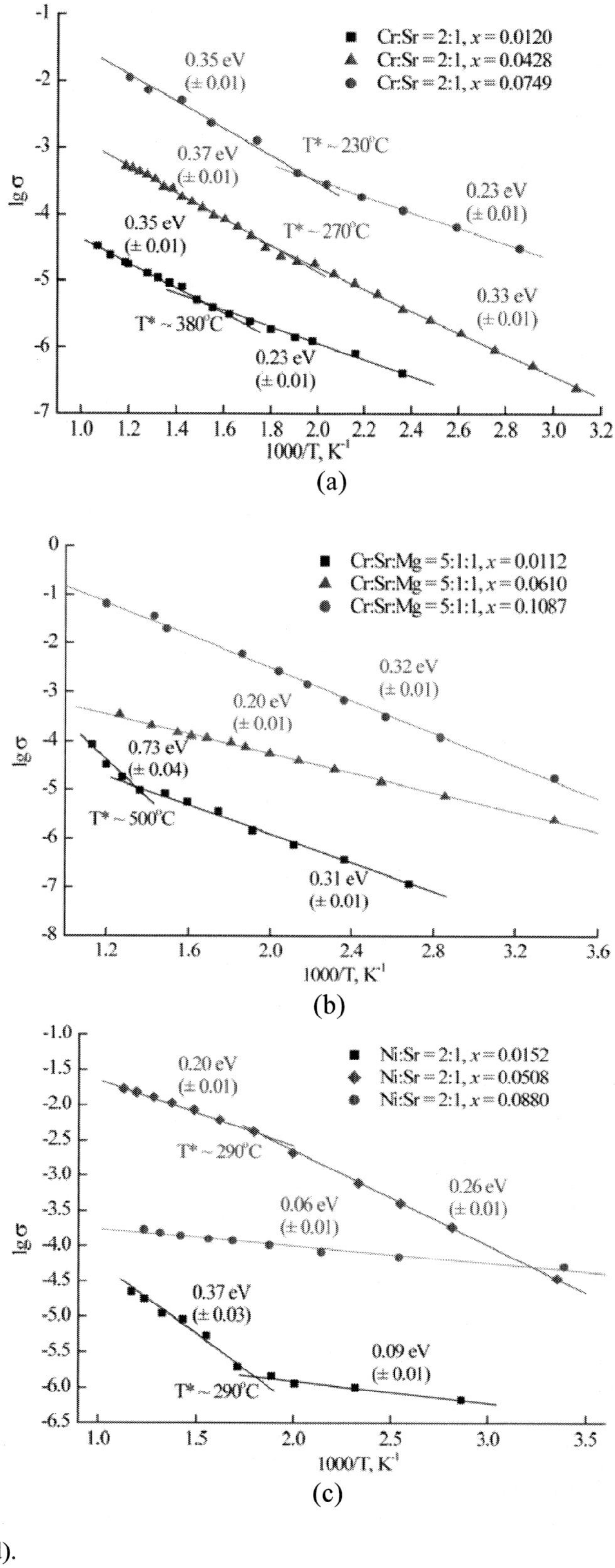

Figure 15. (Continued).

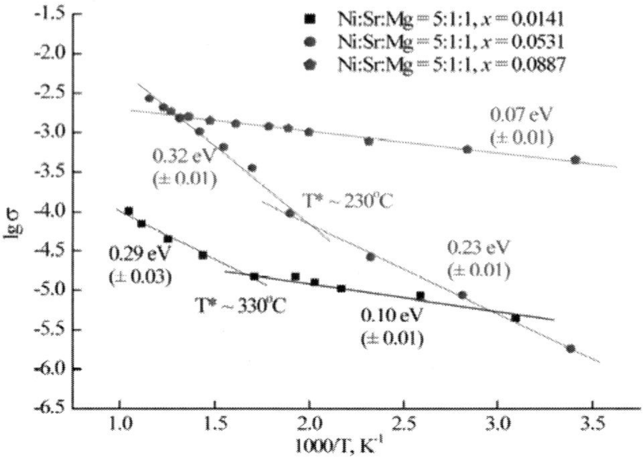

(d)

Figure 15. Plots of lg σ – 1000/T for some systems: (*a*) Cr:Sr = 2:1; (*b*) Cr:Sr:Mg = 5:1:1; (*c*) Ni:Sr = 2:1; (*d*) Ni:Sr:Mg = 5:1:1.

Therefore we found the reason for the structure stabilization upon doping lanthanum gallate with a diamagnetic and paramagnetic elements – it is the formation of nanosize clusters, containing both, transition and diamagnetic elements, and vacancies in the oxygen sublattice Associated with strontium or magnesium. Clustering increases as the content of bivalent element increases, magnesium acting in a slightly different way, since it is in the sites of gallium and Sr – in the sites of lanthanum. In the case of chromium and cobalt these clusters behave as superparamagnetic particles according to our measurements of magnetization. In the case of Ni-containing systems, the exchange between nickel atoms is predominantly antiferromagnetic.

A non-monotony in the changes of magnetic characteristics of the solid solutions containing calcium, strontium, and barium as heterovalent dopant is determined by two factors: an increase in chromium clustering as the size of bivalent atom increases and a decrease in the ferromagnetic component of the exchange. Both factors are associated with the size of the atom and its polarizing action on the orbitals of oxygen atoms. As the result an optimal relationship is attained in the case of strontium, which is used in SOFC as a doping element.

On considering the conductivity as the function of the nature of transition elements we tried to find the reason for nickel atoms giving the highest values of conductivity. The calculation of the band structure of lanthanum gallate and lanthanum gallate, where nickel atoms substitute for a half of gallium atoms ($LaGa_{0.5}Ni_{0.5}O_3$) was carried out in the Institute of Solid State Chemistry (Ekaterinburg, Russia) by the FLAPW-GGA method [44]. The electronic conductivity was shown to arise at the expense of the fact that the band formed by e_g electrons of nickel(III) is located within the energy gap of pure lanthanum gallate. In the diluted solid solutions this band degenerates into an impurity level.

The data of the calculation of the band structure of lanthanum gallate doped with nickel allow us to generalize them to other elements under study and present a scheme (Figure 17). We see the bands of semiconducting $LaGaO_3$. At the level of energy gap there is a *d*-level of a transition element, which is split in octahedral field. Chromium in lanthanum gallate gives three holes in the valence band, thus resulting in small *p*-type conductivity.

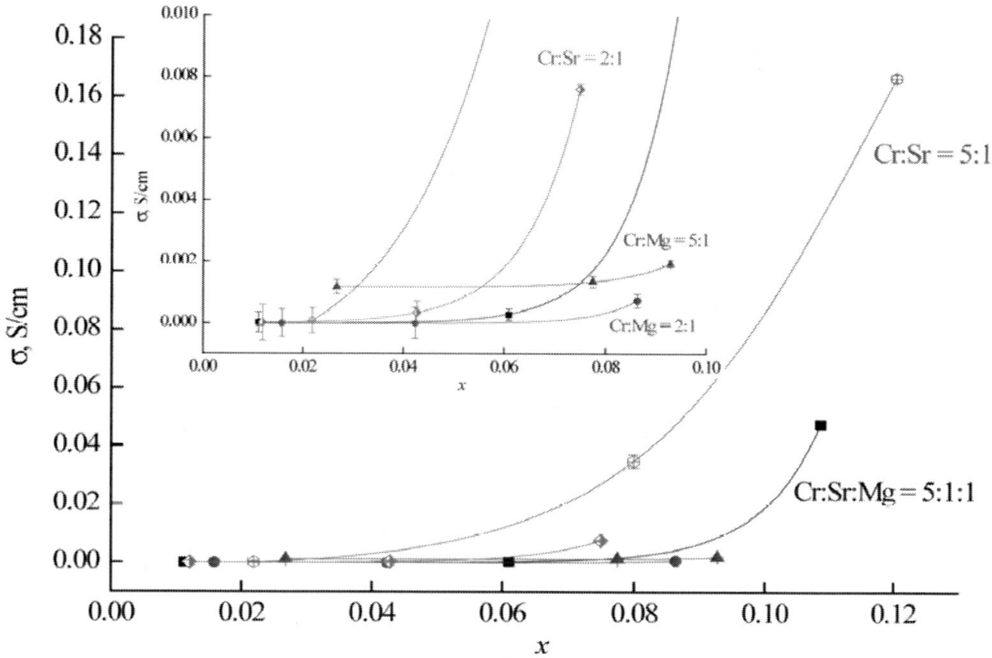

(a)

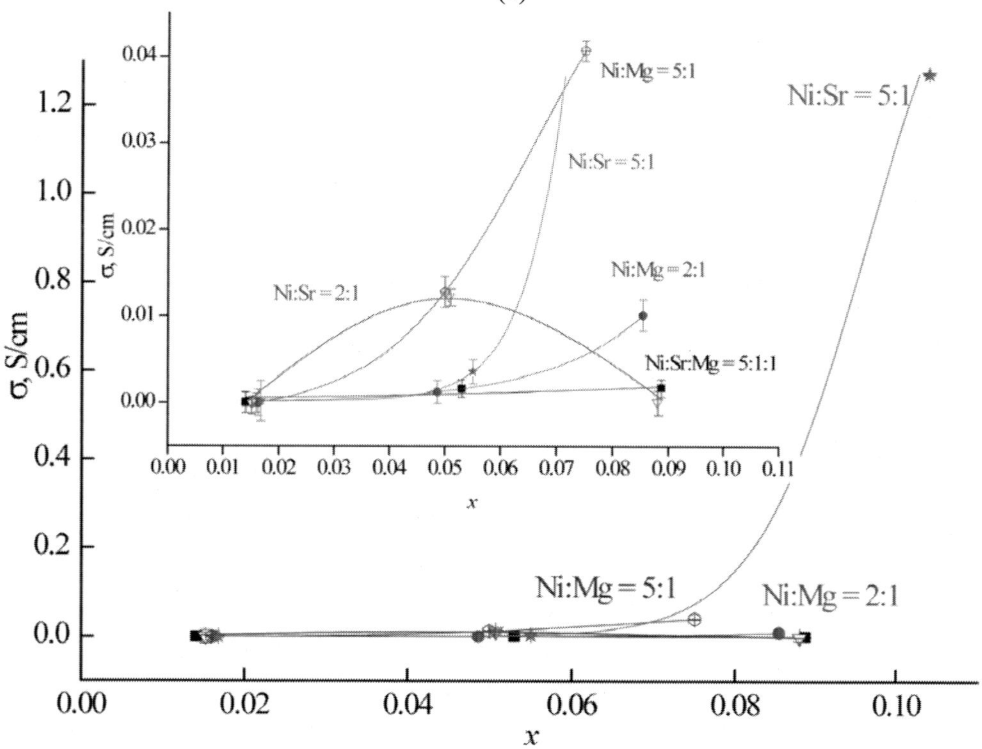

(b)

Figure 16. Plots of conductivity vs concentration of dopants for the systems: (*a*) Cr-containing; (*b*) Ni-containing.

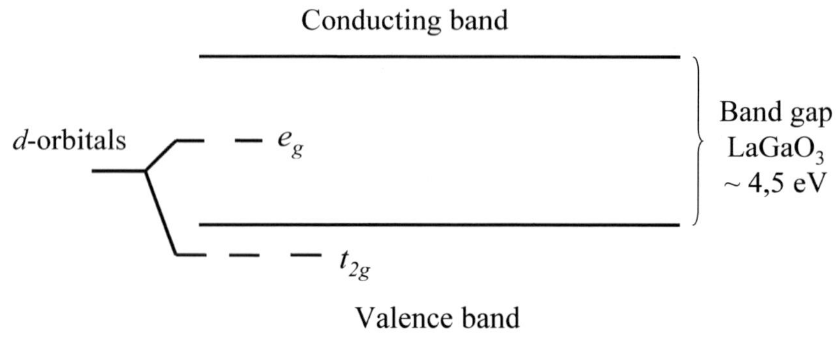

Figure 17. Schematic plot of band structure of doped lanthanum gallate.

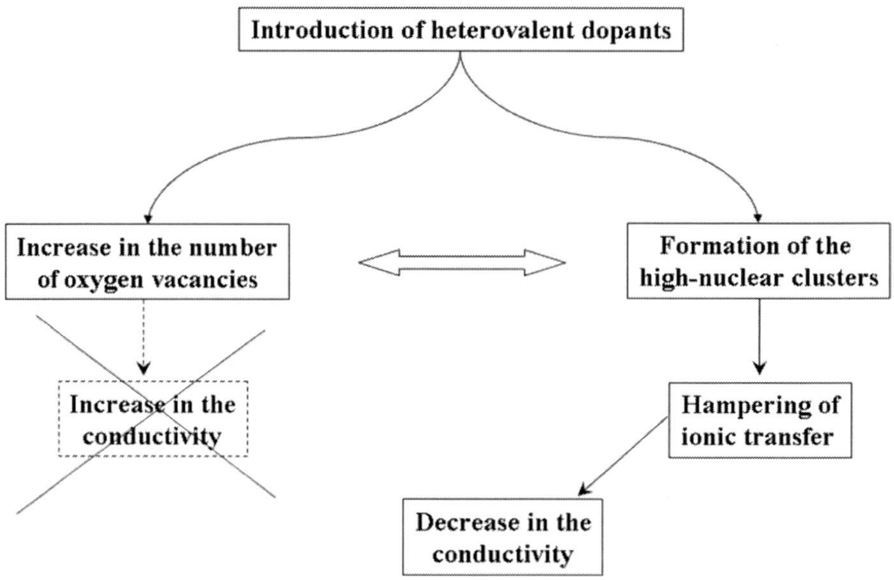

Figure 18. Scheme illustrating the influence of heterovalent substitution in lanthanum gallate on its conductivity.

Cobalt(III) being in two spin states gives electrons to e_g-orbitals only if it is high spin. But by our calculations high spin cobalt is bound in clusters. Therefore chromium and cobalt containing lanthanum gallates may be used as electrolytes and nickel containing – as cathodes for SOFC.

Introduction of greater than 5:1 amount of strontium and introduction of any amount of magnesium results in an increase in clustering, the formation of high nuclear clusters ($n > 20$) of transition metal atoms. The character of the exchange in them is determined by the nature of a transition element.

A correlation between the quantity of bivalent doping element, its nature, and magnetic characteristics accounts for the fact that the clusters include also heterovalent substituents and oxygen vacancies associated with them. The isotherms of conductivity also point to the correlation for all the systems under study – an increase in the concentration of a heterovalent substituent results in a decrease in conductivity.

Conclusion

Therefore we can conclude that clustering of paramagnetic elements in lanthanum gallate is an important factor for their electron structure and conductivity. Clustering is necessary for maintaining the stability of the structure and seems to be inherent for heterovalent doping of complex oxides with perovskite structure. However an increase in clustering following an increase in the bivalent atom concentration results in a decrease in conductivity owing to the fact that the field of negatively charged cluster hampers the mobility of vacancies (Figure 18).

References

[1] Ishihara, T.; Matsuda, H.; Takita, Y. // *J. Am. Chem. Soc.*, 1994, 116, 3801-3803.
[2] Cheng, J.; Navrotsky, A. // *J. Sol. St. Chem.*, 2004, 177, 126-133.
[3] Stevenson, J.W.; Armstrong, T.R.; Pederson, L.R.; Li, J.; Lewinsohn, C.A.; Baskaran, S. // *Solid State Ionics*, 1998, 113–115, 571-583.
[4] Li, Z.-C.; Zhang, H.; Bergman, B.; Zou, X. // J. Eur. Ceram. Soc., 2006, 26, 2357-2364.
[5] Shi, M.; Xu, Y.; Liu, A.; Liu, N.; Wang, C.; Majewski, P.; Aldinger, F. // *Mater. Chem. Phys.*, 2009, 114, 43-46.
[6] Datta, P.; Majewski, P.; Aldinger, F. // Mater. Chem. Phys., 2007, 102, 240-244.
[7] Huang, K.; Goodenough, J.B. // *J. Sol. St. Chem.*, 1998, 136, 274-283.
[8] Ha, S.B.; Choa, Y.H.; Kang, Y.C.; Lee, J.-H. // *J. Eur. Ceram. Soc.*, 2010, 30, 2593-2601.
[9] Hsieh, F.-F.; Okinaka, N.; Akiyama, T. // *J. Alloys Compd.*, 2009, 484(1-2), 747-752.
[10] Trofimenko, N.; Ullmann, H. // *Solid State Ionics*, 1999, 118, 215-227.
[11] Kajitani, M.; Matsuda, M.; Hoshikawa, A.; Oikawa, K.; Torii, S.; Kamiyama, T.; Izumi, F.; Miyake, M. // *Chem. Mater.*, 2003, 15, 3468-3473.
[12] Feng, M.; Goodenough, J.B. // *Eur. J. Sol. St. Inorg. Chem.*, 1994, 31, 663.
[13] Datta, P.; Majewski, P.; Aldinger, F. // *J. Alloys Compd.*, 2007, 438, 232-237.
[14] Zheng, F.; Bordia, R.K.; Pederson, L.R. // *Mater. Res. Bull.*, 2004, 39, 141-155.
[15] Jasiolek, G.; Dabkowska, H. // *J. Less Common Met.*, 1990, 160, 79-84.
[16] Wang, Y.; Liu, X.; Yao, G.-D.; Liebermann, R.C.; Dudley, M. // *Mater. Sci. Eng.*, 1991, A132, 13-21.
[17] Kobayashi, J.; Tazoh, Y.; Sasaura, M.; Miyazawa, S. // *J. Mater. Res.*, 1991, 6(1), 97-100.
[18] Marti, W.; Fischer, P.; Altorfer, F.; Scheel, H.J.; Tadin, M. // *J. Phys.: Condens. Matter*, 1994, 6, 127-135.
[19] Slater, P.R.; Irvine, J.T.S.; Ishihara, T.; Takita, Y. // *Solid State Ionics*, 1998, 107, 319-323.
[20] Slater, P.R.; Irvine, J.T.S.; Ishihara, T.; Takita, Y. // *J. Sol. St. Chem.*, 1998, 139, 135-143.
[21] Majewski, P.; Rozumek, M.; Tas, C.; Aldinger, F. // *J. Electroceram.*, 2002, 8, 65-73.
[22] Wang, W.L.; Lu, H.Y. // *Phys. Chem. Miner.*, 2006, 33(7), 435-444.
[23] Lerch, M.; Boysen, H.; Hansen, T. // *J. Phys. Chem. Solids*, 2001, 62, 445.

[24] Tas, A.C.; Majewski, P. J.; Aldinger, F. // *J. Am. Ceram. Soc.*, 2000, 83(12), 2954-2960.
[25] Khan, M.S.; Islam, M.S.; Bates, D.R. // *J. Phys. Chem. B*, 1998, 102, 3099-3104.
[26] Islam, M.S. // *Solid State Ionics*, 2002, 154–155, 75-85.
[27] Malavasi, L.; Fisher, C.A.J.; Islam, M.S. // *Chem. Soc. Rev.*, 2010, 39, 4370-4387.
[28] N.V. Chezhina, in: A.B. Nikol'skii (Ed.), Physical Methods of Studying Inorganic Substances, Akademia. Moscow, 2006, pp. 212–230. (in Russian).
[29] Chezhina, N.V.; Zolotukhina, N.V.; Bodritskaya, E.V. // *Russ. J. Gen. Chem.* 75(8) (2005) 1233–1236.
[30] Chezhina, N.V.; Piir, I.V.; Zolotukhina, N.V. // *Russ. J. Gen. Chem.* 76(10) (2006) 1585–1587.
[31] Chezhina, N.V.; Korolev, D.A.; Sukharzhevskii, S.M.; Glumov, O.V. // *Russ. J. Gen. Chem.* 80(5) (2010) 745–749.
[32] Korolev, D.A.; Chezhina, N.V. // *Russ. J. Gen. Chem.* 81(10) (2011) 2067–2073.
[33] Chezhina, N.V.; Korolev, D.A. // *The Open Fuel Cells Journal*, 2011, Vol. 4, p. 7-15.
[34] Chezhina, N.V.; Zharikova, E.V.; Knyazev, M.N. // *Rus. J. Gen. Chem.*, 2010, 80(12), 1937–1942.
[35] Chezhina, N.V.; Bodritskaya, E.V.; Zhuk, N.A. // *Rus. J. Gen. Chem.*, 2008, 78(6), 899–904.
[36] Chezhina, N.V.; Korolev, D.A. // *Solid State Ionics*, Article in Press.
[37] Korolev, D.A.; Chezhina, N.V.; Lytkina, Zh.A. // *Russ. J. Gen. Chem.* 2012, 82(3), 354-359
[38] ZView Version 3.1c, Derek Johnson, Scribner Associates, Inc.
[39] Kalinnikov, V.T.; Rakitin, Yu.V. Introduction to Magnetochemistry. Method of Static Magnetic Susceptibility, Moscow: *Nauka*, 1980. (in Russian).
[40] R.M. White, Quantum theory of magnetism, 3rd ed., Springer, Berlin, 2007.
[41] K.P. Belov. Magnetic transformations // *GIFML*, Moscow, 1959. (in Russian).
[42] Kellerman, D.G.; Shalaeva, E.V.; Gusev, A.I. // *Phys. Sol. St.,* 46(9), 2004, 1686-1692].
[43] Brach, B.Ya.; Bobrysheva, N.P.; Reznitskii, L.A.; Chezhina, N.V. Vestnik LGU, 1979, 22, 109-112. (in Russian).
[44] Chezhina, N.V.; Bodritskaya, E.V.; Zhuk, N.A.; Bannikov, V.V.; Shein, I.R.; Ivanovskii, A.L. *Phys. Sol. St.*, 50(11), 2008, 2121-2126.

In: Perovskite
Editors: Jinghua Zhang and Huan Li
ISBN: 978-1-62417-800-9

Chapter 9

Distorted Perovskites: Way to Generate Organic-Inorganic Hybrid Compounds

Prasanna S. Ghalsasi* and Ashok K. Vishwakarma
Department of Chemistry, Faculty of Science,
The M. S. University of Baroda, Vadodara, India

Abstract

The ability to rationally design materials for use in fundamental studies or applications is a key motivation in the field of synthetic chemistry. For this, one needs a basic 'building block' which is flexible for incorporation of organic as well as inorganic counter part.

The cubic AMX_3 perovskite structure will be a 'role' model for the same. AMX_3 pervoskite consists of an extended three-dimensional (3-D) network of corner-sharing MX_6 octahedra, where M is generally a divalent metal and X a halide. With the increase in size of 'A', an organic counter part, one can generate layered perovskite consists where MX_4^{2-} layers of corner-sharing metal halide octahedra alternate with organic cation bilayers (for monofunctional cations) or monolayers (for bifunctional cations). Here two metal centers are connected by halide, which in turn are indulge in hydrogen bonding with the oraganic cations.

That means apart from ionic bonding, hydrogen bonding in designing these type compounds play equally important role, a topic covered in this chapter. These so called, organic-inorganic hybrid perovskites, self-assemble at ambient temperature solutions and are generally found stable up to the decomposition point (typically >250 ºC). But we and other have observed very interesting thermal decomposition behavior in these compounds will be addressed in detail.

Finally, the chapter will discuss technologically important properties explored by these layered perovskites, which include magnetism, thermochromism and structural phase-transitions.

* E-mail: prasanna.ghalsasi@gmail.com.

INTRODUCTION

Organic compounds have structural diversity, ease of processing, efficient luminescence, good optical along with good electrical mobility and/or electrical behavior. Inorganic compounds have a distinct set of advantages, including mechanical and thermal stability, higher structural dimensionality along with interesting electrical and/or magnetic properties. Thus, organic-inorganic hybrid compounds (OIHCs) offer an important opportunity to combine useful properties from these two chemical areas (organic and inorganic) within a single molecule. One of the best strategies to make organic-inorganic hybrid compounds with multifunctional behavior is layered materials based on distortion of perovskite structure [1, 2]. Chemically, synthesize the compounds having multifunction properties (ferromagnetic and ferroelectric) within the single molecule is challenging task. Because as ferroelectricity, in most cases, requires atoms with an empty *d*-orbital [2]. This is in contradiction with the partially filled *d*-orbital's that are required to introduce magnetism.

Materials which have crystal structure similar to calcium titanium oxide ($CaTiO_3$) called perovskite structure compounds [3]. The cubic AMX_3 perovskite structure (Figure 1) consists of an extended three-dimensional (3-D) network of corner-sharing MX_6 octahedra, where A = cation, X = halide and M = divalent metal. The larger A cations fill the 12-fold coordinated holes among the octahedra. For the 3-D perovskites the size of the A cation is limited by the size of the 3-D hole into which it must fit. For a perfectly packed perovskite structure the geometrically imposed condition for the A, M, and X ions to be in close contact is $(R_A + R_X) = t\sqrt{2}\,(R_M + R_X)$, where R_A, R_M, and R_X are the ionic radii for the corresponding ions and the tolerance factor must satisfy $t \approx 1$. Empirically it is found that $0.8 \leq t \leq 0.9$ for most cubic perovskites. Using $t = 1$ and essentially the largest values for R_M and R_X (*e.g.* $R_{Pb} = 1.19$, $R_I = 2.20$ Å), the limit on R_A is found to be approximately 2.6 Å. This calculation, tells us that the size of methyl ammonium or ammonium cation only will be useful for generating three dimensional continuous networked compounds. If inorganic A cation is replace by a suitable organic cation, one uses 'A cation' which is longer, then compounds with the molecular formula A_2MX_4 and AMX_4 [where A = organic ammonium cations (OACs), M = divalent transition metal ion and X = halogen] can be obtained, which is well cited in literature [4]. Organic-inorganic perovskites are widely studied hybrid crystalline systems that exhibit interesting magnetic, optical and electrical properties [5, 6, 7, 8].

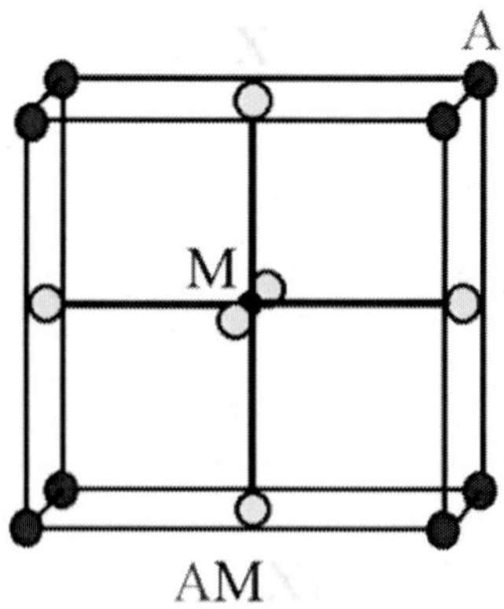

Figure 1. Structure of basic AMX_3 perovskite unit cell.

In A_2MX_4 and AMX_4 class of compounds, the inorganic component consists of more or less regular corner shared octahedral units. This extended MX_4^{2-} layer is exist as quasi two-dimensional (2-D) magnet, and alternating with OACs, bilayers (for monoammonium) or monolayer (for biammonium).

The ammonium groups needs to be able to fit into the "holes" formed by terminal and bridging halogen within the distorted perovskite-sheets, without the rest of the molecule interfering with the sheets. That leads to extreme elongation of one semi-coordinate bond and allowing accommodation of OACs between the inorganic layers. That is why the structure of the MX_4^{2-} layers is characterized by two kinds of metal-halide distances equatorial (D_S) and axial (D_L) distances, a result of Jahn-Teller distortion. In these layered perovskites, Jahn-Teller elongated bonds of adjacent M^{2+} ions lie in the plane of the layer but are oriented at right angles to each other, thus defining an antiferrodistortive structure. The, OACs, are stabilized by the two distinct interactions:

(a) Hydrogen bonding to inorganic sheets at one or both end;
(b) van der Waals interactions amongst their tails

The choice of OACs and hydrogen bonding scheme is important for determining the orientation and conformation of the organic molecule within the layered hybrid structure. For example when we take monofunctional OACs the inorganic layer of metal-halide is separated by bilayers, of organic part, and give staggered form. But when we take bifunctional OACs then inorganic layer of metal-halide is separated by monolayer, of organic part, and gives eclipsed form (Figure 2).

In principle, the ammonium head(s) of the OACs can hydrogen bond to any of the eight halides (*i.e.* four bridging/four terminals) within the holes formed by the corner-sharing MX_6 octahedra (Figure 3). In practice, due to the geometric constraints of the ammonium group and the organic tail, the N-H···X interactions generally form either to two bridging halides and one terminal halide (bridging halide configuration) or to two terminal halides and one bridging halide (terminal halide configuration). That means these ionic bonding, hydrogen bonding design compounds play important role. So these types of compounds are called organic-inorganic hybrid perovskite.

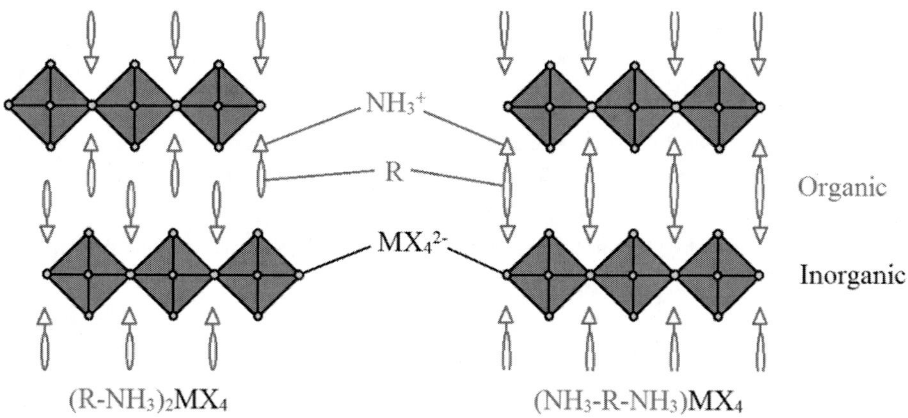

Figure 2. Schematic representation of single-layer perovskite with monoammonium ($R\text{-}NH_3^+$) (a) and biammonium ($^+NH_3\text{-}R\text{-}NH_3^+$) (b), organic cations.

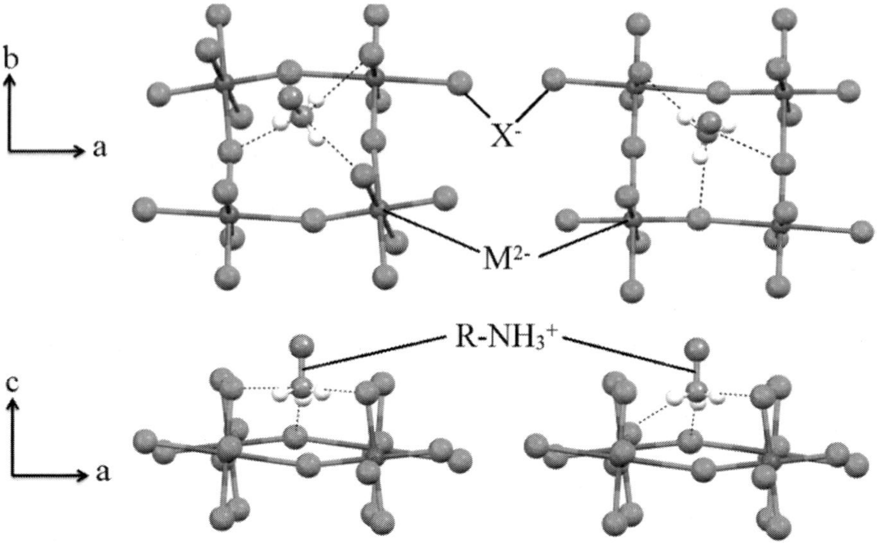

Figure 3. Two hydrogen bonding schemes observed in $(R\text{-}NH_3)_2MX_4$ and $(NH_3\text{-}R\text{-}NH_3)MX_4$ type structures: bridging halide configuration (a) and terminal halide configuration (b).

It is clear that the hydrogen bonding pattern of ammonium group with terminal halide, which always tilts the position of MX_4^{2-} along the axis, and bonding with bridging halide, always forcing the alternating layer halide atoms out of the plane, leading to decreases super exchange pathway through the two halides between interlayer metal. Therefore super exchange interaction through the two halides between interlayer metals normally decreases due to tilted position of MX_4^{2-}. The bridging halide bond angle seems to play an important role in controlling the magnetic coupling to be ferromagnetic or antiferromagnetic. Generally, bent M-X-M bond angle, smaller than 156°, is unfavourable for the ferromagnetic coupling between d_{Z}^2 and d_{XZ} of metal [9].

Furthermore, OACs not only separate the layer of metal-halide but it can also vary the distance between the inorganic sheet and properties of the compounds. Smaller size of OACs decreases the distance of inorganic sheet and larger size of OACs increases the distance of inorganic sheet.

The type of magnetic order ranges from ferromagnetic to antiferromagnetic also depends on the organic component and transition metals that are used. The highest ordering temperatures are observed for the Fe (II) based OIHCs [10]. Moreover, the ordering temperatures do also depend on the organic component. They are highest when short biammonium, R-groups, are used. For example $(NH_3CH_2CH_2NH_3)CuBr_4$ shows long range ferromagnetic ordering below $T_c = 72$ K and the $(CH_3CH_2NH_3)_2CuBr_4$ below $T_c = 11$ K [11]. Thus by changing the OACs and/or metal ions we can get a variety of OIHCs with different properties [12, 13]. These OIHCs are generally stable up to the decomposition point. In a few cases the hybrids melt before decomposition [14].

The crystal structures of $(\text{ammonium})_2CuCl_4$ and $(\text{ethyl ammonium})_2CuCl_4$ were reported by Willett *et. al.* [15, 16]. The structure of $(\text{propyl ammonium})_2CuCl_4$ has been determined by Barendregt and Schenk [17]. The possible mechanisms for magnetic exchange have been a major area of study for decades and numerous empirical and theoretical descriptions have been presented [18]. A variety of pathways are known to propagate the exchange, including single atoms or ions, such as halides, bridging between metal species [19]. Subsequently, Snively *et al.* discovered that magnetic exchange could propagate through non-bonding contacts between halide ions which attached to different metal ions, the pathway, M-X···X-M, described as a double halide bridge [20]. In their studies for $(NH_3CH_2CH_2NH_3)CuCl_4$ consisting of sheets of square planar $CuCl_4^{2-}$ ions linked into square layers by Cu-Cl-Cu bridges, they observed that the antiferromagnetic exchange between the layers. Subsequent work by the same group with a series of alkyl biammonium salts clearly showed that the effect was not unique to $(NH_3CH_2CH_2NH_3)CuCl_4$ and that the strength of the exchange was a function of the distance between the halide ions, rather than the distance between the metal ions [21].

Banks *et al.* report detailed examination of the magnetic structure of anhydrous cupric chloride, $CuCl_2$. Anhydrous $CuCl_2$ shows one-dimensional antiferromagnetic behavior and long range antiferromagnetic ordering below a Neel temperature of 23.9 K [22]. The behavior of magnetic susceptibility of anhydrous chlorides at low temperature (LT) is described by the Starr *et al.* [23]. Thus, by 'incorporating' OACs one can 'tune' the structure of $CuCl_2$ and hence explore technologically important properties such as magnetism.

The structural and calorimetric investigations of the OIHCs showed existence of a large number of phases and phase transitions [24, 25]. These phase transitions are due to OACs. By properly choosing OACs with ferroelectric and/or ferroelastic phases one can generate OIHCs with multifunctionality, along with the magnetism [26]. Recently Polyakov *et al.* have found ferromagnetic order, below 13 K, and ferroelectric order, below 340 K, in OIHCs, $(C_6H_5CH_2CH_2NH_3)_2CuCl_4$ [2]. Thus, it is possible to make multifunctional compounds using this strategy, which a topic of this chapter.

This chapter is focused on exploring role of OACs on OIHCs. With this motive we have synthesized 'novel' OIHCs by using different OACSs and copper (II) chloride. Some selected compounds, as a case study, are discussed below with the help of thermal analysis (TG-DTA, DSC), Raman spectroscopy, thermochromism, single crystal x-ray analysis, and magnetic measurements using SQUID magnetometer.

Results and Discussion

Thermal Analysis: Thermogravimetry-differential thermal analysis (TG/DTA) measurements were carried out by using a SII TG/DTA6300 instrument under nitrogen atmosphere at a scanning rate of 10 K min^{-1}. Samples were taken in aluminum/platinum pan and reference left empty.

The flow was used 100 mL min^{-1} for TG/DTA analysis. While the Differential Scanning Calorimeter (DSC) were carried out by Shimadzu DSC-60 instruments, Al_2O_3 used as a reference material in aluminum pan. The sample were packed in aluminum pan and kept in the flow of nitrogen gas with heating/cooling rate 10 K min^{-1}. The flow was used 50 mL min^{-1} for thermal analysis.

TG/DTA of these layered organic-inorganic hybrid complexes was performed on the powder samples. Figure 4 shows the thermal analysis of $(anilinium)_2CuCl_4$ (**A**), (4-chloro-anilinium)$_2CuCl_4$ (**B**) and (benzilinium)$_2$ $CuCl_4$ (**C**). Complexes **A** and **B** follows a similar degradation pathway as compared to **C**, as tabulated below.

$(C_6H_5NH_3)_2CuCl_4$

Temp. 444.1 K ↓ $C_6H_5NH_2HCl + C_6H_5NH_2$

Loss = 56.57 % (theory)

Loss = 58.10 % (found)

$(H)CuCl_3$

Temp. 776.4 K ↓ $HCl + Cl_2$

Loss = 27.29 % (theory)

Loss = 21.32 % (found)

Cu

16.14 % (theory)

20.58 % (found)

$(4\text{-}Cl\text{-}C_6H_4NH_3)_2CuCl_4$

Temp. 485.1 K ↓ $4\text{-}Cl\text{-}C_6H_4NH_2HCl + 4\text{-}Cl\text{-}C_6H_4NH_2$

Loss = 63.04 % (theory)

Loss = 65.04 % (found)

$(H)CuCl_3$

Temp. 852.2 K ↓ $HCl + Cl_2$

Loss = 23.23 % (theory)

Loss = 17.08 % (found)

Cu

13.74 % (theory)

17.87 % (found)

$(C_6H_5CH_2NH_3)_2CuCl_4$

Temp. 544.0 K ↓ $2HCl + Cl_2$

Loss = 34.17 % (theory)

Loss = 32.36 % (found)

Temp. 568.8 K ↓ $2C_6H_5CH_2NH_2$

Loss = 50.41 % (theory)

Loss = 53.45 % (found)

Cu

15.09 % (theory)

14.49 % (found)

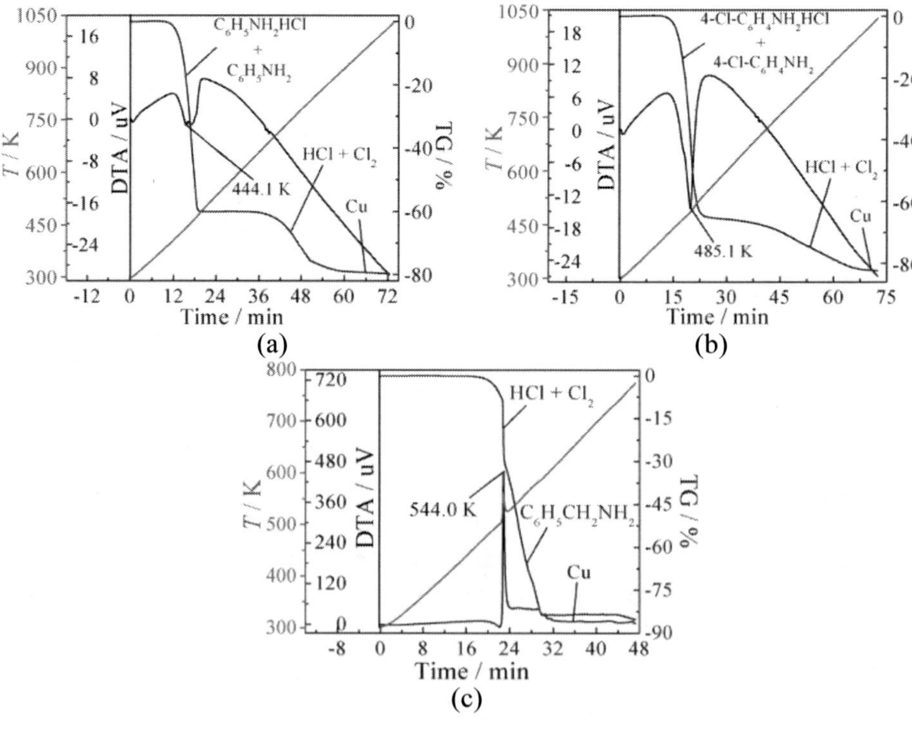

Figure 4. TG/DTA curves of (Anilinium)$_2$CuCl$_4$ (a); (4-chloro anilinium)$_2$CuCl$_4$ (b); and (benzilinium)$_2$CuCl$_4$ (c) showing the decomposition paths.

Aniline being reducing agent a part of $CuCl_2$ converted into CuCl on heating. [15] In case of **B**, DTA shows endothermic peak at 485.1 K due to thermal decomposition. Calculation of first weight loss was observed near to the evaporation of one mole of 4-chloro anilinium chloride along with 4-chloro aniline. After the loses of one mole of 4-chloro anilinium chloride along with 4-chloro aniline, (H)$CuCl_3$ is left. $[CuCl_3]^-$ species are known in literature and prefer to remain in dimeric form if the cation are small. [16] At high temperature, above 773 K, it again loses chlorine molecule in reducing atmosphere of hydrochloric acid so as to convert to Cu. Calculation of second loss come very close to the evaporation of HCl and Cl_2 molecule per formula unit, but it is not completely evaporated. This decomposition path of **B** is very similar to **A**. While in case of **C** the thermal decomposition path is totally different. Here, two moles of hydrochloride are lost along with chlorine molecule in initial heating, instead of benzilinium chloride and benzyl amine, then two molecules of benzyl amine, with subsequent formation of Cu above 573 K. The DTA of **C** shows exothermic peak at 544.0 K. Normally exothermic transition is observed due to crystallization or desolvation/ dehydration. In this case, this exothermic mass loss accounts for Cl_2, a process of dechlorination. Thus, thermal analysis shows different thermo chemical reactions operating these complexes, leading to cleavage of bond or rearrangement of ligand.

Differential scanning calorimetry (DSC) were carried out using Shimadzu DSC-60 instrument. DSC measurements were performed in the temperature range 153 K - 453 K with the scanning rate 10 K min^{-1} under oxygen and nitrogen environments. 4 - 18 mg compounds were used for each test. The α-alumina material was used as reference for DSC. Total flow was maintained at 50 mL min^{-1} for DSC analyses.

To confirm our different degradation pathway and exothermic behavior of DTA we carried out DSC on all OIHCs from 153 K to their degradation temperature. Compounds **B** and **C** shows endothermic peak at 275.83 K (-1.72 J g^{-1}) and 403.95 K (-1.98 J g^{-1}) respectively while heating and exothermic peak at 269.48 K (1.42 J g^{-1}) and 390.07 K (0.93 J g^{-1}) respectively while cooling. These peaks are characteristics for solid-solid phase transition (see Figure 5). These transitions are reversible in nature and can be repeated up to many cycles without any change in peak position. While DSC studies on compound **A** does not show any signature of solid-solid phase transition. In literature for compound **A** phase transitions is observed at 373 K [27]. But TG measurement for compound **A** shows small mass loss at that region which may be due to the low boiling solvent used for the syntheses by the authors. Furthermore, we measured DSC but not observed any change around this region for compound **A**.

Phase transitions observed in compounds **B** and **C** using DSC measurements are related to the thermochromic behavior (see Figure 6). Interestingly many A_2CuCl_4 compounds are known for such structural transitions which lead to the thermochromic behavior. The two thermochromic transitions orange phase to (T = 294 K) yellow phase and to (T = 214 K) green phase are already reported in literature for compound **B** [28]. While no thermochromic records were found for compounds **A** and **C**. To observe thermochromic behavior in OIHCs we heated the compounds near to their transition temperature in open air. As per our expectation we observed, green ↔ yellow ↔ orange for compound **B**, yellow ↔ orange for compound **C**, as shown in Figure 6. This transition is reversible in nature and observed after repeated heating-cooling cycles. But we did not observe any measurable change for compound **A**. This optical behavior is consistent with the transitions observed in the DSC measurements.

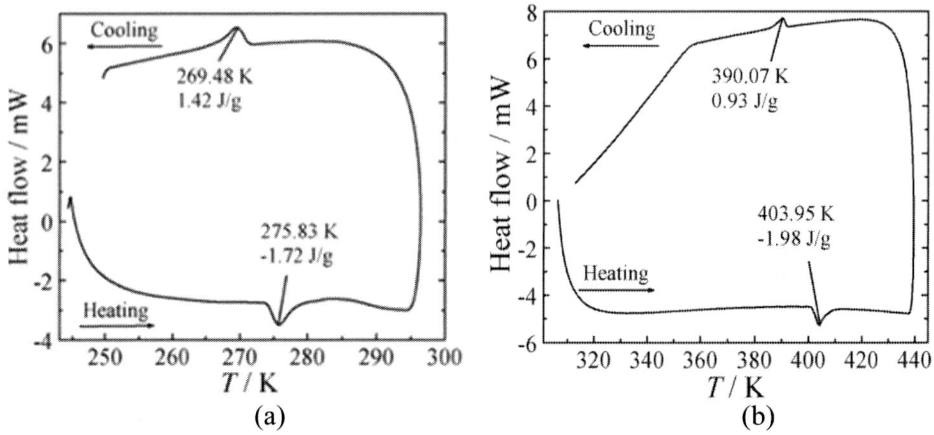

Figure 5. DSC measurement plots for compounds **B** (a) and **C** (b) showing the reversible transitions.

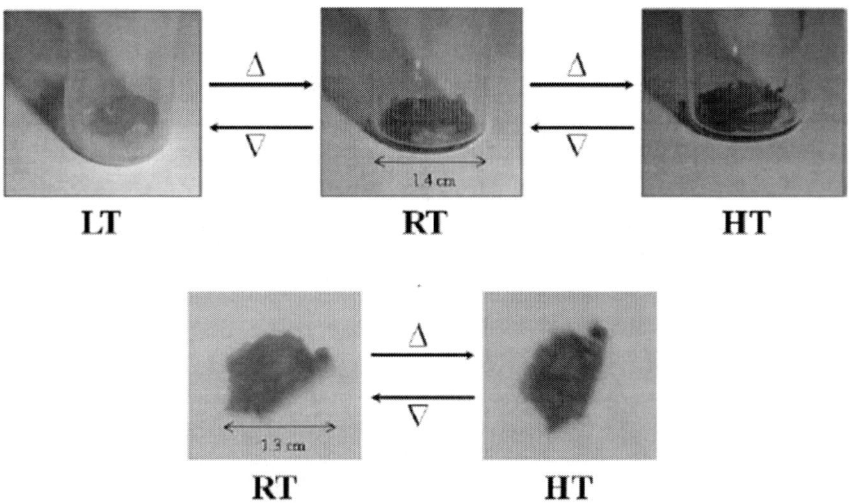

Figure 6. Thermochromic behavior of single crystals for compounds B and C.

Raman Spectra: Figure 7 illustrates temperature dependent Raman study for compound **B** which shows five Raman modes at 73 cm^{-1}, 96 cm^{-1}, 178 cm^{-1}, 257 cm^{-1} and 287 cm^{-1} in the low frequency region below 400 cm^{-1} at room temperature. These Raman modes are intense in lattice mode regions and bring out important information about dynamic of the transition. For instance, with lowering the temperature Raman modes at 73 cm^{-1} and 96 cm^{-1} changes in the background noise, Rayleigh scattering. These Raman modes at 73 cm^{-1} and 96 cm^{-1} increases together to give the new Raman modes at 88 cm^{-1} between the 260 K - 270 K. Compound **B** also shows the clear emergence of Raman mode centered around 130 cm^{-1} between 260 K - 270 K due to change in Rayleigh scattering. This means temperature dependent Raman spectroscopy also shows the solid-solid phase transition at temperature range 260 K - 270 K for compound **B**. This transition is reversible in nature and can be repeated for many cycles without any change.

Crystal Structure: A_2CuCl_4 compounds crystallize in different crystal systems ranging from tetragonal to triclinic ones [29, 30]. During this $CuCl_4^{2-}$ anion changes to all possible geometries tetrahedral, octahedral and isolated square-planar structures, mainly controlled by the ligand. Compounds **A** and **B**, known in the literature, and are crystallized in the monoclinic space group $P2_1/c$ with unit cell dimensions a = 15.050 (2) Å, b = 7.443 (1) Å, c = 7.180 (1) Å, β = 100.71 (1)° and a = 7.3736(3) Å, b = 32.0689(13) Å, c = 7.1611(3) Å, β = 101.51(2)° respectively [31, 32], while at lower temperature compound **B** adopts space group *Pccn* with unit cell dimensions a = 7.3736(3) Å, b = 32.0689(13) Å, c = 7.1611(3) Å, β = 90.00° [33] and resembles very much with compounds **C**, unit cell dimensions a = 31.30 Å, b = 7.59 Å and c = 7.28 Å, β = 90.00° [34].

The structural analysis on the compound shows that it forms the layered structure of copper chloride. The structure of the copper-halide layer is characterized by two kinds of copper-halide distances equatorial (D_S) and axial (D_L) distances as a result of Jahn-Teller distortion see (Figure 8). In these layered perovskites Jahn-Teller elongated bonds on adjacent Cu^{2+} ions lie in the plane of the layer but are oriented at right angles to each other, defining an antiferrodistortive structure [35]. These values suggest small distortions from perfect octahedral geometries for the copper ions.

Surprisingly, single crystal XRD data collected on compound **B** at different temperature, at 296 K (**B**) and 150 K (**B'**), shows structural phase transition from the monoclinic space group $P2_1/c$ at 296 K to the orthorhombic space group *Pccn* at 150 K. Thus reversible peak observed for compound **B**, endothermic peak at 275.83 K (-1.72 J g^{-1}) and exothermic peak at 269.48 K (1.42 J g^{-1}), using DSC analysis confirmed the phase transition observed at around 276 K is due to the structural change.

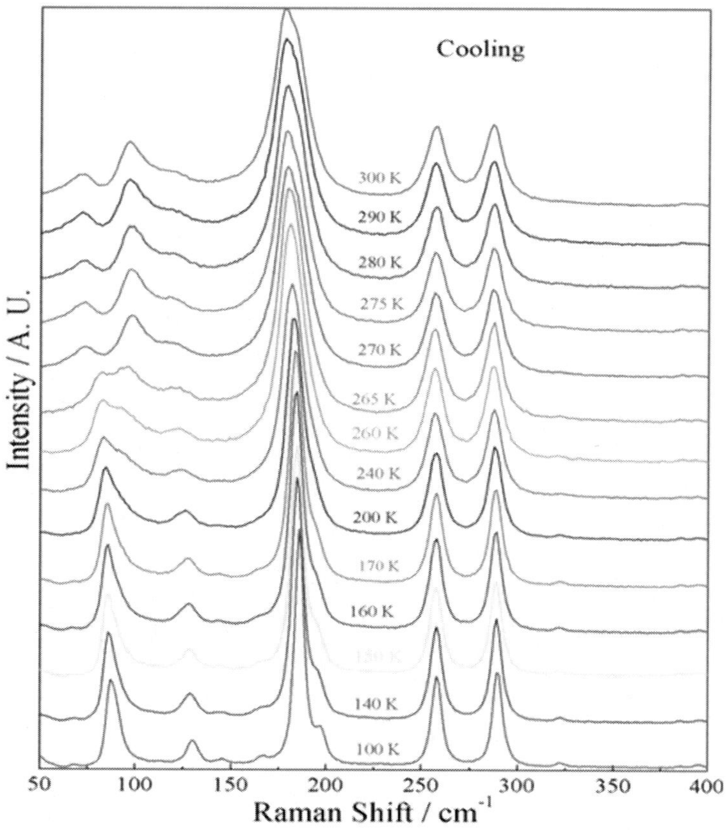

Figure 7. Temperature evolutions of lattice modes in compound **B**.

Though, the basic structure remains constant for both compounds with $(CuCl_4^{2-})_n$ layers sandwiched between 4-chloro anilinium ions moieties. A closer look at the crystal structures on these compounds revealed a few characteristic differences. The 2-D inorganic layers of '$CuCl_4$' anions adopt an eclipsed conformation at 296 K and staggered conformation at 150 K see Figure 9.

According to literature [35], monoammonium organic cation compounds prefer the staggered configuration by displacing each successive perovskite sheets, while diammonium cation compounds should adopt an eclipsed configuration. In the later case the organic cations make hydrogen bond to the inorganic sheets at both ends and thereby removing the van der Waals gap between the layers.

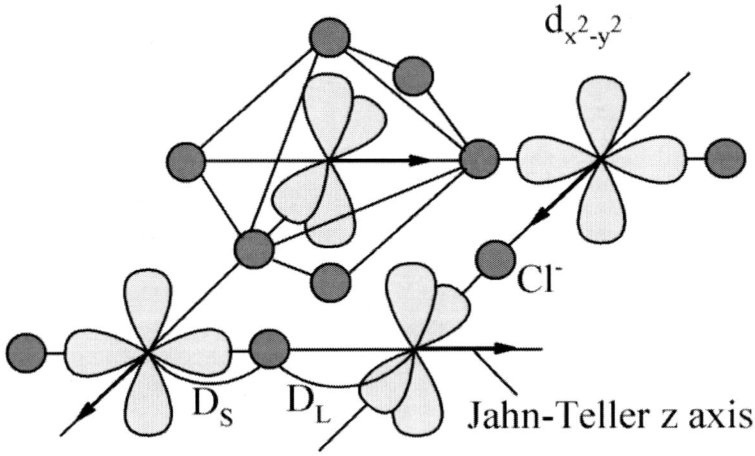

Figure 8. The layered structure of $CuCl_4$.

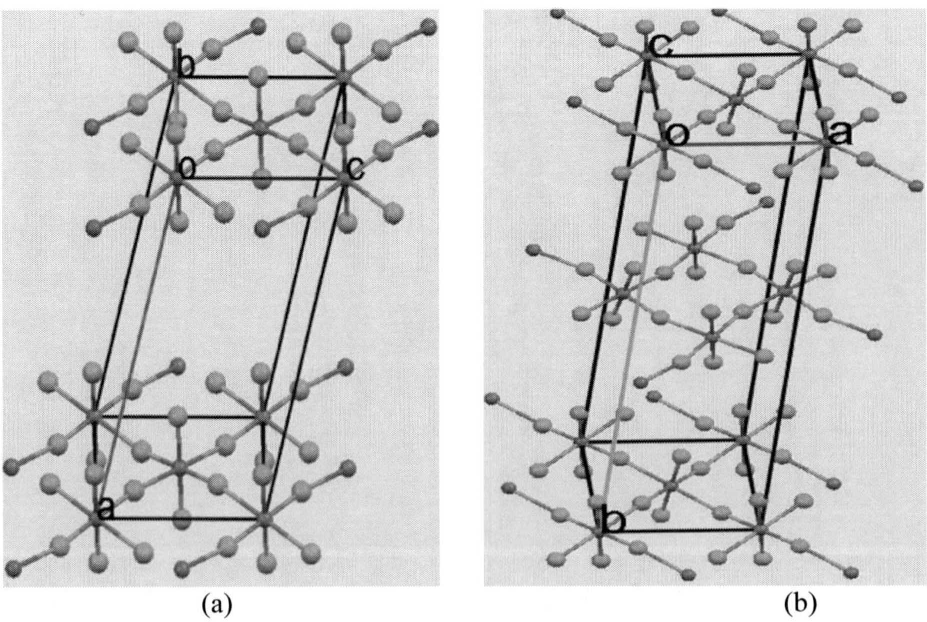

Figure 9. The "eclipsed" and "staggered" conformations of the inorganic 2-D network for compounds **B** (a) and **B'** (b).

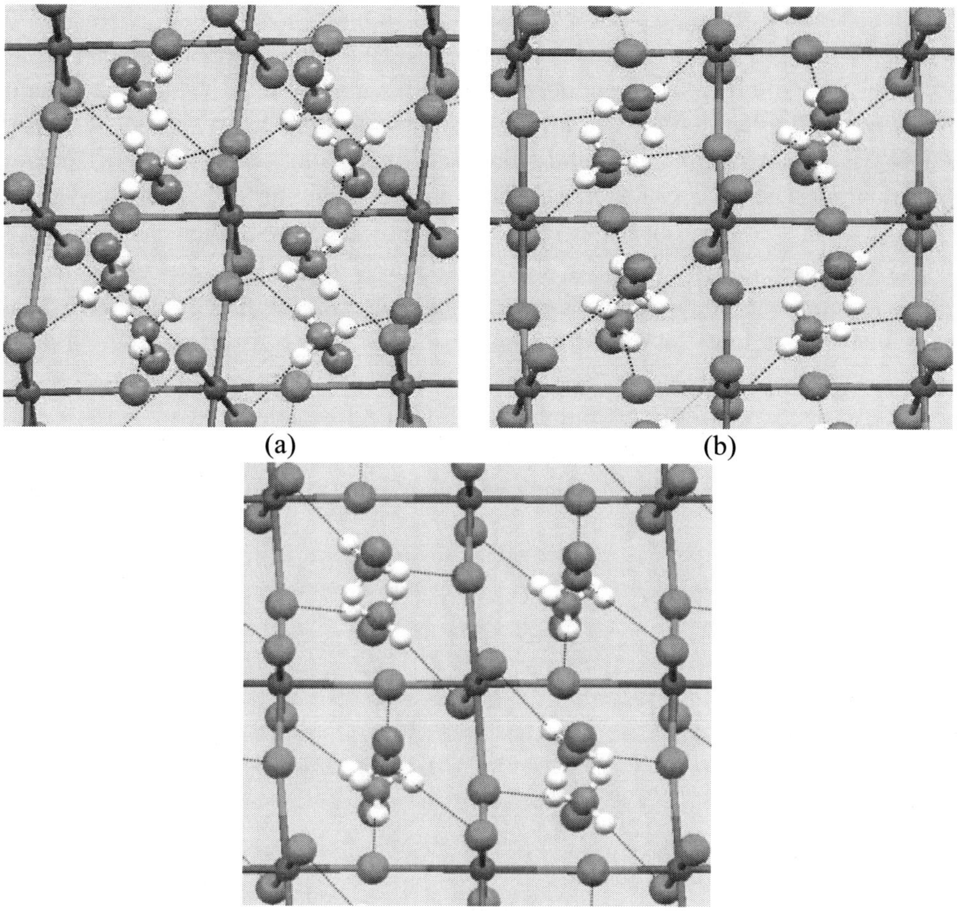

(a) (b)

(c)

Figure 10. N-H···Cl hydrogen bonding interactions for compounds **A** (a), **B** (b) and **B′** (c).

Therefore in the real structures the manner in which organic cations make hydrogen bond to the inorganic sheets influences the alignment of adjacent layers. In the present case as shown in Figure 9, the 2-D inorganic layers of '$CuCl_4$' anions adopt an eclipsed and staggered conformation. Figure 10 represent three hydrogen bonding interactions for compound **A** and two hydrogen bonding (N-H···Cl) interactions for compound **B**, between the ammonium head and Cl from the inorganic layer. In the case of compound **A**, three hydrogen bonding interactions, two with terminal Cl atoms and one with a bridging Cl atom are found. Compound **B** shows two hydrogen bonding interactions, one with a terminal Cl and the other with a bridging Cl atom called bifurcated hydrogen bond.

Magnetic Properties: Magnetic properties of layered perovskite compounds **A** and **C** were reported by R. D. Willett in 1982 [36] and A. Daoud in 1976 [34]. They were found ferromagnetic properties above 200 K for compound **A** and below 4.2 K for compound **C**. We have observed soft ferromagnetic properties with onset long range magnetic ordering below 9 K for compound **B** [33].

Figure 11 illustrate the temperature dependence magnetic susceptibility with temperature (χT) measured under different magnetic fields for compound **B**. The observed χT value for compound **B** is 0.431 cm^3 K mol^{-1} at 300 K, in good agreement with the expected value for one Cu^{2+} ion with $S = ½$ and $g = 2.0023$ (0.375 cm^3 K mol^{-1}). A gradual increase of the χT product upon cooling is observed from 300 K to 20 K. Below this temperature there is a sharp increase of the χT product for compound **B**. A rounded maximum is obtained at around 9 K, which declines on further cooling see Figure 11. These features are typical of ferromagnetic interaction [37]. The temperature dependence susceptibility plot also reflects the spontaneous magnetization present as a consequence of long range ordering of Cu^{2+} ions within the layers below the 16 K as shown an inset in Figure 11. Figure 12 shows a plot of χ^{-1} versus T (K), which obey the Curie-Weiss law above 30 K with positive θ values of 51 K for compounds **B**.

Figure 13 represent the field dependence of the magnetization for compound **B**. It reveals the sharp increase at low fields and tends to saturate below 15 kOe. The saturation of magnetization reached approx 1.05 μ_B, less than expected 1.73 for Cu^{2+} ion. The slightly lower than expected spin-only saturation value is due to the interlayer antiferromagnetic interaction. The hysteresis effects observed are characteristic of soft ferromagnet. Inset in Figure 13 shows the hysteresis loop in the low-field region, for clearly showing soft magnetic behavior.

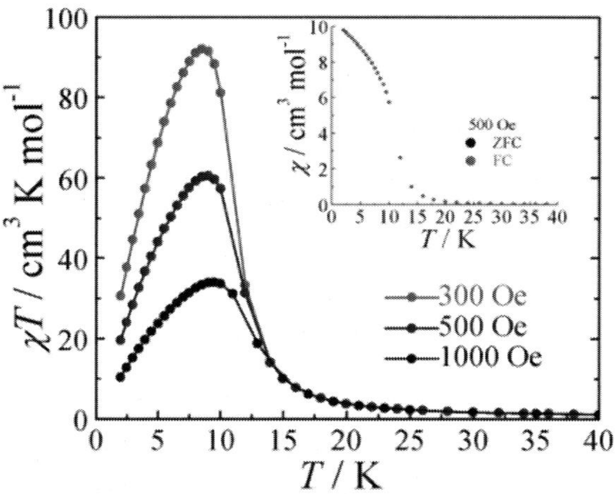

Figure 11. Temperature dependence χT with different applied fields for compounds **B**.

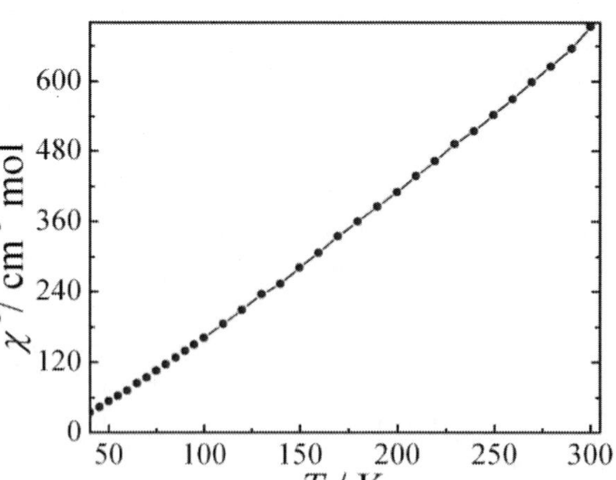

Figure 12. The temperature dependence χ^{-1} plot for compounds **B** obeying the Curie-Weiss law above the 30 K.

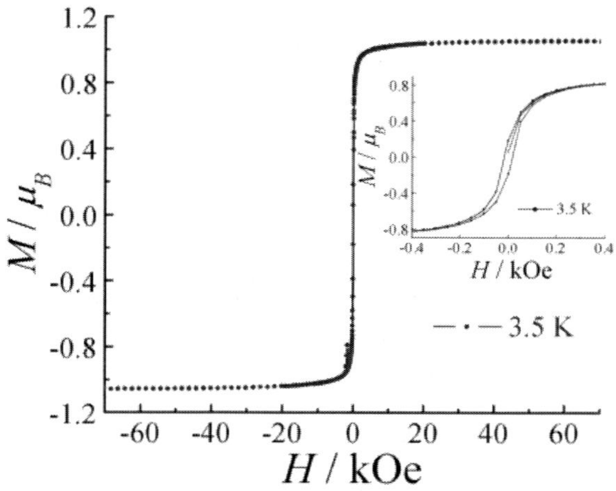

Figure 13. The *M-H* plot for compounds **B** measured at 3.5 K. The inset shows the hysteresis loop in the low field region.

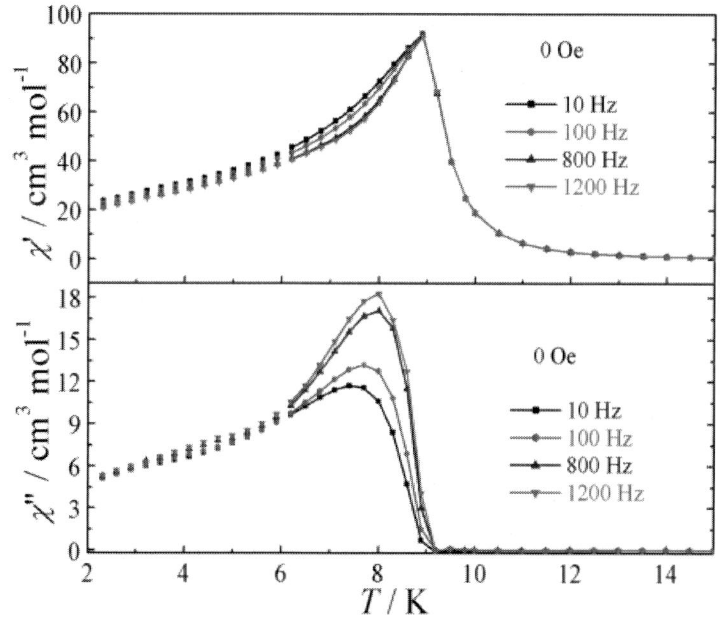

Figure 14. Temperature dependence of the ac-susceptibility for compounds **B**.

Ac-susceptibility measurements for compound **B** was performed to confirm this ferromagnetic ordering and to determine precisely the critical temperature (T_C). Figure 14 illustrates the magnetic susceptibility real (in-phase) and imaginary (out-of-phase) components as a function of temperature for selected frequencies of 10 Hz, 100 Hz, 800 Hz and 1200 Hz at 0 Oe. The in-phase ac-magnetic susceptibility (χ') exhibits a rounded maximum at 11 K for compound **B**. The out-of-phase ac susceptibility (χ'') shows onset of a long range magnetic ordering at 9.0 K for compounds **B**. The weak frequency dependent behavior suggests that the long range magnetic ordering appears below a critical temperature. Although the structural changes observed in compound **B** with a variation of temperature does not lead to any observable changes in the magnetic behavior.

Thus, preset chapter outlines importance of OACs in the designing of OIHCs with layered perovskite structure. The structural phase transitions inherent in OACs remain, or change during its transformation into perovskite structure directed by inorganic metal halide network. These phase transitions are 'driven' by non-covalent interactions such as hydrogen bonding, and result in thermochromic, ferroelectric, and/or magnetic behaviour. Thus, these strategies of making OIHCs for multifunctional or multiferroics compounds have already showed great promise and will be act as a viable alternative for future generation electronics.

ACKNOWLEDGMENT

Author thanks Mr. Hemant Mande for his help. PSG thanks to Department of Science and Technology New Delhi (SR/S1/IC-43/2009) and UGC-DAE Consortium for Scientific Research, Kalpakkam node (CRS-K-04/19) for financial support for this research activity.

REFERENCES

[1] D. B. Mitzi, *Prog. Inorg. Chem.* 1999, *48*, 1-121.

[2] J. Wang, J. B. Neaton, H. Zheng, V. Nagarajan, S. B. Ogale, B. Jiu, D. Viehland, V. Vaithyanathan, D. G. Schlom, U. V. Waghmare, N. A. Spaldin, K. M. Rabe, M. Wutting, R. Ramesh, *Science* 2003, *299*, 1719-1722.; A. O. Polyakov, A. H. Arkenbout, J. Baas, G. R. Blake, A. Meetsma, A. Caretta, P. H. M. van Loosdrecht, T. T. M. Palstra, *Chem. Mater.* 2012, *24*, 133-139.

[3] M. F. C. Ladd, *Structure and Bonding in Solid State Chemistry*, Ellis Horwood Ltd. Publishers, England, 1981, 1-326.

[4] R. D. Shannon, *Acta Cryst. A* 1976, *32*, 751-767.

[5] B. C. Watson, V. N. Kotov, M. W. Meisel, D. W. Hall, D. A. Jensen, R. Backov, M. A. Petruska, G. E. Fanucci, D. R. Nagler, *Phys. Rev. Lett.* 2001, *86*, 5168-5171.

[6] T. Ishihara, J. Takahashi, T. Goto, *Phys. Rev. B* 1990, *42*, 11099-11107.

[7] D. B. Mitzi, S. Wang, C. A. Field, C. A. Chess, A. M. Guloy, *Science* 1995, *267*, 1473-1476.

[8] D. B. Mitzi, L. Kosbar, C. E. Murray, M. Copel, A. Afzali, *Nature* 2004, *428*, 299-303.

[9] H. R. Wen, Y. Z. Tang, C. M. Liu, J. L. Chen, C. L. Yu, *Inorg. Chem.* 2009, *48*, 10177-10185.

[10] M. F. Mostafa, A. S. Atallah, R. Emrick, *J. Appl. Phys.* 1997, *81*, 4134-4136.

[11] G. V. Rubenacker, S. Waplak, S. L. Hutton, D. N. Haines, J. E. Drumheller, *J. Appl. Phys.* 1985, *57*, 3341-3342.

[12] M. M. Turnbull, C. P. Landee, B. M. Wells, *Coord. Chem. Rev.* 2005, *249*, 2567-2576.

[13] D. W. Smith, *Coord. Chem. Rev.* 1976, *21*, 93-158.

[14] M. J. M. Van Oort, M. A. White, *J. Solid State Chem.* 1988, *75*,113-123.

[15] R. D. Willett, *J. Chem. Phys.* 1964, *41*, 2243-2244.

[16] J. P. Steadman, R. D. Willett, *Inorg. Chem. Acta* 1970, *4*, 367-371.

[17] F. Barendregt, H. Schenk, *Physica* 1970, *49*, 465-468.

[18] O. Kahn, *Molecular Magnetism*, VCH Publishers, New York, 1993, 1-380.
[19] W. E. Marsh, E. J. Valente, D. J. Hodgson, *Inorg. Chem. Acta* 1981, *51*, 49-53.
[20] L. O. Snively, P. L. Seifert, K. Emerson, J. E. Drumheller, *Phys. Rev. B* 1979, *20*, 2101-2104.
[21] L. O. Snively, G. F. Tuthill, J. E. Drumheller, *Phys. Rev. B* 1981, *24*, 5349-5355.
[22] M. G. Banks, R. K. Kremer, C. Hoch, A. Simon, B. Ouladdiaf, J. -M. Broto, H. Rakoto, C. Lee, M. -H. Whangbo, *Phys. Rev. B* 2009, *80*, 0244041-02440415.
[23] C. Starr, F. Bitter, A. R. Kaufmann, *Phys. Rev.* 1940, *58*, 977-983.
[24] T. Maris, N. B. Chanh, J. -C. Bissey, N. Filloleau, S. Flandrois, R. Zouari, A. Daoud, *Phase Transitions* 1998, *66*, 81-98.
[25] M. F. Mostafa, S. A. El-Hakim, *Phase Transitions* 2003, *76*, 587-599.
[26] S. Kalużа, M. Suchańska, R. Belka, S. Leśniewski, *Ferroelctrics* 2002, *273*, 143-148.
[27] A. Purkayashtha, J. B. Baruah, *Thermochim. Acta* 2002, *390*, 187-193.
[28] M. F. Mostafa, M. M. Abdel-Kader, S. S. Arafat, E. M. Kandeel, *Physica Scripta.* 1991, *43*, 627-629.
[29] R. Belka, *J. Mater. Sci.* 2007, *42*, 10090-10097.
[30] D. A. Tucker, P. S. White, K. L. Trojan, M. L. Kirk, W. E. Hatfield, *Inorg. Chem.* 1991, *30*, 823-826.
[31] K. P. Larsen, *Acta Chem. Scand. A* 1974, *28*, 194-200.
[32] M. M. Zhao, P. P. Shi, *Acta Cryst. E* 2010, *66*, m656.
[33] A. K. Vishwakarma, P. S. Ghalsasi, A. Navamoney, Y. Lan, A. K. Powell, *Polyhedron* 2011, *30*, 1565-1570.
[34] A. Dupas, K. Le Dang, J. -P. Renard, P. Veillet, A. Daoud, R. Perret, *J. Chem. Phys.* 1976, *65*, 4099-4102.
[35] D. B. Mitzi, *J. Chem. Soc., Dalton Trans.* 2001, 1-12.
[36] R. D. Willett, F. Waldner, *J. Appl. Phys.* 1982, *53*, 2680-2682.
[37] D. J. Price, F. Lionti, R. Ballou, P. T. Wood, A. K. Powell, *Phil. Trans. R. Soc. Lond. A* 1999, *357*, 3099-3118.

INDEX

D

E

F

G

H

I

N

O

P

Q

R

S

T